RENEWALS: 691-4574

DATE DUE

FEB 17			
AUG 1 3			
DEC 17			

Demco, Inc. 38-293

HARDWARE DESCRIPTION LANGUAGES

Advances in CAD for VLSI

Volume 7

Series Editor:

T. OHTSUKI
*Department of Electronics
and Communication Engineering
School of Science and Engineering
Waseda University
Tokyo, Japan*

NORTH-HOLLAND
AMSTERDAM · NEW YORK · OXFORD · TOKYO

Hardware
Description Languages

Edited by

R. W. HARTENSTEIN
Department of Informatics
University of Kaiserslautern
Kaiserslautern, F.R.G.

1987

NORTH-HOLLAND
AMSTERDAM · NEW YORK · OXFORD · TOKYO

ISBN: 0 444 87897 1
ISBN set: 0 444 87890 4

Publishers:

ELSEVIER SCIENCE PUBLISHERS B.V.
P.O. Box 1991
1000 BZ Amsterdam
The Netherlands

Sole distributors for the U.S.A. and Canada:

ELSEVIER SCIENCE PUBLISHING COMPANY, INC.
52 Vanderbilt Avenue
New York, N.Y. 10017
U.S.A.

Library of Congress Cataloging-in-Publication Data

Hardware description languages / edited by R.W. Hartenstein.
      p.   cm. -- (Advances in CAD for VLSI ; v. 7)
   Includes index.
   ISBN 0-444-87897-1 (Elsevier Sci. Pub. Co.)
   1. Computer hardware description languages.   I. Hartenstein,
Reiner.   II. Series.
TK7885.7.H37 1987
621.3--dc19   87-24494
                                                          CIP

PRINTED IN THE NETHERLANDS

INTRODUCTION TO THE SERIES

VLSI technology has matured to the extent that hundreds of thousands or even millions of transistors can be integrated in a single silicon chip, and VLSIs are now the key to the design of efficient electronic systems. It therefore follows that the problems of designing integrated circuits are also becoming increasingly complex. A wide variety of topics on computer aided design (CAD) have emerged. This is a period when no-one can be a specialist in all of the topics in CAD for VLSI, and the whole area is beyond the scope of a single volume. The requirement for information and communication is increasing.

In 1982, Dr. R. Morel, at that time Acquisition Editor at North-Holland, conceived a range of projects to relieve this problem. Initially this resulted in "INTEGRATION, the VLSI journal", a quarterly aimed at speeding up communication among VLSI designers. It was further decided to launch a book series in the field, an idea enthusiastically supported by Dr. L. Spaanenburg, Editor-in-Chief of Integration. Dr. Morel approached me to edit this series.

It was agreed that the book series should be aimed towards a comprehensive reference for those already active in areas of VLSI CAD. Each volume editor was asked to compile a present state of the art, scattered in many journals. The book series therefore should help CAD specialists to get a better understanding of the problems in neighbouring areas by reading particular volumes. At the same time, it should give novices a foothold for doing research in areas of VLSI CAD, although a basic knowledge of VLSI technologies and design methods will aid understanding.

The book series, entitled "Advances in CAD for VLSI" consists of the following seven volumes:

Vol. 1	Process and Device Modeling	W. Engl (RTWH Aachen, FRG)
Vol. 2	Logic Design and Simulation	E. Hörbst (Siemens, Munich, FRG)
Vol. 3	Circuit Analysis, Simulation and Design (Part I and II)	A. Ruehli (IBM, Yorktown Heights, USA)
Vol. 4	Layout Design and Verification	T. Ohtsuki (Waseda University, Tokyo, Japan)
Vol. 5	VLSI Testing	T. Williams (IBM, Boulder, USA)
Vol. 6	Design Methodologies	S. Goto (NEC, Kawasaki, Japan)
Vol. 7	Hardware Description Languages	R. Hartenstein (Univ. of Kaiserslautern, FRG)

The first five volumes deal with major phases of VLSI design separately. The following two volumes are devoted to recent approaches which span the whole design phase. The integrated approach combining system design and VLSI design is also treated in these volumes, and it will suggest a major trend in the future. Each volume is reasonably self-contained so that it can be read independently.

All of the volumes were intended to include up-to-date results, and the latest developments with a good balance between theory and practice. Moreover, emphasis was placed on basic techniques, methods and algorithms rather than on descriptions of existing design tools. This, I hope, will prevent obsolescence at the time of publication. Only the readers however, can judge to what extent our intentions were successful.

Selecting the volume editors was not an easy task. In order to produce a quality book series, it was necessary to utilise authorities well known in their respective fields who in turn would attract outstanding contributors. Those active in VLSI design had other commitments. I was relying on their volunteering spirit, and they in turn faced the same dilemma with their authors. It was a great pleasure to me that ultimately we were able to attract such an excellent team of editors and authors.

The Series Editor wishes to thank all of the volume editors and authors for the time and effort they put into the book series, and especially Prof. W. Engl for his assistance in arranging other volume editors. I also thank Dr. E. Fredriksson, Dr. J. Julianus and Dr. R. Morel of North-Holland for their continuing effort to bring the book series from the initial planning stage to final publication.

Waseda University
Tokyo, Japan
1985
                                                        Tatsuo Ohtsuki

PREFACE

The area of Hardware Description Languages exhibits some indications of forthcoming maturity. Meanwhile a little bit of common sense is beginning to take shape with respect to terminology, systematic treatment, and a clear partitioning of the field into subfields. It is the intention to contribute to the progress of this development by providing a handbook giving a survey, and also covering recent developments and future trends. The objective of the book is to give a systematic representation of the discipline and its implications: fundamentals, implementation techniques, applications, and, its embedding into the area of CAD for VLSI. The book serves as a navigation aid in a Babylonian World having produced hundreds of languages.

Reiner W. Hartenstein

TABLE OF CONTENTS

Introduction to the Series   v

Preface   vii

1. Introduction
    R.W. HARTENSTEIN   3

2. The Classification of Hardware Description Languages
    R.W. HARTENSTEIN   15

3. Application-oriented HDL Specializations

    3.1. Data Path Descriptions
          K. LEMMERT   51

    3.2. Control Part Descriptions and Languages for Microprogramming
          P. PRINETTO   75

    3.3. Interface and I/O Protocol Descriptions
          A.C. PARKER, N. PARK   111

    3.4. Graphic Hardware Description Languages
          U. WELTERS   137

    3.5. Languages for Simulator Activation and Tester Operation
          A. MAVRIDIS   163

4. HDL Applications

    4.1. Conceptual Design Based on HDL Use
          K. LEMMERT, W. NEBEL   197

    4.2. Silicon Compilation from HDL and Similar Sources
          M. GLESNER, H. JOEPEN, J. SCHUCK, N. WEHN   227

4.3. Test Pattern Generation from HDL and Similar Sources
      S.Y.H. SU, T. LIN   253

4.4. Hardware Verification Based on HDL Sources
      M. FUJITA   283

4.5. RT Languages in Goal-oriented CAD Algorithms
      A. WODTKO   313

4.6. Fault Simulators at Functional Level
      M. MELGARA   337

5. HDL Subsystem of an Integrated CAD System
    G. GIRARDI, S. GIORCELLI, G. GIANDONATO   375

6. Implementation Techniques for Multi-level Hardware Description Languages
    D. BORRIONE, C. LE FAOU   409

7. Standardization Efforts

7.1. The CONLAN Project: Concepts, Implementations, and Applications
      R. PILOTY, D. BORRIONE   441

7.2. The Electronic Design Interchange Format (EDIF) – Proposed Standard
      P.H. STANFORD   463

Author Index   495

1. Introduction

HARDWARE DESCRIPTION LANGUAGES
R.W. Hartenstein (Editor)
© Elsevier Science Publishers B.V. (North-Holland), 1987

1. INTRODUCTION

Reiner W. HARTENSTEIN

Department of Informatics, University of Kaiserslautern
Postfach 3049, D-6750 Kaiserslautern, F.R.G.
phone: (xx49 - 631) 205 - 2606 or (xx49 - 7251) 3575

WHY HARDWARE DESCRIPTION LANGUAGES (HDLs) ?

A single VLSI Chip may include hundreds of thousands of transistors or more than a hundred thousand gates. That's why to cope with complexity is the most difficult problem in VLSI design. Design methodologies based on the gate level are no longer sufficient. The next higher abstraction level above gate level is called register transfer level (RT level), since its primitives are registers, register arrays, and data transfer paths, such as e.g. operators, buses, multiplexers, and others. Hardware descriptive notations at RT level (and sometimes above) are called *(Computer) Hardware Description Languages (HDLs)*. Some of them have a mnemonics which is similar to that of the Pascal programming language. To cope with complexity made possible by the progress of IC technology there is an urgent need for a break-through of HDL use in design practice.

HDLs may be used to feed higher level simulators, to feed silicon compilers (at least future ones), for automatic test pattern generation, for hardware specification, and for documentation, for more concise teaching the principles of digital hardware, use as a design calculus and for many other applications. Let us look more closely onto all these applications. All modern HDLs are hierarchical and thus have structural description capabilities, so that this should not be used as a classification criterium. We may distinguish two major classes of HDLs: non-procedural languages and algorithmic languages. For a more detailed discussion on HDL classes see chapter 2 in this volume.

Substantial Reduction of Complexity. What are the benefits in using such HDLs, compared to traditional abstraction levels, such as e.g. the gate level ? The most important benefit is the reduction of notational complexity. Gate level notations, such as e.g. Boolean equations, do not yield a substantial reduction of complexity, compared to circuit diagrams. The average number of transistors per gate is the quotient of complexity reduction: this is only about 3 to 5, in CMOS and some other circuit techniques only about 4 to 8. In using HDLs this quotient may by much higher, sometimes up to several hundreds.

One reason for reduced complexity is the fact, that HDLs use to bundle vectors of bits into words, like in high level programming languages. In describing a 32 bit data path, for instance, its data values are kept in a single word to be processed at once within the simulator and other tools. A second reason is the fact, that at RT level more powerful operators are available, such as e.g. multiplication, an equivalent to up to hundreds of gates, and many others. A third reason for reduced complexity is found in the flexibility of RT language use. Modern CHDLs feature capabilities to describe a particular hardware in different levels of abstraction. That's why in using the same language a design process may start with a very high level specification of very low complexity, and after several steps of refinement it may end up with a more complex and more detailed description of a solution concept.

Training. So it is sure, that the complexity problem of VLSI design can be solved only by using HDLs. Later in this paper a number of additional benefits will be illustrated, such as support of design for testability, early test pattern development, structured VLSI design, experimenting with alternative architectures before starting logic design, and many others. Why does a majority in industry hesitate to introduce HDLs? Some quite interesting tools are available. However, most universities do not teach using HDLs, which would be an effort taking about less than half of the time needed to introduce Pascal. Another problem is the lack of methodology. At gate level a very elaborate and formal design methodology has grown, mainly within the last 30 years, due to contributions of thousands of scientists throughout the world. At register transfer level, however, most contributions are more of narrative character and of analytical nature, rather than being a design methodology. A generally and widely accepted formal notation - comparable to Boolean algebra at gate level - has not yet been established in most application areas at RT level. So design tends to be more a trial and error procedure, or, to use one of a few popular concepts, such as e.g. systolic arrays and others.

EARLY PHASES OF THE DESIGN PROCESS

CHDLs are an important opportunity to avoid expensive redesigns needed to correct errors, such as e.g. bad testability, bad topology and bad structure of the circuit, too much area consumption, or, bad (VLSI-) architecture, missing the requirements, and others. Many of such errors could be avoided, if design concepts would be decided at a very early phase of the design process, definitely before the costly logic design procedure has been started.

Specification and Design Problem Capture. One important role of HDL use could be design problem capture. To be sure to meet the requirements the design problem has to be pinned down the correct way. A concise notation has to be used to express the design problem. A description of a design problem using such a notation is called a specification. To be sure to capture the design problem correctly the specification has to be checked against the requirements.

Specification Verification by Simulation. If an HDL implementation including a simulator is available, the requirements could be simulated after the specification has been accepted by the tool. So the HDL system may serve for design problem capture. Such an HDL system could be also used as a communication medium between customer and design center, or, if within the same company: between product planning division and design division. The customer uses the HDL system, such as e.g. an HDL compiler and simulator, for design problem capture. Reacting to simulation results the customer successively debugs the specifications. Finally the verified and debugged specifications are handed over to the design center. Positive practical experiences have been made about this kind of HDL use.

Experimenting with alternative Architectures. Bugs in specifications are not the only possible reasons for missing the requirements. Sometimes the principles of a design concept are critical with respect to real-time performance, to design cost, or other important aspects. Often several possible solutions have to be considered and analyzed, so that experimenting with alternative architectures is needed. Of course such experiments should be carried out at the highest possible level of abstraction to avoid incomprehensibility of descriptions and high labour cost because of high complexity. The use of an HDL is the only systematic way to work at the high level needed.

Design for Testability. This section should illustrate, why HDL use could enhance design for testability. Testing VLSI circuits currently is a major disaster area in industry, since the technology of testing and test pattern generation is far behind the possibilities of manufacturing technology and design capabilities. For mass production very often the time needed for testing is too long. Desirable would be around a second or less. For automatic test pattern development often an excessive amount of CPU time is needed. A very critical aspect is the fact, that for a given set of test patterns often the test coverage is much too low. This means, that the percentage of circuit faults, which will be detected by using a given set of test patterns, is far below 100%. This issue is critical, since it severely affects product quality. In some applications, such as where malfunction of circuits could be a danger to human life (process control, aerospace, some modern automotive, medical applications, etc.), or, could make the entire mission fail (aerospace applications, etc.) this quality aspect is one of the most important objectives at all.

Very early Test Pattern Development. In many cases the design is the reason, why a circuit's fault coverage is low. In such a case the best possible test patterns could not achieve high fault coverage. That's because of properties of the design important inner subcircuits cannot be reached by a sufficient percentage of stimuli. Nor a sufficiently high percentage of its responses could be observed from outside the circuit. Only an expensive redesign of the circuit, which takes testability aspects into account, could solve such a problem. The product development schedule could slip for months or more.

All this illustrates that design for testability is an important ingredient of the VLSI design process. Not only testability per se, but also the length of the test needed is a very important objective in design for testability. The best solution is to carry out test pattern development in very early phases of the design process, at least before logic design has been started. The most desirable time would be, when the specifications are ready, concisely expressed by means of a suitable HDL. Instead of being part of the logical design, and thus being highly expensive, testability would be a subject of early design planning and partitioning definition. The designer could fully concentrate on the testability architecture of a circuit and could experiment with alternative architectures. However, this would require, that the test patterns are available at such an early time that testability data and test length data of different version architectures are available. Otherwise the designer would not know, which alternative to decide, and, whether the design for testability efforts have to be continued or not.

Functional Testing. All this is feasible, since fortunately for production testing of integrated circuits only a *go / no go* test (sometimes called a *functional test*) is needed. That's because integrated circuits are not repaired, so that fault locating is not required. A functional test can also be developed without any structural knowledge about the circuit. (A test also exhibiting fault location diagnostics would be called a *structural test*.) That's why a functional test can already be developed from the functional description of a circuit, i.e. from its specification by means of an HDL.

Integrating Simulation and Test Development. Although the area of functional test pattern generators currently is rather immature, it is useful to have it available already along with the circuit specificaiton for designing for testability. For large circuits an exhaustive simulation is not possible, unless an accelerator is available which runs the simulator, or, a physical model extension is used. So the user will have to select subsets of the test patterns in a clever way to run the simulator. This is not the only reason, why there is a need to integrate high level simulation and test pattern development. An HDL-based simulator could be an efficient basis of such an integration, if its activation language would be capable to express test patterns concisely and in a highly comprehensible way.

Integration of Simulation and Testing. In production testing, where normally fully automatic test equipment is used for go / no go tests only, the integration of the testing process itself with simulation would not make sense. However, for low production volumes in the area of ASICs it might be interesting. But prototype testing which normally has the goal of design debugging is a highly interactive process using tracing and breakpointing strategies for fault location. The procedure is quite similar to interactive simulation normally carried out to reach the same goals. In connection with a simulator featuring a physical hardware extension (section 3.5 within this volume) a simulator activation language which not only supports test pattern description, but also interactive testing as well as interactive simulation, may be a very efficient aid to achieve synergisms between testing and simulation. One part of the synergism is achieved by the fact, that under notational integration under a common activation language the same hardware device may be used as a tester and (for simulator acceleration) as a physical model extension. Such a language has been implemented in 1987 (see section 3.5 in chapter 3 within this volume).

Structured VLSI Design. The term of *structured VLSI design* has been coined by the Mead-and-Conway scene. It stands for a method to implement algorithms directly onto the planar surface of silicon in a way, which attempts that most cells of the design are connected by abutment. This means, that by means of port matching between neighbour cells no routing area between these cells is needed. This in many application problems is a highly efficient way to save chip area, since routing areas tend to eat up very much more chip area (sometimes up to about 95% of the chip) than active cells. The best way to use this method is it, to try to plan the chip in a way, that most of it is made up by arrays of abuttable cells. The success of such a solution highly depends on the cleverness in planning the shape and the topology of *key cells,* being efficiently abuttable. Often a successful key cell design is possible only, when a clever partitioning and placement strategy has been used in chip floor planning. All this means, that layout considerations are needed at very early phases of the design process, about when the specification is formulated.

Supporting the Innovative Power of VLSI Design. Structured VLSI Design as a design style has an innovative power. The success of structured VLSI design efforts depends on selection of the best possible task realization algorithm for a VLSI solution. Sometimes the smart memory approach is a good solution (this is shown in tutorial-like explaining the design of a simple sorter chip example in [22] of chapter 2). Also this illustrates the benefit of very early chip planning. To provide means for design plan verification at this early phase, an HDL is needed, which can express such partitioning and topological features already at specification level.

FEATURES OF HDLs

There is a strong tendency toward multi-level HDLs. Quite a number of them have been proposed. This has good reasons: sometimes small pieces of a description cannot be expressed at RT level, so that this 'remainder' can only be presented in using gate level primitives. Another important reason is the bus, being an important architectural resource at high levels of system description: it is a switch level concept (see chapter 2 [26]). This multi-level paradigm has more advantages: in using the same language as a top-down design medium a specification can be successively refined to a more hardware-near detailed concept.

Comprehensibility. Also when being multi-level and multi-paradigm languages HDLs are substantially more easy to learn than high level programming languages, such as e.g. Pascal. Most languages use reserved words with rich mnemonics designed skillfully, so that HDL descriptions are self-documenting and reach a high degree of readability and comprehensiblilty. Comprehensibility is even more enhanced when an interactive gráphic user interface is available. One such graphic interface, which in fact is the implementation of a graphic HDL, has been implemented in 1986.

Language Primitives. Power and flexibility of an HDL are determined by the repertory of its primitives. We may distinguish structural primitives from functional ones. Structural primitives within an HDL normally are uniform throughout all abstraction levels: they are all the same, no matter whether being used at RT level, gate level, or, at switch level, and even at circuit and symbolic layout levels. Structural primitives may be subdivided into module definition features (such as functions, modules and cells), and, into notations to specify interconnect. This eases interfacing of HDLs and HDL-related tools to other tools, which deal with more silicon-near methodological levels. The repertory of functional primitives, however, highly depends on the methodological levels covered by the HDL. For a classification of those levels see chapter 2 within this volume. Within HDLs these levels may reach down to switching level. At this level, for instance, a method for modelling bus systems has been implemented which is also capable to model systematically the general circuit principles of generalized symbolic layout.

Cell Modules with Floor Plan Capability. At least one HDL efficiently supports structured VLSI design and the integration of HDL-based CAD tools into physical design by cell definition and instantiation features similar to floor planning features. For this purpose an HDL needs topological cell declaration attributes to support automatic interconnect generation for cell abutments. With operators to express horizontal and vertical abutment, as well as rotate and mirror transforms, a very powerful topological instantiation sublanguage for abutment expressions has been implemented, capable of describing very concisely the structure of rather complex cell clusters including many nesting levels.

Concise Description of Routing Patterns. Normally wiring patterns are subject of other disciplines within design sciences: schematics entry systems and routing algorithms. Of course, any arbitrary wiring can be expressed in an HDL by user-defined descriptions, and, by user-defined routing cells. However, particular classes of wiring patterns are concisely predefined by primitives of some HDLs. Widely spread examples from HDLs are wiring descriptions for: changing path width (juxtaposition of paths to create a wider path, as well as for subscripting to split up a path), and for routing pattern which preserve data path width. The latter ones may be split up into subgroups: direct connections which do not affect the sequence of bits, user-defined routing boxes, and, *wiring operators* which rearrange the sequence of bits algorithmically, such as shift, shuffle, reflect, and butterfly operators. Within an HDL such operators are very powerful. With only the operator name, the path width specification, and sometimes an additional single integer parameter, the entire complex wiring pattern may be expressed. To be used within the simulator the HDL compiler derives this wiring pattern algorithmically from such few information. For more details see section 3.1 in chapter 3 within this volume.

For instance, shuffle and butterfly patterns have many applications, such as for example in digital signal processing, micro processor and micro computer interconnect networks and many other areas. For wider data paths the butterfly pattern usually is shaped by the superposition of several elementary butterflies. The array format versions of these wiring operators are useful for concise description of interconnect patterns in switch boxes, like banyan networks, etc., for parallel signal processing circuits, such as e.g. for fast fourier transform. Shuffle operators are also useful for multi level HDLs and their simulation as an adapting interface between single-slice level and the next higher level, where operators have been formed by iterative slice instantiation.

Step-wise refinement using an HDL. Refinement capability is an important language property to achieve its use as a design language. So you do not need to change the language in top-down planning from a purely functional specification to a more detailed description which then may be used to enter logic design and physical design. So you may use the same language and the same tools to analyze and synthesize a conceptual description being a structural / topological / functional notation which carries along all the clever architectural ideas for testability, for structured design, etc. over to the silicon implementation team. Fig. 8 in chapter 2 within this volume illustrates stepwise refinement by an example circuit, a simple 2-way multiplexer.

HDL Use as a Design Calculus. Section 3.4 in chapter 2 of this volume has illustrated the consistency of a particular HDL and its capability for step-wise refinement, so that top-down design can be carried out without changing the language. This is a requirement if it is desired to use the language as a design calculus in order to reach design goals by means of a sequence of algebraic manipulations. However, such a language can only be a medium to express such algebraic rules, however, it cannot be this algebra itself. This medium, however, is powerful, suitable for effective exploration of many areas of application by experimenting with alternative architectures and structures. In section 3.1 (in chapter 3) an example is described, where a regularly structured integer multiplier layout has been developed from the algorithmic description by a sequence of successive algebraic manipulations. Also for other particular kinds of digital circuits a number of approaches to a general algebraic schematics development have been published, using examples, such as binary-to-BCD, BCD-to-binary code converters, and universal shifter examples.

CHDL-BASED DESIGN ENVIRONMENTS

Not only an HDL itself, but also other languages used within an HDL simulation system (the simulator activation language, and the simulator-executable intermediate form generated by the HDL compiler) in more than one case have become an interchange format to interface a wide variety of CAD tools. Therefore these languages have turned into quasi-standard interfaces to a number of tools within a larger user group. For more details see sections 4.1, 4.4 and 4.5 in chapter 4 within this volume. For instance, this has been the opportunity, that for a particular HDL three different simulators have been implemented: a heuristic simulator, a fault simulator, and a fast event-driven simulator (also see section 4.6 in chapter 4).

Using an Interactive Graphic HDL Editor. In 1986 an interactive graphic HDL editor has been implemented which is the user interface to an HDL system (see section 3.4). DOMINO notation is used to show symbolic abutment (of abstract boxes) in architectural diagrams, and, to show physical abutment as well (when boxes reflect the shape of real cells) in case of a partitioning derived from a chip floor plan. To be efficient such an editor features modern menue techniques and windowing and also includes an on-line graphical syntax check, which immediately diagnoses illegal design manipulations. This accelerates working with an HDL considerably, since most of the diagnostics are already interactively available. This is much more efficient than using such an editor as a front end to an HDL compiler. Compared to this batch mode the syntax directed graphic editor improves design turn-around time by about a factor of 5. See section 3.5 within this volume. This creates a situation, where an HDL turns from a user interface into a tool makers' and tool integrators' intermediate language, at least partly.

Parameterized cell descriptions. By means of precompilers or interpreters generating HDL source descriptions the parameterized documentation of cell descriptions is possible. An example is the HDL description of a cell using iterative logic such, that the data path width is a formal parameter. Another example would be the description of a memory array of 2^n memory locations, such that n is a formal parameter. This way a larger family of cells may be represented by just a single HDL description. One simple way to implement such a parameterization feature is to have an HDL extension accepted by a precompiler or filter program for translation into HDL code. Such a filter program has been implemented for the generation of cell arrays under control of the HDL extension's array size parameters and data path width parameters: one-dimensional cell arrays, bit node arrays, and word node arrays, two-dimensional cell arrays with linear growth, exponential growth, node arrays with exponential growth, as well as recursively defined two-dimensional cell arrays.

HDL Extraction from Layout. Extraction is inverse to synthesis. That's why it could be an important ingredient of verification strategies which compare the specification with its extracted version. Not only circuit extractors are available. Also an HDL extractor has been implemented so that verification may be carried out at higher levels. In section 4.1 such an extractor is dealt which directly generates HDL descriptions from Layout. It is rule-driven, and thus technology-adaptable.

Behavioural Extraction from Symbolic Layout. Another application of such an HDL extension is the extraction of behavioural descriptions from symbolic layout. This extension needs to have features to accept personality matrixes of all types, as well as for technology-independence, to accept switch-level-to-logic relations. Such an HDL extension has been implemented which features a rule-driven, and thus technology-adaptable algorithm for translation of personality matrixes into HDL functional descriptions for a wide variety of matrix-oriented logic (MOL) circuit techniques, such as e.g. PLAs, folded PLAs, Weinberger arrays, folded Weinberger arrays, Lopez/Law dense gate matrix layout, KOLTE arrays, and others. The advantages of this type of symbolic extraction are the following ones. Extraction from symbolic layout is more efficient than from geometric layout. The level of symbolic layout is the ideal basis to manipulate MOLs to meet topological constraints such as needed to benefit from structured VLSI design. This level gives a very efficient support to ASIC design, since many efficient MOL-oriented layout generators are available.

CHDL-based CAD Tools. Several HDLs have been implemented which became some sort of integrating core of a set of CAD tools, such as for silicon compilation, for interactive synthesis of microprograms, for RT level verification, for test pattern development and testability analysis, for assembly of tester-specific test programs and for many other purposes. For more information see chapters 3 through 5 within this volume. An important aspect of HDL to tool relations is the question whether an HDL and its implementation is more suitable for semi-custom ASIC design support, for full-custom circuit design. This mainly depends on the type of simulator being implemented for the particular HDL. It turns out, that library-oriented simulators are better for semi-custom circuit design, whereas calculus-oriented (language-oriented) simulators better fit for full-custom design environments. One problem of library-oriented simulators is the high cost of library maintenance. On the other side, calculus-oriented simulation environments need higher user skills than library-oriented ones. Another important aspect is, that HDL simulators normally run in the state domain. So it seems to be rather wise, to do the timing simulation with other tools, such as e.g. separate timing simulators or logic and timing simulators.

CONCLUSIONS

This paper has illustrated the many benefits which may be obtained from using HDLs, HDL-related CAD tools, and CAD systems having been integrated under the umbrella of an HDL implementation. Up to now still silicon-near VLSI design tools are dominant. To fully exploit the innovational power behind the steady progress of modern IC fabrication technology, the use of system-near conceptual HDL-based design tools is urgently needed. It is a cronic disease of the HDL area, that proposing new languages is a hobby widely spread all over the world. Already 10 years ago an analysis of Rome Air Development Center, Dayton, Ohio, has counted much more than 200 languages. It also highlights the quality of program committees of conferences and workshops, that all these languages are published, independent of whether they have been implemented or not, whether they are clones of other languages or not, whether they have new features or not. What is really needed is to practice using such languages instead of continuously inventing new ones. What is needed is a design staff being skilled to use such languages and related design tools. One major problem is the training problem. Universities do not teach the use of HDLs sufficiently. That's why industry lacks man power being qualified to use conceptual design tools and methods. This problem is highlighted drastically by a Dataquest prognosis saying, that by 1990 worldwide 400,000 digital system designers would be needed, where 100,000 of them would have to have IC design skills. The in-house continuing training effort would be to large: a one-week's course would by far not be sufficient. So it turns out, that problems of training and engineering education are a major reason, why the break-through so urgently needed is proceeding so awfully slow.

2. The Classification of Hardware Description Languages

HARDWARE DESCRIPTION LANGUAGES
R.W. Hartenstein (Editor)
© Elsevier Science Publishers B.V. (North-Holland), 1987

2. THE CLASSIFICATION OF HARDWARE DESCRIPTION LANGUAGES

Reiner W. HARTENSTEIN

Department of Informatics, University of Kaiserslautern
Postfach 3049, D-6750 Kaiserslautern, F.R.G.

Abstract

This paper gives a survey on languages and similar notations used for documentation and design of hardware, and, mixed hardware / software systems as well. It also introduces the terminology of the area, and, it tries to clarify, where the current terminology is confusing. Classification schemes of language domains are introduced with respect to abstraction levels, and, with respect to particular application areas.

1. Introduction

Several hundreds of languages for hardware-related applications have been published, and still about a dozen of them is added every year. Designing and publishing such languages seems to be a widely spread "Volks-Sport" hobby, obviously supported by a majority of editors and reviewers of conference proceedings, professional magazines and other publications. Although most of these languages do not survive, since not having been implemented, or, for some other reasons. When the candidate has received the Ph.D. degree, the "one man show" normally is shut down. Some papers, having been abstracted from the dissertation, and being submitted to various conferences and periodicals, are some sort of abortion dump. The scene is confusing, and almost each new language is another voice from the tower of Babylon.

There is a need for navigation aids to enter this area without getting lost. This paper tries to be such an aid to some extent by a clear presentation of terminology, and, by providing transparent classification schemes. It tries to be a guide in moving through the more than a dozen abstraction levels, and, though the various application areas of such languages.

However, before talking about language classification we have to solve another problem: here is no uniform use of the term '(computer) hardware description language'. The CHDL community being organized in groups like the IFIP working group 10.2 tends to use this term in a more special sense. It is another problem, that the terminology in the field is not used in an uniform manner.

PL   procedural language (programming language)

HL   hardware language

DHL   digital hardware language (HL for digital hardware only)

Procedural Languages are notations to express procedures or algorithms by means of control structure and data structure of desired or real sequences of actions or instructions. Hardware Languages (HLs) are graphic and/or textual notations, which may be used for the description of the structure, the behaviour, and the morphology (geometry and/or topology) of hardware at one or more abstraction levels. Since there are languages and applications (e.g. design procedures) which require the representation of procedures and their underlying hardware at the same time, I would like to introduce the term *HSL* for hardware / software languages.

HSL   hardware / software language (merged PL and HL)

DHSL   digital HSL (HSL for digital systems only)

In most cases DHSLs are used in the field of computer structures, which is a subarea of hardware engineering. The reason for this is, that the merging of PL and HL into the same description is a typical requirement for an application to processor-based digital systems. If such a language is specialized for an application to such processor-based systems only, we may give a different name to this class of languages:

CHL   computer hardware language

CHDL   computer hardware *design* language

(Instead of CHL also the term 'CHDL' (here: Computer Hardware *Descriptive* Language) is often used in literature.) A computer is a special kind of hardware, normally partitioned into processor, memory, and i/o interface, where the structure of the processor is characterized by its dichotomy into a control part and an operative part. That's why CHLs provide special language features to model this dichotomy. For more details see below.

DL   design language (multi-level language with refinement capability)

SL   specification language (descriptive language)

HDL   hardware design language (DL for hardware)

DHDL   digital hardware design language (HDL for digital hardware only)

The term 'Computer Hardware Language' (CHL) is only used in the field of digital hardware. It very often is used to focus on higher methodological levels, such as e. g. functional level, or, behavioural level. However, the term 'Hardware Description Language' (HDL, we propose: 'HL') has only a less widely accepted definition. So I would like to propose the following more general definition.

2. Language Classification Criteria

For the classification of the languages under consideration in this paper, there is quite a number of criteria. For classification of languages used for specification of hardware and hardware/software systems, and for manipulation of descriptions of such systems, there are the following four main ciriteria:

- the *level*   (abstraction level)
- the *area*   (application area)
- the *dimensions* of notation
- the *source medium* used

the methodological level or *abstraction level* for which the language has been designed (henceforth called *'level'*, also see figures 1 and 3 for illustration of level examples, for more details see chapter 3), its intended *application area* (henceforth called *'area'* , for more details see chapter 5), its notational *dimension,* which determines the general class of objects being subject of a description (see chapter 4), as well as the *source medium* having been used (graphic or textual). For more details see section 6.

There is about a dozen of levels (see chapter 3). Depending on its intended application a language may cover more than one particular level. Possible areas are: synthesis, analysis, specification, documentation and others. The dimensions of a language are determined by its capabilities of expressing structure, behaviour, and morphology (geometry and/or topology). For more details see chapter 4. The source medium may be a graphical (i. e. when a diagram language is used), or, textual (if strings of characters are used as a notation).

3. Abstraction Levels

In computer science and engineering much more abstraction levels are needed for an exhaustive description of systems and subsystems, than in any other scientific discipline. When we use a top-down strategy to walk through the levels, and, when we start at the world of computer applications, we encounter open networks, local area networks, multi processor systems, processors, control part and data part of a processor, etc. until finally we arrive at a level, where individual transistors are described in detail. More than a dozen of levels of abstraction may be identified all this way down. That's why it is quite useful to use a hierarchical classification scheme. We may distinguish the following classes of abstraction levels:

- transactional levels
- procedural levels
- semi procedural level
- non-procedural levels

At *non-procedural levels* of abstraction (for illustration see fig. 3) only the hardware is described. The term 'non-procedural' indicates, that software is not included in such descriptions, so that the specification of sequencing (control structures) is not subject of these levels. Non-procedural descriptions are mainly combinational. Also the assignment of values to registers may be subject of non-procedural descriptions. Also the data path network of a sequencer may be described at these levels, as far as no control structures are specified.

The *semi procedural level* is an intermediate level between procedural and non-procedural levels. To a description which is primarily a non-procedural one, the concept of the finite state machine and features to specify its state sequence are added. However, program-like procedural notations are not used at this level.

Procedural levels (for illustration see rows 2 thru 6 in fig. 1) are those levels, where hardware and software, or, hardware and control structures, respectively, are described simultaneously. During a top-down design of a multi processor system or a multi computer system we encouter the following levels of abstraction: the level of *concurrent processes* (showing their synchronisation and coordination, for example within a concurrent program running in parallel on several processors), the level where a single *sequential process* is described (for example by a sequential program running on one processor), the *micro program level,* where the implementation of a processor described algorithmically (for example by a micro program), and finally the *dichotomic level,* where the data part / control part cooperation within a processor is described in detail. The *'guarded micro program level'* may be used as a level for the representation of an intermediate result in designing a processor from a micro program level specification.

The *transactional levels* (network levels) describe communication relations and communication links between computers within a network. Subject of these high abstraction levels are: the network structure description, the description of communication channels, the communication protocols used, and the data structure of messages to be transmitted.

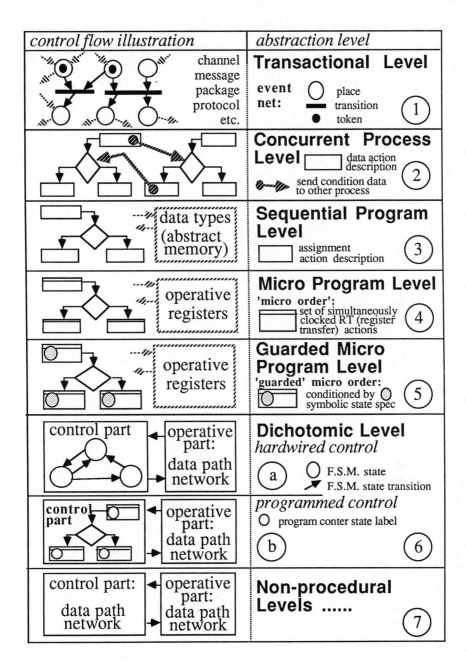

Fig. 1.

3.1   Higher Abstraction Levels

Those levels of abstraction, which are above the group of non-procedural levels, we would like to call *higher abstraction levels*. The scope of the higher levels of abstraction is illustrated by rows 1 thru 6 in figure 1. Higher levels of abstraction may be subdivided into the following subclasses:

- transactional level (network level)

- procedural levels
  - concurrent process level
  - sequential process level

- semi procedural level

At *procedural levels* the objects being described are typically treated like computer structures, where software (programs, which are *procedure descriptions*) runs on a processor (sequential process level), or, in parallel on several processors simultaneously (concurrent processes level). The model of the processor used is a generalized von-Neumann machine, suitable to model machines with CISC (complex instruction set), RISC (reduced instruction set), HISC (high level language instruction set), SISC (specialized instruction set), or other machines with von-Neumann-like sequencing and accessing. An interesting HISC machine example is the SAL machine [1].

All other higher levels (transactional level and semi procedural levels) do not include procedure descriptions. At transactional level the communication of networks of computers and computerised systems is modelled without showing the details of implementation via programs. At semi procedural level only *state sequences* of finite state controllers, or, *transition firing sequences* of event net modelled controllers are modelled, however, programs are not visible.

3.1.1   Transactional Levels

The *transactional levels* (network levels) describe communication relations and communication links between computers within a network. Subject of these high abstraction levels are: the network structure description, the description of communication channels, the communication protocols used and the data structure of messages to be transmitted. A universal hardware / software language at that level should specify the structure of networks, where a node could be a computer (processor, multi processor system, (local) computer network), or, a non-von-Neumann special hardware, or, a mixture of those. An arc could be any high level specification of a communication channel, such as for example a bus, a queue, or, a mail box memory. A module frame definition at this level should feature *procedural ports* including the specification of several levels of hand shake (bit hand shake, word hand shake, package level hand shake, etc.).

Some high level programming languages (such as for instance Ada) include some primitives and some other features which are useful for these purposes. However, in the field of hardware / software languages, only very few research has been carried on at this level of abstraction. One approach into this direction, for example, is the channel concept of the CVS_BK mixed level language of the CVS project within ESPRIT [2]. A relatively old example is the PMS notation [3], which, however, is only a structural notation, which does not specify any transactions nor other actions of communication. PMS describes nets of processors, memories, and 'switches' (which are I/O interfaces and other interfaces).

3.1.2   Procedural Levels

At transactional level network configurations and their communication relations are specified in terms of message formats and channels, and of protocols [31]. At this level no data processing algorithms are specified. The specification of algorithmic data processes is subject of the procedural levels. Programs and algorithms are typical examples of procedural notations. The term 'procedural' reminds to the fact, that in these levels it is shown how a process proceeds instruction step by instruction step. This means that at procedural levels primarily sequences of actions (desired actions or 'real' actions) are specified. At procedural levels we may distinguish two fundamentally different kinds of descriptions:

- procedure descriptions (software spec only)
- procedural hardware descriptions

To illustrate the difference between these two kinds of descriptions, let us consider the following two different kinds of problems:

- execute a program written for a high level language
  interpreter running on an existing target machine

- simulate the execution of a program during simulation
  of its target machine, which has not yet been implemented

In the first case a program, written in a high level programming language is only a procedure description. It does not specify any information about the resources it will be running on, since this kind of information is implemented (and hidden) inside the language interpreter used. In the latter case, however, we need two things to activate the program: we need the program itself, and, at the same time we also need an executable description of the target machine. This combination of program and hardware description I would like to call a *'procedural hardware description'*. But now let us come back to procedure descriptions: *procedural programming languages* are suitable notations to express these. At algorithmic level (above RT level) we may distinguish the following kinds of languages:

- concurrent programming languages
- sequential programming languages

Sequential programming languages are used to express a sequential program, i.e. a single sequence of instructions to be executed on a single computer. Traditional *sequential programming languages* are von-Neumann-type languages, since the abstract machine model behind such languages use a *processor,* some sort of abstract CPU (central processing unit), which can carry out only one action at a time. So these languages are called procedural languages, since execution proceeds step by step.

Concurrent programming languages are used to express at the same time several instruction execution processes, which are running concurrently. Compared to sequential programming languages, extra features are needed, such as e.g. for the coordination of concurrent processes. Such features, for example, are needed to make one process wait for completion of another process, or, to start a process by a particular instruction in another one. For example by a co-begin statement we may start several processes. By a co-end we can specifiy a rendezvous coordination, which means, that all these processes started at co-begin have to wait for completion of the slowest one of them (also see [32]). Traditional *concurrent programming languages* are notations to specify the coordinated actions of ensembles of cooperating von-Neumann-type machines.

Concurrent languages normally are extended sequential programming languages. Sequential and concurrent programming languages having been discussed so far are located in the group of *algorithmic levels* of abstraction. This will be sufficient for procedure descriptions. However, for procedural hardware descriptions, since being used as an input to a design process, a further refinement into a more detailed description is needed. So we get another level of abstraction, the *RT level,* which is below the algorithmic level. Our language classification scheme has to be extended:

- procedural languages
  - programming languages

    (algorithmic level)
  - micro programming languages

    (register transfer level (RT level))

A program describes sequences of actions to be carried out by a von Neumann device. It describes, *what the processor is doing*, such, that the processor appears to be a black box featuring a repertory of data manipulation operations. Under guidance by the control structure of the program the device scans the primary (data) memory in a read/modify/write mode. A program is an *algorithmic level* description, which describes an algorithm which is written in a language being accepted by a particular von-Neumann-like device.

A micro program, however, describes, *how the processor is doing it*. It shows, how a particular instruction of the processor has been implemented. The processor is no longer a black box. Instead, the data path network inside a processor is shown, along with the sequence of path activations needed to carry out a particular data manipulation operation of the processor. The execution of a single machine instruction, such as seen from a higher level of abstraction, has been broken down into a sequence of data transfers between temporary registers. That's why a micro program is also called an *RT level (register transfer level)* description: it shows the sequences of register-to-register transfers inside the processor. It shows, how the execution of an algorithmic level instruction is broken down into consecutive actions inside the processor. At this level instead of the processor's memory only its ports are visible, which communicate with the memory's ports.

In introducing concurrent programming languages we mentioned one type of parallelism, which we call *concurrent parallelism*. This kind of parallelism is typical for the algorithmic level. Like in racing each process proceeds independently in its own lane. Only at relatively few coordination points the progress depends on signals from other processes. At RT level, however, there is also another kind of parallelism, so that we now may distinguish two kinds of parallelism:

- concurrent parallelism
- synchronous parallelism

After having been started together, may be by a co-begin, concurrent processes may run independently and at individual clock speeds without any mutual interference for long sequences of instruction executions, until finally the rendezvous point is reached. In synchronous parallelism, however, a much more tight coordination is used: individual instructions are executed in parallel under a common clock and during the same clock cycle. Let us illustrate this by using an ISPS notation, where <action-1> ';' <action-2> means, that both actions are carried out at exactly the same clock cycle of the same clock, and, where <action-1> 'next' <action-2> means, that <action-2> is carried out one clock step later than <action-1>. The example is:

next
        <action-1> ; <action-2> ; <action-3> ; <action-4>
next
        <action-5> ; <action-6> ; <action-7> ; <action-8>
next

This example specifies, that the statements for actions no. 1 thru 4 are carried out simultaneously at time step no. n, whereas the statements for actions no. 5 thru 8 are carried out simultaneously during time step no. n+1.

The Guarded Micro Program is an important intermediate result during synthesis of processor-based hardware [3, 4]. In such a micro program state labels are needed not only for jumps, but for every statement. It helps to define the state assignment needed later at dichotomic level (also see next section).

3.1.3   Dichotomic Level (Semi Procedural Level)

In algorithmic level descriptions, i.e. in programs or micro programs, we may distinguish two parts: a declaration part and a statement part. The statement part defines the control flow and specifies the desired sequence of instructions, whereas the declaration part with its data type declarations is modelling some sort of structured data memory, where the values of variables used by the statement part are stored. The data paths inside the processor used are not visible, so that the processor is some sort of black box. The repertory of operations provided by the processor is specified only by the language accepted by it. So at algorithmic level only data processes are specified.

To proceed in a hardware / software design process this information is not sufficient. More details have to be created to provide a specification for designing the processor needed. That's why at next lower abstraction levels the resources are specified, which are capable to carry out programs. Instead of the statement part of a program a behavioural description of a controller is needed. Instead of a declaration part of a program the behavioural description of the operation part (the data manipulation unit) is needed. In the literature this level of abstraction sometimes is called 'behavioural level'. This is not a good terminology, since in all-day language 'behavioural' is a much more general term. We propose to reserve this term to characterize the dimensions of a description or language (see chapter 4).

Note: *'behavioural'* and *'structural'*

are **dimensions -**

but **no** (methodological) **levels of abstraction**

We prefer to call this level of abstraction the *dichotomic level,* since it clearly specifies a dichotomic hardware subsystem, divided into a *control part* and an *operative part*. This level is no more procedural in a way, such as known from programs. The instruction part has been replaced by lower level notations. Fig. 2 a and d (von-Neumann machine or 'Princeton machine'), and b (Harvard machine) illustrate this familiar partitioning scheme from a global point of view. Fig. 2 c shows the usual partitioning of the processor itself (for more details also see [5-8]).

We could say, that the dichotomic level is of *semi procedural* nature. In my opinion, also the term 'semi procedural' would be better than 'behavioural', since at least parts of descriptions of this level (e.g. the operative part) are described by non-procedural notations. At this semi procedural level we may distinguish two different kinds of implementations of the control part:

- hardwired control
- programmed control

If *programmed control* is used (see fig. 2 f), the control part may be specified by the non-procedural description of the program store and the data paths inside the sequencer. To describe a processor with *hardwired control* normally a finite state machine model primitive is used (fig. 2 e), such as provided by a number of hardware languages.

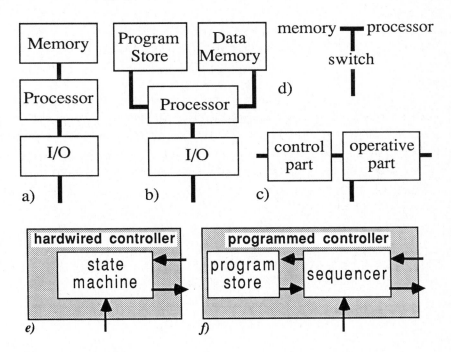

Fig. 2

3.2   Non-Procedural Abstraction Levels

Procedural levels have been subject of section 3.1. Now we are going to explain non-procedural
levels. First let us identify the most important difference between these two kinds of levels.
Procedures describe step-by-step behaviour, so that the time sequence of the execution of instructions
primarily depends on the spatial sequence of these instructions within the procedure text (procedures
are von-Neumann-type).

The semantics of non-procedural objects is quite different. In 'executing' (e.g. simulating) a
non-procedural description, the order of execution follows a fundamentally different pattern: During
each execution cycle ('label cycle') always every statement is scanned to find out, whether it should be
executed now. More precisely: each instruction has a label (like 'guarded commands' proposed by
Dijkstra) and each label expression (referencing clocks, enable signals etc.) is checked for its truth
value during each label cycle. Whenever the label is true, the instruction will be executed. All those
descriptions are called 'non-procedural' ones, which are at lower levels, than procedural descriptions
(see section 3.1.2). Possible non-procedural abstraction levels of HDL use may be, but are not
restricted to:

- functional level
- gate level
- switching level
- circuit level
- symbolic layout level
- physical layout level

Fig. 3 gives a survey, which illustrates these levels by showing a simple description example and a
repertory of elements used in each particular level. The same circuit example used for all levels: it is a
two input multiplexer. Rows d thru f show, that this example is implemented in using depletion load
silicon gate NMOS technology.

Physical layout level (compare row f in fig. 3). At this level the real geometry of the layout of
materials of integrated circuits or other forms of hardware implementation is specified. Fig. 3 f shows
a simple example of graphical layout representation. Many textual layout formats are in use, such as
the CIF format [9], which is used as an interchange format. CIF stands for 'Caltech Intermediate
Form'.
The layout specification includes the shape, size and material of rectangular boxes, polygons and other
geometrical objects. Such a layout is a purely morphological description. Layout data may also be
organized hierarchically, such that its modules (called 'cells') may be partitioned into subcells, and,
may be part of a higher level cell (of its mother cell). When such a hierarchical organisation is used, a
layout is a mix of morphological and structural description. Layout does not include any behavioural

specifications, so that it cannot be used directly by any kind of simulator.

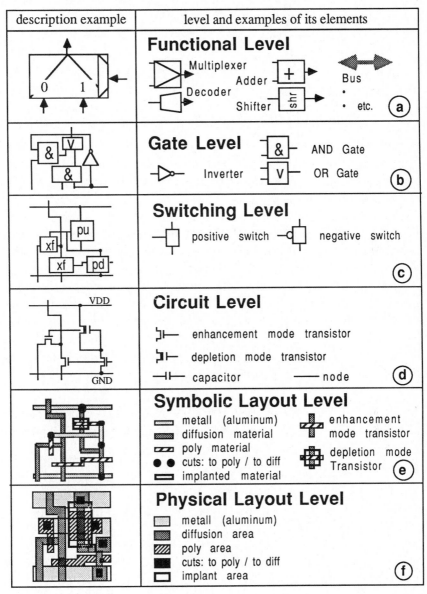

Fig. 3.

Symbolic layout level (see row e in fig. 3). Symbolic layout is an abstraction of physical layout, such that geometry has been stripped off. At this level layout is represented in a way, that it does not specify physical geometry. A rectangular box, for instance, may be presented by a thin line only, the colour or shading of which may indicate the material. That's why graphical presentations of symbolic layout are sometimes called *stick diagrams*. This sticks notation has been invented and implemented at first time by the M.I.T. For an introduction to stick diagram use see [9].

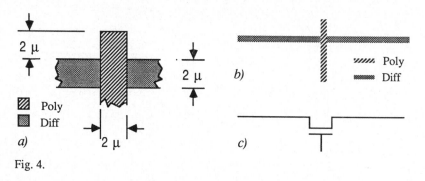

Fig. 4.

For comparison fig. 4 shows an NMOS transistor at 3 different levels of abstraction: fig. a shows the *layout* of the transistor, fig. b shows its *stick diagram,* and, fig. c shows its graphic symbol used at *circuit level*. Row e in fig. 3 shows another simple example of a stick diagram (a multiplexer). The 'sticks' are a little more thick than 'thin lines', so that it is more easy to show 'shaded sticks'. By symbolic layout only the relative placement of different objects (relative to each other) is shown. However, the physical (i.e. geometrical) distance between these objects is not specified. Also symbolic layout data may be organized hierarchically, so that it is a mix of topological and structural description. Symbolic layout directly does not include any behavioural descriptive elements, so that it cannot be used directly by any kind of simulator.

Circuit level (illustrated by row d in fig. 3). The notation at this level is the circuit diagram, which shows the components (transistors, resistors, capacitors, etc.) and the interconnection between these

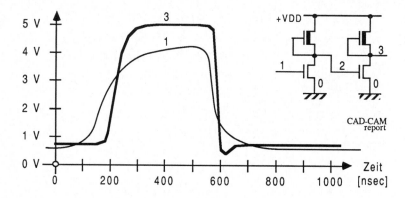

Fig. 5.

elements. That's why circuit diagrams are structural descriptions. Fig. 3 shows the graphic version of a simple circuit diagram example (a two-input NMOS multiplexer). The textual version of a circuit diagram is called a circuit level net list. At circuit level signal values are analog values. Fig. 5 shows the waveforms resulting from using a ciruit simulator for a simple circuit example. Many analog domain details are visible, such as on-voltage, off-voltage, rise time, fall time, and even the little undershoot at the output signal.

A circuit diagram is not only a structural description. At the same time it is a model of the analog behaviour of the circuit. That's why circuit simulators, such as SPICE [10] use such circuit level net list input. In some sort of declaration part of such a format the model parameters of the individual transistors are specified, so that a net list specifies the interconnection between single transistor models. That's why a net list is a structural and behavioural description at the same time. Also hierarchical circuit diagrams and hierarchical net list formats are known, where each circuit or subcircuit may be decomposed into subcircuits. For instance, the most recent versions of SPICE and other circuit simulators accept hierarchical net lists.

Switching level (compare row c in fig. 3). At switching level, and, at all levels above it, only digital signal values are distinguished, such as, for example, {0, 1}, {on, off}, {high, low}, or

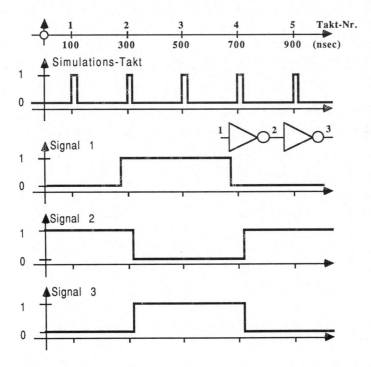

Fig. 6.

others. Fig. 6 illustrates this by showing the resulting waveforms of a simple logic simulation example. At switching level the same net list is used, which is known from circuit level. However, a much more simple transistor model is used, which is sufficient, when only the digital behaviour of a system or subsystem is modelled. Instead of a precise analog model only a simple switch model is used for a transistor. For a transistor this model distinguishes an 'on state' (low impedance) and an 'off state' (high impedance). Transistors with different low impedance values may be distinguished, so that switches may have different degrees of softness. Fig. 3 (c) shows a simple switching level diagram example using a pull-up transistor (pu), a pull-down transistor (pd), and, two pass transistors (xf). A number of switch level simulators has been implemented, such as MOSSIM [11], and others. Meanwhile the switch level is beginning to become a scientific discipline, sometimes also called CSA level [12, 13], since the elements of its formal methods are Connections, Switches, and Attenuators.

Gate level (see row b in fig. 3). This level is the abstraction level, where logic design is carried out. Its notations are net lists of gates and flipflops. Fig. 3 b shows the graphic presentation, i.e. the logic diagram, of a simple example with two AND gates, an OR gate, and an inverter). A set of Boolean equations may be interpreted as a special form of a gate level net list. Gate level descriptions are structural and behavioural at the same time. A number of logic simulators has been implemented [14]. Normally also timing simulation is performed at gate level by using so-called logic-and-timing simulators [15].

Level of Abstraction	Degree of Detailedness	Kind of Description	Complexity for Example	Primitives
RT Level	lowest	RT Level Specification	ca. 100	RT elements
		refined RT Description	ca. 500	RT elements
Gate Level		Logic Diagram	ca. 10.000	gates
Circuit Level		Circuit Schematics	ca. 30.000	transistors
Physical Layout	highest	Layout	ca. 200.000	rectangles

Fig. 7.

Functional level (row a in fig. 3). For highly complex circuits, such as VLSI circuits, very often logic diagrams are too complex for comprehensibility by human beings. An abstraction level above gate level, called 'functional level', may result in much less complex descriptions by using much more powerful primitive operators, such as (arithmetic:) '+', '-', '*', '/', mod, rem, (relational:) '>', '<',

'=', '≠', '≤', '≥', ':=', '.=', (miscellaneous:) <u>decode</u>, <u>encode</u>, <u>priority</u>, <u>msb</u>, <u>lsb</u>, <u>msw</u>, <u>lsw</u>, etc., (for clocking expressions:) <u>at</u>, <u>on</u>, <u>while</u>, (for multiplexers:) <u>if</u>, <u>then</u>, <u>else</u>, <u>case</u>, (for wiring patterns:) <u>shift</u>, <u>circulate</u>, <u>fold</u>, <u>merge</u>, <u>shuffle</u>, <u>butterfly</u>, <u>reflect</u>, and others.

Not only more powerful functions result in less complex descriptions, but also the bundling of bits into words, and, the bundling of single bit signal paths into word-formatted data paths helps a lot in reduction of notational complexity. Fig. 7 illustrates possible benefits in notational complexity when migrating a non-procedural description up to functional level. Data path formats are specified by declarations, which look similar to type declarations in programming languages. Also the mnemonics and part of the syntax (especially the expression syntax) of some functional hardware languages very much look like those of high level programming languages. However, functional hardware languages are non-procedural, whereas programming languages are procedural languages.

Mixed level descriptions. In many cases it would be more convenient to describe hardware at several abstraction levels, than in a single level only (simple mixed level description examples are intoduced in [9], for example). Sometimes mixed level descriptions of circuit principles are more comprehensible, than equivalent single level descriptions. In attempts to describe a circuit at a high level, such as for example at functional level, sometimes there are 'remainders' left over, which can be described only at a lower abstraction level, such as, for instance, at gate level. That's the reason, why most implementations of functional languages also include gate level primitives, and, sometimes also switch level primitives.

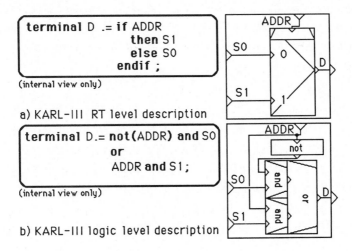

Fig. 8. a) - b)

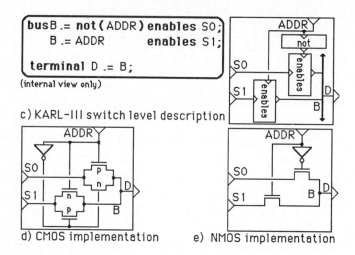

c) KARL-III switch level description

d) CMOS implementation   e) NMOS implementation

Fig. 8. c) - e)

An example is KARL-III [16], which is a multi level language covering functional level, gate level, and switching level. Normally design languages need to be mixed level languages, so that a top-down design by successive refinement can be carried out all the way down without switching over to another language. Fig. 8 shows an example, represented in KARL and its graphic version ABL (except figures d and e): a mulitplexer at functional level (fig. a), at gate level (fig. b), at switching level (fig. c), and circuit level implementations in CMOS technology (fig. d) and NMOS technology (fig. e). The ABL-like diagram is shown as a mixed-level representation (gate level and circuit level). ABL [17, 18] is the graphic version of the textual non-procedural language KARL-3, such that the scope of ABL is slightly extended, compared to that of KARL.

3.2.1   Using Procedural Languages for Non-Procedural Descriptions

Although programming languages are normally procedural languages, also hardware at non-procedural levels may be described in using such languages. More than once there have been attempts and proposals (e.g. [19]) to use programming languages for hardware description. Also the design of (in this case: so-called) hardware languages, which have very close similarity to programming languages (e.g. [20, 21]), may belong into that category. However, this use of such languages is rather inefficient. A few (but not exhaustive) experiments we made have indicated, that by using Pascal instead of KARL to describe a clocked LSI circuit, about 3 to 5 times as many lines of code are needed.

Fig. 9 illustrates the efficiency relationship between the level of description (target level) and the level of the notation used for it (language level). The larger the gap between target level and language level, the more sophisticated the description will be, no matter, whether the language level is too high, or, too low. If the language level is too high, low level primitives needed are missing. It is very difficult to describe low level constructs in using high level primitives, since sometimes a lot of artefacts have to be formulated to replace a single missing primitive. It also is inefficient to use a language having a level which is too low. In such a case missing high level primitives force the user to replace them by extensive complexes based on low level primitives. This is like being forced to implement subsystems by yourself instead of picking them from stock. A data path network may be described much more concisely, if a functional language is used instead of using a logic diagram or a circuit diagram.

There is also another good reason, not to use programming languages for hardware description. Such notations are harder to read, since the procedural semantics obscure of the intended non-procedural meanings of the description. This is the consequence of the intended misinterpretation resulting from the unusual language application: misinterpretation has been made principle of notation. For similar reasons programming languages also are in most cases no suitable notations for source input to silicon compilers and other hardware synthesis algorithms. Exceptions are generators for strictly processor-based solutions.

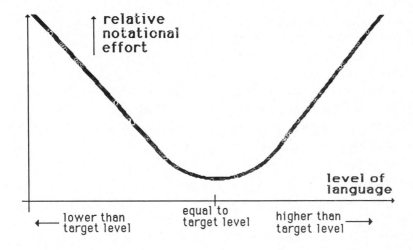

Fig. 9.

Also for integrated hardware / software design (compare section 3.1.2), even where procedural notations are needed, using a programming language is not a good idea. For such kinds of design problems a HSL is much better to use, although a HSL sometimes may be slightly more complex, than a high level programming language. Such a HSL needed here is a multi-paradigm multi-level language, where procedural paradigms and non-procedural paradigms are well interfaced to each other, and where the block structure is well taylored for the partitioning requirements of processor-oriented hardware / software design. A recent example is the CVS_BK language [2], which is a KARL-3 extension featuring strict upward compatibility. Multi-level languages also provide more freedom in minimization of notational effort, than single-level languages. There is no danger that such languages become too baroque.

3.3   Levels of Innovative Computer Structures

Computer structures are special structures - a subset of digital hardware structures. The area of computer structures deals with sequential algorithms and their implementation on top of a von-Neumann-like processor, or, on top of ensembles of von-Neumann-like processors. A typical and principal characteristic of computer structures is the dichotomy of control structures and data structures.

Most solutions having been implemented on computer structures can be converted into other implementations which do not use any von-Neumann mechanisms. For example, instead of a sequential algorithm, a *VLSI algorithm* may be used. Quite a number of sorter chips have been proposed or designed, which replace sequential sorting algorithms, and, which do not need any program code for operation. One example is the *Shuffle Sort* algorithm [22], which is a VLSI algorithm having been derived from the Bubble Sort algorithm (a sequential algorithm). Within this implementation there is no sequencer, except the clock. Since procedure descriptions are no more needed, the sequence of abstraction levels is different from those of computer structures:

- • transactional level
- (•) *normally not needed:* procedural levels
- (•) *normally not needed:* semi procedural level
- • non-procedural levels

Sometimes it is useful to have a uniform partitioning for both, (von-Neumann) computer structures, and, non-von-Neumann structures. So the non-von-Neumann may have trivial implementations of familiar blocks: e.g. a sequencer, which contains nothing else but a clock, an operative part, which is extremely simple, a memory which contains only one or two registers, a 'switch' (the term used here like in PMS), which contains only wires, or a fragment of a bus. This philosophy is explained in more detail in [5].

4.   Dimensions of Notation

Except the level of abstraction, there are other important characteristics of descriptions, which are rather independent of the abstraction level(s) of its application. Those important characteristics are the dimensions of the information it conveys. The three possible dimensions of information are the following:

- structural information

- behavioural information

- morphologic information

These three dimensions are illustrated by the Gajski diagram (fig. 10 a): a program macro (no hierarchy) is a behavioural notation, a schematics entry system only handles structural information, whereas physical layout is only geometric (which is morphologic). Fig. 10 b illustrates, that KARL-3 (and its graphic version ABL) is a three-dimensional language. With respect to these dimensions, a notation or description may be one-dimensional (such as e.g. the formats of schematics entry systems), two-dimensional (such as, for example, structural and behavioural at the same time), or, three-dimensional. The following sections explain these three dimensions in detail.

4.1 Behaviour

The behaviour of a system, subsystem, or component is characterized by its input / output relationship, or, (what is almost the same) by its stimuli / responses relations. The way, how behaviour is expressed the most important ingredient of a description, in order to determine, to which abstraction level it belongs to. The form of behavioural information is very much dependent on the

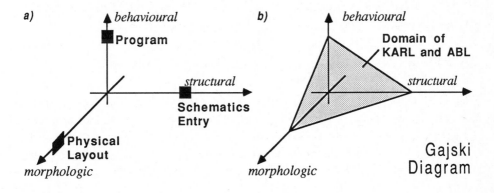

Fig. 10.

abstraction level, whereas structure and morphology are relatively level-independent.So we may distinguish the following forms of behaviour (also compare chapter 3):

- Procedural Behaviour

  - Parallel Behaviour
    - synchronous (synchronously parallel actions)
    - concurrent: (asynchronously parallel procedures)
  - Sequential Behaviour

- Semi-Procedural Behaviour

  - dichotomic behaviour (RT behaviour including state sequences)

- Non-Procedural Behaviour

  - combinational behaviour (combinational switching networks)
  - RT behaviour (data path networks including registers)

These forms of behaviour have been used as classification criteria, so that subclasses of levels of abstraction are formed. Notations for concurrent and *sequential behaviourr* are, what we find in programming languages. Notations for *synchronously parallel behaviour* are found in mirco-programming languages. The term *dichotomic behaviour* is, what we propose to use instead of the traditional, but misleading term 'behavioural level'. Some authors also use the traditional term 'CHDL' (Computer hardware description language), or, 'behavioural CHDL' for those languages, which cover this level.

4.2   Structure

Structural description is one of three dimensions of information in a general description of complex systems. (The three kinds altogether are: structure, behaviour, and morphology.) The form of the structural ingredient of a given description at a particular level of abstraction is almost independant of this level. This will be explained later. Structural descriptions may be subdivided into the descriptions of:

- Partitioning
- Interfacing
- Interconnect

Partitioning defines how a given system, subsystem, or module is broken up into submodules.

Interfacing specifies the external view of a module, which defines the arrangements of ports, prepared for its connection to the module's outside world. Interconnect description specifies the connections between different objects of a structural description, such as e.g. wiring patterns. Let us look more closely at these three kinds of specifications.

Partitioning. Partitioning defines a hierarchical structure: it defines how a given system, subsystem, or module is broken up into submodules. In the area of VLSI design such modules are often called *cells*. Each submodule or cell again may be partitioned, so that we may have a module hierarchy or cell hierarchy with many nesting levels. This hierarchy we will call a *structural hierarchy* for easy distinction from other kinds of hierarchy. Those cells, which are not broken further down into subcells are called *leaf cells*, since in the tree diagram illustrating a hierarchy those cells are found as leaves of the tree.

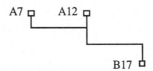

Fig. 11.

Interfacing. Interfacing specifies the *external view* of a module, which defines the arrangements of ports, prepared for its connection to the module's outside world. It specifies name (and / or pin number), format, and location of the ports of a given module. If the module is a cell, also the topology of the set of all ports is an important part of interfacing information: in case of a rectangular cell it specifies, at which of the four sides of the cell a particular port is located. It also specifies the relative location of a port within the row of ports at a particular side of the cell.

Interconnect. Let us start with a few definitions. A *net* is a single wiring pattern which directly connects two or more ports or other 'pins', and, which carries the same signal. Fig. 11 illustrates a simple example of a net. A *net list* is a list, which specifies all nets of a given routing area or similar wiring pattern. Interconnect wiring may be specified by net lists, or by graphic notations, between modules and components at same nesting level, between a module's peripheral ports and its internal modules and components. (To avoid confusion, a set of several nets should be called a *network*. It never should be called 'net'.) If wiring by abutment is used to interconnect two adjacent cells, no routing area is needed: the interconnect pattern is defined implicitly and no net list is needed. A structural module description which completely shows the modules peripheral ports, its internal components and submodules, as well as all its internal nets, is called the *internal view* of the module.

Structural ingredients of descriptions are almost independent of the current level of abstraction. Only a very limited number of changes are possible, during migration from one level of abstraction to another one. One difference could be for instance, whether bits are named individually, or, whether bits are bundled to words of predeclared size and format. (This bundling of bits often happens, during a translation of a gate level description into a functional level description.) The relative stability of structural descriptions in migrations through levels of abstraction may help to simplify the implementation of multi level interactive graphic schematics editors such as the MLED editor. We get almost complete orthogonality between two dimensions of hierarchy: between the hierarchy of levels of abstraction, and, the structural hierarchy (the hierarchy of nesting levels).

4.3   Morphology

If a graphic notation is used for a description, or, if physical geometry or physical topology has to be conveyed by a description, we need - in addition to structure and behaviour - a third kind of ingredient: morphology information to specify the following properties of objects:

- Shape and Colour

- Exact (geometric) Placement

- Relative Placement (Topology)

We here use the term 'morphology' to talk about geometry **and** topology. Topologic descriptions are sometimes abstractions of geometric information. For example, symbolic layout may be the topology-only abstraction of physical layout (see section 3.2). We have to distinguish symbolic morphology from physical morphology. If, in a graphic presentation, the shape and location of objects have only the purpose of setting up a nice drawing, the geometry is called *symbolic geometry*. If, however, the geometric specification conveys information about physical objects, the geometry is called *physical geometry.*

Of course at physical layout level and symbolic layout level descriptions without morphology information would not make sense, since geometry or topology is the most important subject of descriptions at these levels. This morpologic information is also found in textual versions of such descriptions. However, also at higher levels of abstraction morphologic information sometimes is important - also in textual descriptions. To support structured VLSI design approaches it is very useful to convey topologic information also at functional level.

4.4   Structural versus behavioural descriptions

To prepare a simulation means to model the object to be simulated in terms of the primitives of the language used. In modelling we may observe, that behavioural descriptions sometimes very much look like structural descriptions. Before explaining this phenomenon let us define the term of *Modelling*. When we have a language implementation which includes a simulator (like KARL [16], for instance), then the semantics of this language is based on a set of *primitive models* implemented inside the simulator. If, for instance, the '+' function is referenced, inside the simulator a primitive model of '+' is activated, which models the behaviour of '+'. For each language primitive, such a primitive model is inside the simulator.

When, for simulation, we set up a description of a complex module, using such primitives of a description language, we are *modelling* this complex module such, that we construct a network of the primitive models available. The primitive models are our *modelling components*. When this construct is completed, it (hopefully) is an executable *model* of our complex system, which, during simulation, should behave like the module. We have written a behavioural description, which is translated into a behavioural model.

To construct this model we have been forced to use structural language elements to establish links between these primitive models. That's why behavioural descriptions look like structural descriptions. What we have done is, that we have used a *pseudo-structural* notation. That's why behavioural descriptions do not look like black boxes. That's why G. Jack Lipovski talked about *grey box* descriptions. This is surprising, since we once have learnt, that a behaviorual description be a black box description. But it is a joke: our grey box is only a model, but normally it does not show the real structure of the physical module we are going to design. So it is a *black box,* with respect to the physical realisation, if the model's structure is not isomorphic to the structure of the real physical module. If the latter would be the case, then the behavioural description would be equivalent to the structural description.

5. The Source Medium of a Language

An important characteristic of a description or a language is it, whether we use a *textual* form, or, a *graphic* form for presentation. We would like to call this property the type of *source medium* (or 'presentation medium'). If a language is used for man / machine interaction, such as using a functional language in designing a piece of hardware, a graphic version of the language would substantially increase the designer's productivity, compared to its textual version. Experiences with KARL (a non-procedural textual multi-level language) and its graphic version ABL [17, 18] report a productivity improvement of about a factor of five [23, 24].

One reason for this success of interactive graphic language implementations is based on the fact, that the visual data input channel of human beings is much faster, much more powerful, and much more intelligent, than his (or her) text string input channel. But in the field of VLSI design, this is not the only reason. Because of the shortage of planar chip surface needed, topologic and geometric problems are much more important than in classical methods of hardware design. Of course physical layout in textual form is completely incomprehensible to human beings, so that check plots cannot be avoided. However, also at much higher levels of abstraction graphic presentations are inevitable.

The design style called *structured VLSI design* successfully advocates chip floor planning considerations already at very early phases of the design process [25]. This design style requires experiments with different alternatives of combined behavioural / topologic descriptions of key cells and cell arrays. Key cells are those cells, which are essential for an optimum structured design concept solution. A very important method is the application of *wiring by abutment* to eliminate the need for routing areas, whenever possible. We will try to illustrate this: Fig. 12 c shows the external view of a full adder slice FA. We are going to abut horizontally two instants FA&1 and FA&2 (fig. 12 d, the '&' separates slice name from its instant number). The result is seen in fig. 12 f. This has been an example of horizontal abutment. Figure 12 g illustrates vertical abutment.

The use of abutment has a substantial impact on graphic notations. Traditional structural notations, which are not suitable to present wiring by abutment, show a net between the ports to be connected (see fig. 11 and right side of figure 13). New notational ideas are needed, since the ports to be interconnected overlap in presentation (see fig. 12 f, m, and p). The left side of fig. 13 illustrates the *Domino Notation* [26] having been developed for this within the ABL language. We use little triangles or rhombs to show the port and its direction of data flow (fig. 12 b). One problem during an interactive design process using Domino notation is, that the interconnect is invisible (you do not see, whether there is one, or not). Missing interconnect could happen, if a port is available only at one of the abutting cells. Also this missing of a mate to this port is invisible because of the overlap in presentation. In ABLED this problem is solved by turning the port from white to black, as soon as the interconect is effective. If it does not turn black in abutment, then this indicates a port mismatch.

The need to present cell hierarchies has an impact on the Domino notation. Fig.12 h illustrates the traditional way to show interconnect between a mother cell and its internal daughter cell. Fig. 12 i shows the Domino notation version of this structure, which reflects the topology of such a cell hierarchy. KARL-3 which is the textual version of ABL, has a textual version of the Domino notation. Fig. 12 n shows the text for the abutment of two instants (fig. 12 m) of a cell X (declaration: fig. 12 l). This text also presents mother / daughter abutment (the mother is unnamed). Ports of the mother (in parentheses) are: L, R, and B. Fig. 12 q shows another example of a KARL-3 make expression. The equivalent Domino notation is found in fig. 12 p.

6.   Application Area and Language Scope

Many languages are highly goal-oriented and the question for the class of the language has never been asked. Such languages are often implementations of a philosophy or methodology for synthesis,

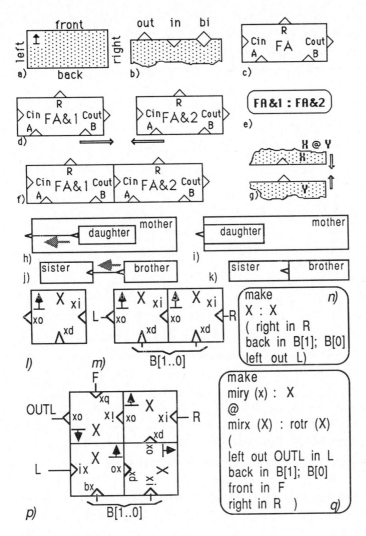

Fig. 12.

analysis, extraction, simulation, deductive verification, specification, design problem capture, schematics capture, and other applications. From application requirements we may distinguish single-level applications (specification, rules check, optimization, single-level simulation, timing verification, model description, activation pattern description, etc.), and multi-level applications. The latter ones are required for tools translating a description from one level into another one (e.g. synthesis, extraction, deductive verification, modelling), if not a separate language is provided for input (source language) and output (object language). So we have the alternative between multi-level language and a family of separate languages. The optimum choice sometimes also depends on the intended application.

Let us illustrate this by looking at simulation systems. In the source languages of circuit simulators, such as e.g. SPICE, many applications are mixed into a single language notation for a single string of source text. If the ingredients are separated, we may isolate a *NDL (net description language)* which describes the interconnect of transistors and other circuit level components, a *MDL (model definiton*

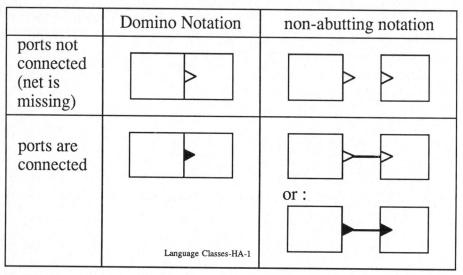

	Domino Notation	non-abutting notation
ports not connected (net is missing)		
ports are connected	Language Classes-HA-1	or :

Fig. 13.

language) to describe the characteristics of transistors and other components, as well as an *activation description language (ADL)* to describe abstract stimuli generators, their waveform and pulse shapes, and connection to the circuit under simulation. ADL descriptions also define simulator output, such as e.g. the response signals of which nodes will be outputted for which period of time.

We may distinguish different types of ADLs. Also a *test description language (TDL)* is an ADL, since it describes the activation patterns (test patterns) and the desired responses (output) of a piece of real hardware. The main difference between a hardware ADL (a TDL) and a simulator ADL is, that a simulator has some extra features which real hardware does not have. The cooperation between a simulator and a simulation model is similar to the cooperation between a testing device (Tester) and the device unter test (DUT) being tested by this tester. This is the reason why a simulator ADL is a combination of a hardware ADL and a *Tester Control Language (TCL)*. A TDL is a subset of a TCL, since command features to set-up and control the tester operation and output have to be added to the test patterns.

CAD tools using several sublanguages may be organized in two different ways. All sublanguages may be mixed into one compound language, fed into a single input of the system. An example is SPICE, where hardware description, model parameters, stimuli description and simulator commands are mixed into a single input string (fig. 14 a). However, also different organisation is possible, which uses separate languages, accepted by separate inputs of the system, and/or generated by separate modules using separate outputs of the system. An example is the KARL system which uses 3 different input languages (fig. 14 b): a HDL description language KARL (for the source input of the KARL compiler), an ADL/TCL simulator activation language SCIL [27] which may be used in batch mode or interactively to activate the simulator, as well as an intermediate form HDL language RTcode [28] which is the executable code generated by the KARL compiler. Fig. 14 shows the difference between both types of systems.

7. A Few Language Examples

Figure 15 illustrates the location of a number of languages and similar descriptive notations within our classification schemes, especially within our levels of abstraction. There are single-level languages as well as multi-level languages, which cover more than one level. For instance, most RT languages also include the boolean primitives of the gate level. However, also baroque languages are possible, which cover several levels. Pascal is a single level language, covering sequential procedures only. Modula-2

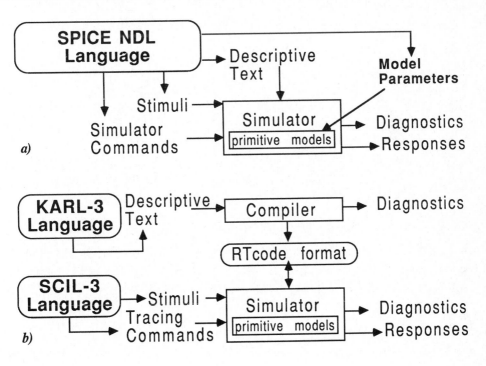

Fig. 14.

is a two-level language, since covering sequential and concurrent processes. Ada is a three-level language, since in addition it also covers - at least partially - the transactional level.

Now let us look at hardware languages. Sticks and CIF (more data formats than languages) are single-level notations: geometric, or, topologic, respectively. The SPICE NDL (SPICE net definition language) is a four-level language, since it expresses transistor area size and shape (layout level) up to structural information at circuit level and switch level. KARL is as three-level language covering switch level, gate level, and functional level. CVS_BK being implemented for the ESPRIT project is a KARL extension into semi-procedural and procedural levels. It has been a surprise, that CVS_BK is not much more complex than KARL-3, although covering much more abstraction levels: its grammar has only about 30% more production rules. It is still much more simple than Ada.

This is a proof which indicates, that multi-level languages are not necessarily baroque languages. ISP [29, 30] is a typical processor-oriented hardware language. That's why it stresses procedural levels. DDL and AHPL controller specifications are primarily restricted to finite state notations, such that really procedural levels are excluded. To our surprise the classification scheme revealed, that the most recent version (7.2, December 1986) of VHDL, the so-called 'hardware description language' of the DoD, is no hardware language at all: it is a procedural / transactional language (see fig. 15).

8. Summary and Acknowledgements

We have clarified terminology and have set up new language classification criteria for hardware languages and hardware / software languages: levels of abstraction, dimension of notation, source medium, and application area. Based on these criteria we presented in detail a new improved version of the definitions of levels of abstraction from transactional level over procedural levels and semi-procedural levels down to non-procedural levels. By means of a few examples we also discussed the language space defined by the four sets of classification criteria. The scheme classifies the area of programming languages as a subset of the area of digital system description languages.

We found the following three dimensions of description (fig.15 a): morphologic, structural, behavioural, where morphologic descriptions distinguish geometric and topologic descriptions. Descriptions or languages may also have mixes of 2 or 3 of these dimensions. Behavioural descriptions may present procedural behaviour and / or non-procedural behaviour. Procedures may be sequential, concurrent (asynchronously parallel), or synchronous (clocked parallelism). Descriptions may be physical (presenting real systems), or symbolic (presenting models) (see fig. 16 b).

Part of the work reported here has been partially funded by the Commission of the European Communities. The work on KARL has been partially funded within the CVT project. The work on CVS_BK is currently being partially funded as part of the CVS project within ESPRIT. Some other

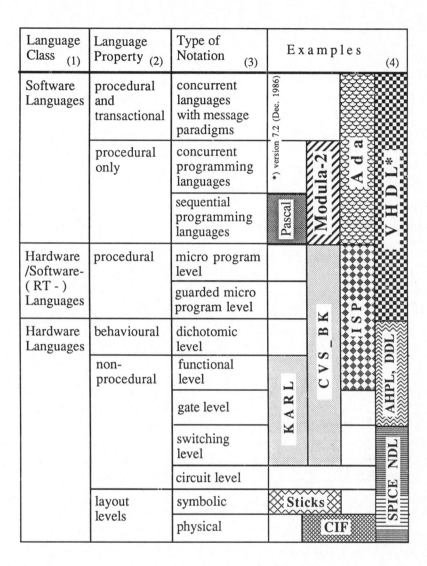

Language Class (1)	Language Property (2)	Type of Notation (3)	Examples (4)
Software Languages	procedural and transactional	concurrent languages with message paradigms	
	procedural only	concurrent programming languages	
		sequential programming languages	
Hardware /Software- (RT -) Languages	procedural	micro program level	
		guarded micro program level	
Hardware Languages	behavioural	dichotomic level	
	non- procedural	functional level	
		gate level	
		switching level	
		circuit level	
	layout levels	symbolic	
		physical	

Fig. 15.

KARL-related work (e.g. automatic extraction of KARL-descriptions from Layout) has been funded by the German Minister of Research and Technology within the Multi University E.I.S. Project. We gratiously acknowledge the fruitful cooperation especially with Guglielmo Girardi and his crew (CSELT, Torino, Italy) and with colleagues from the VLSI group at GMD (Gesellschaft für Mathematik und Datenverarbeitung, Birlinghoven near Bonn), as well as with a number of KARL users throughout Europe. Without these activities and without ideas from the colleagues involved it would have been impossible to develop this classification scheme for languages.

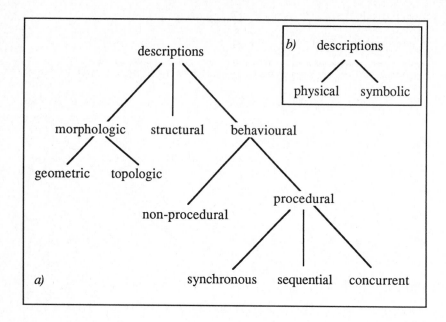

Fig. 16.

9. Literature

[1] S. Wang: On the Direct Implementation of Algorithmic Scientific Computer languages; Ph.D. dissertation; University of Minnesota, 1968

[2] A. Bonomo, G. Girardi, R. Hartenstein, R. Hauck, L. Lavagno: Specification of CVS_BK (CVS Behavioural KARL); CVS report; Kaiserslautern / Torino; March 1987

[3] C. G. Bell, A. Newell: Computer Structures: Readings and Examples; MacGraw-Hill, New York, 1971

[4] J. A. Nestor, D. E. Thomas: Behavioural Synthesis with Interfaces; Proc. IEEE ICCD, 1986

[5] R. Hartenstein: Computer Structure Partitioning Schemes; rep. FB Informatik, Kaiserslautern, F.R.G., 1987

[6] R. Hartenstein: Hierarchies of Interpreters for Modelling Complex Digital Systems; Proc. GI 3rd Ann. Conference, Hamburg, 1973; Springer-Verlag, Berlin / Heidelberg / New York, 1973

[7] R. Hartenstein: Increasing Hardware Complexity - A Challenge to Computer Architecture Education; ISCA 1 (1973); ACM, New York, 1973

[8] R. Hartenstein: Microprogramming Concepts - a Step towards Structured Hardware Design; Micro 7 (1974), ACM New York, 1974

[9] C. Mead, L. Conway: Introduction to VLSI Systems; Addison-Wesley; 1986

[10] L. W. Nagel, D. O. Pederson: SPICE (Simulation Program with Integrated Circuit Emphasis), University of California, Berkeley, Electronics Research Laboratory; Memorandum ERL-M 382, April 12, 1973

[11] R. E. Bryant: MOSSIM: A Switch-Level Simulator for MOS-LSI; Proc. 18th Design Automation Conf., July 1981, pp 786-790

[12] J. P. Hayes: Pseudo-Boolean Logic Circuits; IEEE Tans. C-32 (July 86), pp. 602-612

[13] J. P. Hayes: A Unified Switching Theory with Applications to VLSI; Proc. IEEE 70 (Oct. 1982), pp 1140-1152

[14] S. A. Szygenda: TEGAS - Anatomy of a General Purpose Test Generation and Simulation System for Digital Logic; Proc. 10th Design Automation Workshop; Portland, Oregon, June 1973, pp 116-127

[15] T. M. McWilliams, L. C. Widdows: SCALD: Structured Computer-Aided Logic Design; Digital Research Lab, Techn. Report No. 152, Stanford University, March 1978

[16] N. N.: KARL Manual; Kaiserslautern University; Kaiserslautern 1986

[17] G. Girardi, R. Hartenstein, U. Welters: ABLED: a RT level Schematics Editor and Simulator Interface; Proc. EUROMICRO Symposium Brussels, Belgium, 1985, (ed.: K. Waldschmidt), North Holland Publ. Co., Amsterdam/New York, 1985

[18] A. M. Biraghi, A. Bonomo, G. Girardi: ABLEDitor: User Manual; CSELT, Torino, Italy, July 1986

[19] M. Barbacci, S. Grout, G. Lindstrom, M. Maloney, E. Organick, D. Rudisill: ADA as a Hardware Description Language: An Initial Report; Proc. 7th Int'l Symposium on Computer Hardware Description Languages; Tokyo, Japan, 1985

[20] D. D. Hill: ADLIB: A Modular, Strongly Typed Computer Design Language; Proc. 4th Int'l Symp. on Computer Hardware Description Languages; Palo Alto; 1979

[21] D. Coelho, W.M. vanCleemput: HELIX, A Tool for Multi-Level Simulation of VLSI Systems; Int'l Semi-Custom IC Conference; London, Nov. 1983

[22] W. Bastian, R. Hartenstein, W. Nebel: VLSI-Algorithmen: innovative Schaltungstechnik statt Software; VDI/VDE-GMR-Tag. Mikroelektronik i.d. Automatisierungstechnik; VDI-Ber. Nr. 550, Düsseldorf, 1985

[23] G. Balboni, V. Vercellone: Experiences in Using KARL-III in Designing a CMOS Circuit for Packetswitches Networks; ABAKUS (ABL and KARL user group) workshop, Passau, F.R.G., 1986

[24] D. Farelly: The Practical Application of ABLED and KARL in Designing a Speech Synthesizer; ABAKUS (ABL&KARL user group) workshop, Passau, F.R.G., 1986

[25] R. Hartenstein: Higher Level Simulation and CHDLs; Proc. (Eds.: W. Fichtner, M. Morf) IFIP Summer School on VLSI Design; Beatenberg, Switzerland, 1986; Kluwer Academic Publishers; Boston et al., 1986

[26] R. Hartenstein: Fundamentals of Structured Hardware Design - A Design Language Approach at Register Transfer Level; North Holland / American Elsevier; Amsterdam / New York, 1977

[27] N. N.: KARL Manual; Kaiserslautern University; Kaiserslautern 1986

[28] R. Hartenstein, A. Mavridis: RTcode Instant for KARL-III, second edition; report, Fachbereich Informatik, Kaiserslautern University, Kaiserslautern, F.R.G., 1986

[29] M. Barbacci: Instruction Set Processor Specification (ISPS): The Notation and Its Applications; IEEE-Trans.-C, January 1981

[30] M. Barbacci, G. E. Barnes, R. G. Cattell, D. P. Siewiorek: The ISPS Computer Description Language; CMU-CS-79-137, Department of Computer Science, Carnegie- Mellon-University, August 1979

[31] H. Zimmermann: OSI Reference Model, the OSI Model of the Architecture for Open Systems Interconnection; IEEE-Trans-COM-28, pp 425-432, (April 1980)

[32] P. B. Hanson: The Architecture of Concurrent Programs; Prentice-Hall, Englewood Cliffs, New Jersey, 1977

[33] U. Welters: Specification of the MLED Multi-Level Editor; internal report, Fachbereich Informatik, Universität Kaiserslautern, Kaiserslautern, F.R.G., 1986

3. Application-oriented HDL Specializations

HARDWARE DESCRIPTION LANGUAGES
R.W. Hartenstein (Editor)
© Elsevier Science Publishers B.V. (North-Holland), 1987

3.1. DATA PATH DESCRIPTIONS

Karin LEMMERT

Department of Computer Science, University of Kaiserslautern
Postfach 3049, D-6750 Kaiserslautern, F.R.G.

This chapter presents a short survey of data path descriptions in HDL's at the register transfer level. Particular data path description concepts necessary to describe structure and topology of regularly structured hardware designs and regular wiring pattern are examined closer. Their use will be illustrated by means of a multi-stage interconnection network and other simple application examples.

0. INTRODUCTION

It is the goal of new design strategies, such as e. g. the structured VLSI design approach advocated by the Mead and Conway [4] scene, to manage the complexity and the number of components on an integrated circuit by decreasing the design cost. Structured VLSI design is based on the conception of keys cells, so that a regular placement and routing pattern can be formed using multiple instantiations of such key cells. The most simple pattern in this context is the synthesis of arrays just by iterative abutment of instants of the key cells. However, not all design problems lead to such a simple solution of regularity: routing pattern outside the key cells are needed in such cases. It is the goal of this chapter to give an introduction on regular wiring pattern other than by cell abutment. The wiring pattern considered here may be called 'parameterized algorithmic wiring pattern'. One wellkown example is the Perfect-Shuffle [32].

This chapter stresses data path descriptions at Register Transfer (RT) level. It specifies hardware in form of multiplexers, registers, operators etc.. The goal of a hardware description language at the RT level is to facilitate the design process of integrated circuits. In the last few years RT level languages have been developed which support such structured VLSI design methods. Those languages are capable to express the behavior as well in a highly concise way the routing and placement structure and topological hardware attributes, like in floor plans. This chapter does not deal with the structural implementation of arithmetic, relational, and logic operators, since these are subject of abstraction levels lower than RT. There is a lot of literature available, e.g. see [2] [4] [5].

1. DATA PATH DESCRIPTIONS

This chapter will not discuss in detail the different data path descriptions at RT level. It is the author's intention to give only a short illustration, how the data path descriptions most widely used are implemented in the several RT languages and to focus some particular data path descriptions needed for concise RT level descriptions. Therefore the paper gives a short survey of regularly structured hardware designs to demonstrate the need of additional or extended concepts to describe data flow at the RT level in a concise and simple way. Furtheron it will present in more detail some particular wiring schemes and illustrate their use by means of a multistage interconnection network and other simple application examples.

1.1 DATA PATH DESCRIPTIONS MOST WIDELY USED

Traditional RT languages like AHPL [13], CDL [8], ISP [18] etc. provide primitives for

- multiplexer expressions,
- catenations (juxtaposition of paths),
- forks (fanout),
- subscripting (forming sub paths),
- simple connection,
- arithmetic, relational, and logical operations

to describe data paths at RT level. The tables 1-4 give a short summary how the data path descriptions most widely used are implemented in several RT languages. Free places in the tables indicate, that the detail is either not available or not known to the author. More informations about data path descriptions can be found in surveys of HDL's [3], [7], [17] and in the several language descriptions [8], [9], [11], [13], [15], [18] and [19].

	Data path selection	Catenation of		Connection	
		words	arrays	static	dynamic
AHPL	IF ... THEN ... ELSE	,	!	<-	<-
CDL	IF ... THEN ... ELSE	-	-	<-	<-
ISP	=> 1)	□	□	<-	<-
FDL	IF .... THEN .... ELSE CASE <EXPR.> OF <EXPR.>	.COMB.	.COMB.	=	=
HARPA	IF .... THEN .... ELSE CASE <EXPR.> OF <EXPR.>	&	&	:=	:=
KARL-III	IF .... THEN .... ELSE CASE <EXPR.> OF <EXPR.>	.			:=	:=

1) corresponds to the IF .. THEN ...; case of ALGOL

Table 1: Selecting and structuring data path primitives

	Arithmetic Operations			
	add	subtract	multiply	divide
AHPL	+	−	x	:
CDL	. ADD.	. SUB.	no	no
ISP	+	−	*	:
FDL	. ADD.	. SUB.	. MUL.	. DIV.
HARPA	+	−	*	
KARL−III	+	−	*	/

Information not known or not available to the author

Table 2: Arithmetic data path primitives

	Relational Operations					
	equal	not equal	greater than	greater or equal	less than	less or equal
AHPL	=	no	>	>=	<	<=
CDL	. EQ .	. NE .	. GT.	. GE .	. LT.	. LE.
ISP	=	<>	>	>=	<	<=
FDL	. EQ .	. NE .	. GT.	. GE .	. LT.	. LE.
HARPA			>	>=	<	<=
KARL−III	=	<>	>	>=	<	<=

Information not known or not available to the author

Table 3: Relational data path primitives

	Logical Operations				
	and	or	not	exclusive or	equivalance
AHPL	∧	∨	–	⊕	░░░░░░
CDL	. and .	. or.	'	. ERA .	░░░░░░
ISP	∧	∨	–	⊕	≡
FDL	. and .	. or.	. not .	. xor .	. coin .
HARPA 1)	and	or	not	░░░░░░	░░░░░░
KARL-III	and	or	not 2)	exor	. coin .

░░░░ Information not known or not available to the author

1)   functional notation of logical operators in HARPA
2)   standard function in KARL-III

Table 4: Logical data path primitives

1.2 PARTICULAR DATA PATH OPERATORS

Structured VLSI designs can be synthesized:

- automatically, such as e.g. by the application of problem specific silicon compilers.

- by the conception of keys cells with array capability, which can be abutted without additional wiring nets.

- by the conception of key cells, which are connected by regularly structured wiring pattern (see Fig. 19), such e.g. shift, shuffle, and butterfly wiring schemes. These wiring pattern may be called 'parameterized algorithmic wiring pattern'. If e.g. in a shift operator the shift parameter k specifies the number of bit positions by which the bits are shifted, the wiring pattern is specified by k and the path width w of the input data path (see Fig. 1).

VLSI designs composed of abutted key cells without additional wiring are iterative and systolic arrays. Systolic arrays are lattices of simple processing units connected in regular structures. The processing units only communicate with their nearest neighbour cells. These characteristics make systolic arrays suitable for a regularly structured VLSI implementation. Problems solved by systolic arrays are for example Finite Impulse Filters (FIR) and the Discrete Fourier Transformation. It is beyond the scope of this chapter to discuss exhaustively the application aspects in the design of systolic arrays. Surveys can be found in [36] ... [41].

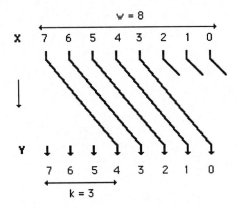

Fig. 1: Computation of the shift wiring pattern

Shuffle and butterfly wiring pattern are applied to interconnect the different stages of multistage interconnection networks. Well known networks of this class are the indirect binary n-cube [29], omega [31] and FLIP network [31]. With the availability of VLSI technology this class of special purpose networks to interconnect parallel processing units is becoming more attractive and has a number of interesting applications, such as airborne communication processors, image processors, Fast Fourier Transform processors, matrix multiplication arrays, and sorting networks. Detailed discussions of multistage interconnection networks can be found in [20] ... [32].

Regular implementations of arithmetic operations, like multiplication and division, or conversion functions, such as for binary-to-BCD and BCD-to-binary mostly use shift operators. See [33] for a regular multiplier design using shift operators and [34] [35] for binary-to-BCD conversion implementation descriptions based on shifter wiring operators.

In describing structures like systolic arrays, multistage interconnection networks, and shift application networks we need some more or improved concepts to describe

- topology of cell interconnections,
- floor plan topology,
- regular wiring pattern

in a systematic way.

HDL's supporting hierarchical hardware descriptions are suitable to describe concisely iterative and systolic arrays. Such languages are AHPL [13], FDL [15], HARPA [19], µFP [39]...[41] and KARL-III [11]..[12]. µFP is developed for the special purpose to design and verify systolic arrays. In the past systolic arrays have been designed using ad hoc techniques. µFP a starting point to formalize such designs based on the functional programming language FP [42]. The language provides high order functions (hofs) to plug smaller circuits, like logical gates, full adders, multiplexers, ALUs etc. together. Fig. 2 gives two examples of high order functions. To transform µFP circuit descriptions the language provides laws which guarantee that the final circuit has the same behavior as the original one. Fig. 3 illustrates one of these laws.

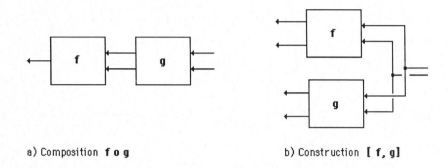

a) Composition **f o g**   b) Construction **[f, g]**

Fig. 2: Examples of µFP high order functions (hofs)

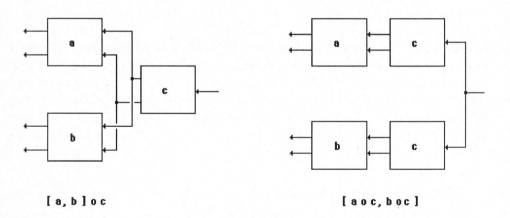

[a, b] o c   **[a o c, b o c]**

Fig. 3: Example of an µFP algebraic law **[a, b] o c = [a o c, b o c]**

FDL [15], HARPA [19] and KARL-III [11] are structural behavioral languages capable to describe modularized hardware. They use input and output attributes to specify input and output ports of the module interfaces. Additionally KARL-III provides concepts to describe the topology of ports and floor plans. The cell concept of KARL-III distinguishes input, output, and bidirectional ports at all four sides of a rectangular hardware cell by the prefixes **'front'**, **'back'**, **'left'**, and **'right'**. Floor plan topology can be specified by the slice abutment operator ':' and the layer abutment operator '@'. The ':' operator abuts cells in horizontal direction to layers and the '@' operator abuts layers or single cells in vertical direction.

All RT level languages considered in this paper provide features to describe shift wiring patterns. The following table gives a little survey:

	shift				
	destructive	nondestructive	constructive	circular	starring
AHPL	YES	NO	NO	YES	NO
CDL	YES	NO	YES	YES	NO
DDL	YES	NO	YES	YES	NO
KARL-III	YES	YES	YES	YES	YES

Table 5: Shift operators in HDL's

The meanings of the various shift operators are explained in the following section 1.2.1.1.

In all of the traditional languages at RT level, like AHPL [13], CDL [8], ISP[18]], shuffle wiring patterns can be described only by individual specification of a large number of connection assignments or by complex catenation expressions. KARL-III [11] is the only language featuring standard functions which may be used like primitives for specifications of such wiring shemes. The function **'fold'** rearranges the positions of bits of a vector and **'merge'** rearranges words within an array. To describe for example a shuffle wiring pattern with the destination step width m=4 (see Fig. 4) you may write the following assignment in KARL-III:

$$\text{terminal Y .= fold\&4(X);}$$

Only the reserved word **fold** the 'destination step width parameter m' (m=4 in this example) and the input path width (defined in a declaration part) need to be specified to completely define the wiring pattern. A detailed definition of the shuffle operator is given in subchapter 1.2.1.2.

None of the design languages considered in this paper provide primitives to describe butterfly wiring pattern (e.g. see Fig. 19) in a compact way instead of long net lists. For a detailed description of the butterfly wiring pattern see chapter 1.2.1.3.

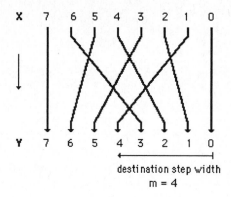

destination step width
m = 4

Fig. 4: Computation of the shuffle wiring pattern

1.2.1 BASIC DEFINITIONS

To have a closer look at shift and shuffle operations as well as at butterfly wiring patterns first some notational conventions will be introduced:

- X, Y   are vectors of N bits or arrays of N bitvectors numbered from 0 (least significant bit) through N-1 (most significant bit).
- w(X)   indicates the path width N of X.
- x_i is the i-th. bit, if X is a single bit vector or the i-th. word of the array X.
- k   shift parameter: the number of bit positions by which the vector is shifted.
- m   shuffle destination step width
- K   catenation operator

1.2.1.1 SHIFT OPERATORS

Considering e.g. a shift right operation by k steps at the right end k bits are discarded (we say 'discarded bits'). At the left end however, k output bits are left undefined (we say 'shift-in bits'), since no values are provided from the input source of the shift operator, that's why for the shift-in bits another value source is needed. Depending on the shift direction (e.g. 'shift left' or 'shift right') a shift operation by k steps applied to a vector of bits or to an array of words can be formally defined in the following way:

DEFINITION: 1) shift right by k steps:

$$Y = k \ \mathbf{shr}(X)$$

$$= \mathop{\mathbf{K}}_{i=N-1}^{N-k} c \ . \ \mathop{\mathbf{K}}_{i=N-k-1}^{0} x_{i+k}$$

2) shift left by k steps:

$$Y = k \, \textbf{shl}(X)$$

$$= \mathop{K}_{i=N-1}^{k} x_{i-k} \cdot \mathop{K}_{i=k-1}^{0} C$$

Where C stands for the shift-in values. For example in KARL-III depending on the shift type C has the following values:

shift operator	KARL primitive		shift-in bits output		
				C value	
	word :	array	source	bit :	word
starring shift	shl shr	push pop	internal constant	* STAR bit	(*)k STAR word
destructive shift	dshl dshr	dpush dpop	internal constant	0 ZERO bit	(0)k ZERO word
constructive shift	cshl cshr	cpush cpop	internal constant	1 ONE bit	(1)k ONES word
nondestructive shift	nshl nshr	npush npop	extra wiring	connected from input of the same bit position	
circular shift	cirshl : cirshr :	cirpush cirpop	extra wiring	connected from discarded bit positions	

(x)k : means k times x

Table 6: Shift-in bit/word computation

Considering the nondestructive shift operation in Fig. 5, it is obvious that a nondestructive shift by k steps is different to k sequential nondestructive shifts.

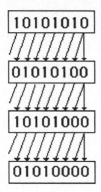 

a) sequential nondestructive shift k times   b) nondestructive shift by k steps

Fig. 5: Nondestructive shift operator

1.2.1.2 SHUFFLE OPERATORS

An m-shuffle of a vector of N bits or an array of N words numbered from 0 to N-1 can be illustrated by means of a card game example:

- Given N cards.
- Divide the N cards into m equal piles of N/m cards.
- Pick the first card of the first pile, the first card of the second pile and so on to form a new first pile of m cards.

    Continue this process to build a second new pile and so on until N/m new piles each of m cards exist.

This has been a procedural definition, which will also be used to illustrate the synthesis of the shuffle wiring pattern. In that case the arrows at the wires of Fig. 6 show the movements of the cards. The number m of cards of each original pile is identical to the destination step width m. Formally the m-shuffle operation can be defined as follows:

DEFINITION: An m-shuffle executed on a vector X with N bits or an
             array of N words is given by

$$Y = \prod_{i=0}^{m-1} \left(\prod_{k=0}^{w(X)/(m-1)} x_{w(x)-1-i-k*m} \right)$$

Fig.6 illustrates a 2-shuffle operation of the vector X, 8 bit wide. In literature of multistage interconnection networks the 2-shuffle operation is well known as the so called Perfect-Shuffle [32]. One further application of shuffle operators occurs by the refinement of RT descriptions from word to bit level (slice pathing) e.g. the shuffle&2 for dyadic, the shuffle&3 for triadic operators etc..

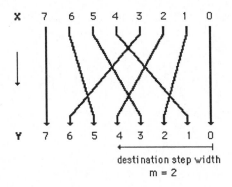

Fig. 6: Perfect-Shuffle operator

1.2.1.3 BUTTERFLY WIRING PATTERN

The butterfly wiring shape is defined for N = 2, 4, 8, 16 ... bits of a vector or words of an array. The basic shape of the butterfly wiring sheme with two inputs is shown in Fig. 7a). Fig. 7b) presents an example of the butterfly wiring pattern with 8 inputs and Fig. 7c) shows the detailed representation of it, which can be achieved by duplicating each input wire. In chapter 2 the reader will find rules to describe the butterfly wiring pattern by shuffle operators. In chapter 3.3 these rules are applied to describe a multistage interconnection network in KARL-III.

a) Butterfly shape

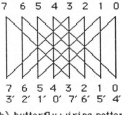

b) butterfly wiring pattern
with 8 inputs

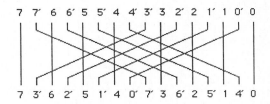

c) Detailed illustration of the butterfly wiring pattern

Fig. 7: Butterfly wiring pattern

(1) $2^1 * X$ = i shift_left(X);

(2) i shift_left(X) = (i mod n) shift_left(X)
i shift_right(X) = (i mod n) shift_right(X)

(3) i shift_left(X) = (n - i) shift_right(X)
i shift_right(X) = (n - i) shift_left(X)

(4) i shift_left (j shift_left(X)) = (i + j) shift_left(X)
i shift_right (j shift_right(X)) = (i + j) shift_right(X)

(5) i shift_left(X) + i shift_left(Z) = i shift_left(X+Z)
i shift_right(X) + i shift_right(Z) = i shift_right(X+Z)

Derived Rules:

(6) n shift_left(X) = X   with (2)
n shift_right(X) = X   with (2)

(7) i shift_left(X) = j shift_left (i - j shift_left(X))
i shift_right(X) = j shift_right (i - j shift_right(X))
with (3) and (4)

(8) i shift_left(X) = j shift_left (j - i shift_right(X))
i shift_right(X) = j shift_right (j - i shift_left(X))
(3) applied to (7)

Notation:

(X) input vector (x_{n-1} , x_{n-2} , ..., x_1 , x_0)

n path width of the vector (X)

i, j shift step width

Fig. 8: Rules of a shifter algebra

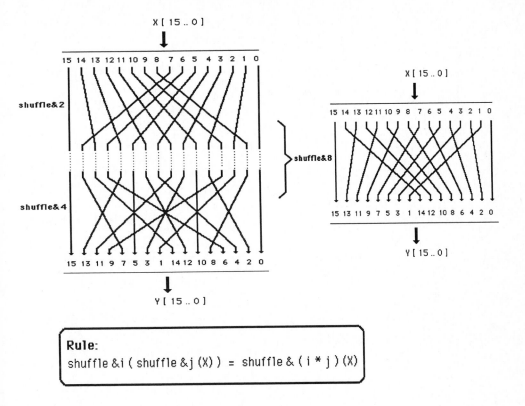

Fig. 9: Composition of the shuffle operator

2. APPROACHES TOWARDS A DESIGN CALCULUS

As mentioned in chapter 1.2 structured VLSI design is based on the conception of key cells which can be either connected to arrays by wiring by abutment or by additional regularly structured wiring cells outside the key cells. To facilitate the design of the basic cells and of the regularly structured wiring cells the designer needs something like a design calculus which considers not only one level of the design processs. The design calculus should be based on an algebra in similar way as the boolean algebra can be used as a calculus to describe the logic of boolean networks and to transform the networks. The design algebra should provide rules to transform the circuit structure without changing its originally defined behavior. As an approach towards such a design calculus the author will present some rules concerning shift, shuffle, and butterfly wiring patterns.

The rules of the shifter algebra (see Fig. 8) describe transformationes of a general shifter model. It is assumed to the shift model that e.g. for a shift right operation no bits are left at the right end. The basic rules (1) and (2) illustrate a shift right or a shift left operation. Rule number (3) describes a shift left operation by a corresponding shift right and rule number (4) specifies the composition of the shift operation. Rule number (7) can be derived with rule number (3) and (4). Applying rule number (7) we

7 6 5 4 3 2 1 0

7 3 5 1 6 2 4 0

a) Butterfly wiring pattern with 4 inputs

Notation:

X : input vector of n bits $(x_{n-1}, .., x_0)$

n = 4, 8, 16, 32 ....

i : $n/2$

7 6 5 4 3 2 1 0

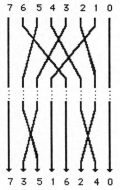

shuffle&4 $(x_7, .., x_0)$

shuffle&(n/2) $(x_{n-1}, .., x_0)$

shuffle&2 $(x_7, .., x_4)$.shuffle&2 $(x_3, .., x_0)$

shuffle&2 $(x_{n-1}, .., x_i)$. shuffle&2 $(x_{i-1}, .., x_0)$

7 3 5 1 6 2 4 0

b) Version 1: butterfly wiring pattern described by shuffle operators

7 6 5 4 3 2 1 0

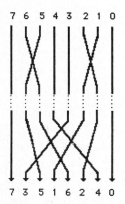

shuffle&2 $(x_7, .., x_4)$.shuffle&2 $(x_3, .., x_0)$

shuffle& (n/4) $(x_{n-1}, .., x_i)$. shuffle& (n/4) $(x_{i-1}, .., x_0)$

shuffle&2 $(x_7, .., x_0)$

shuffle& 2 $(x_{n-1}, .., x_0)$

7 3 5 1 6 2 4 0

c) Version 2: butterfly wiring pattern described by shuffle operators

Fig. 10: Butterfly wiring pattern described by shuffle operators

get rule number (8) which will be fundamentally to the regularly structured multiplier design of chapter 3.1. Fig. 9 shows the general rule of the composition of the shuffle operator and its illustration by means of an example. Rules to describe butterfly wiring shemes by shuffle operators are given in Fig. 10. They are applied to describe the multistage connection network of chapter 3.3 in KARL-III.

3. EXAMPLES

In the following the author will give examples to demonstrate the need of particular wiring primitives for concise descriptions of regular wiring pattern instead of long net lists. For the descriptions the hardware description language KARL-III [11] has been chosen because of the authors familiarity with the language.

3.1 REGULARLY STRUCTURED MULTIPLIER USING SHIFT OPERATORS

This example shows the regularly structured design of a multiplier using shift operators. The design will be described using a graphical block diagram notation similar to ABL [10], because the presentation of the whole textual KARL description would spread the scope of this paper. ABL is the graphical companion to the textual language KARL. It illustrates each KARL primitive by a rectangular box with in, out, and bidirectional ports. The function of a primitive is indicated by its name or a corresponding symbol standing within the box.

The product P of the data vectors X and Y each 4 bits wide can be expressed in the following way:

$$P = \sum_{i=0}^{3} 2^i * y_i * X$$

Applying rule number (1) of the shifter algebra of Fig. 8 and then rule number (8) we achieve the following equation which is illustrated in Fig. 11a):

$$P = \sum_{i=0}^{3} i \; shl(y_i * X) \overset{(8)}{=} 4 \; shl \sum_{i=0}^{3} (4-i) \; shr(y_i * X)$$

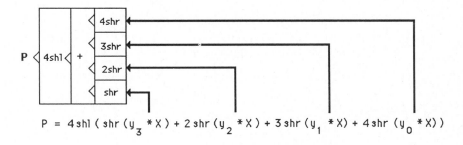

$$P = 4 \, shl \, (\, shr \, (y_3 * X) + 2 \, shr \, (y_2 * X) + 3 \, shr \, (y_1 * X) + 4 \, shr \, (y_0 * X) \,)$$

Fig. 11: Irregularly structured multiplier

Up to now the design of the multiplier has not a regular structure. However if we use rule number (4) which specifies the composition of a shift operation we can cascade the shift right operations in the following way getting a regular structure:

$$P = 4\,shl\,(shr(\,y_z * X \;+\; shr\,(\,y_2 * X \;+\; shr\,(\,y_1 * X \;+\; shr\,(\,y_0 * X \;+\; K\,)\,)\,)\,)\,)$$

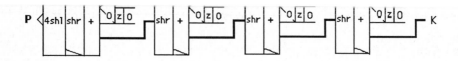

Fig. 12: Regularly structured multiplier

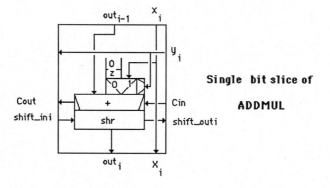

Single bit slice of

ADDMUL

Fig. 13: Single bit slice of regularly structured mutliplier

After the design of a single bit slice (see Fig. 13) the floor plan of the 4 x 4 bit mutliplier can be generated (see Fig. 14).

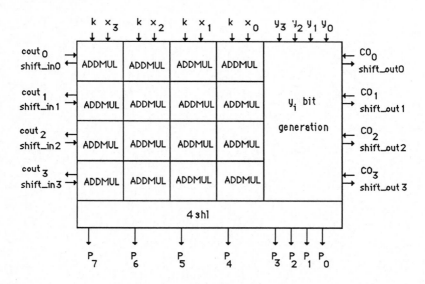

Fig. 14: Multiplier floor plan

What we still have to design is the generation of the y_i bits and the connections of the 'shift_out$_i$ bits' with the corresponding one of the product vector P. Realizing the wiring by simple connections we get a highly irregular structure (see Fig. 15a). Stretching the two overlapping wiring pattern in horizontal direction we achieve a regular structure with shift wiring shemes (see Fig. 15b). The basic cell of the regular wiring structure can be described by concise wiring primitives (see Fig. 15c). The primitive '**fork**' duplicates its input data path and the standard function '**lsb**' selects the least significant bit of its input data path.

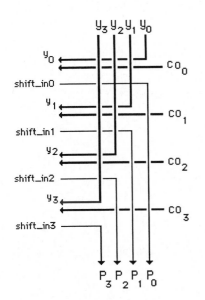

a) irregular structure

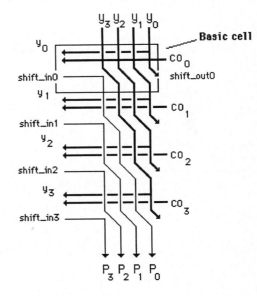

c) regular structure

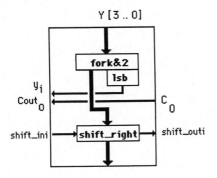

c) basic cell described by RT wiring functions

Fig. 15: y_i bit generation

3.2 APPLICATION EXAMPLE OF THE STARRING SHIFT OPERATOR

the data vector SHIFT_CODE.

```
cell S_R_SHIFT (back in INPUT [7..0]
               front out OUT_SHR [7..0]
               left  in SHIFT_CODE [2..0]);

node OUT_SHR_1, OUT_SHR_2, OUT_SHR [7..0];

begin
terminal
  OUT_SHR_1 .= if SHIFT_CODE[0]
               then shr(INPUT) (* shift by 1 step *)
               else INPUT
               endif;

  OUT_SHR_2 .= if SHIFT_CODE[1]
               then shr&2(OUT_SHR_1) (* shift by 2 steps *)
               else OUT_SHR_1
               endif;

  OUT_SHR  .= if SHIFT_CODE[2]
               then shr&4(OUT_SHR_2) (* shift by 4 steps *)
               else OUT_SHR_2
               endif;
end.
```

Fig. 16: KARL-III description of starring shift right cell

Two starring shift right cells S_R_SHIFT can be used to design a circular shift right cell C_R_SHIFT (see Fig. 17). One instantiation of the cell S_R_SHIFT is used for the shift right operation and the second together with reflect wiring operators to perform a shift left operation. The reflect operators reverse the order of its input data vector. The step width of the shift left operation is the difference: SHIFT_MAX (path width of INPUT) - SHIFTCODE. It is implemented by the function SHIFTPARAM. The outputs OUT_SHR and OUT_L_SHIFT of the two shifter operations are connected to the bus OUT_CSHR by tristate drivers to form the result of the circular shifter (wired AND function).

```
cell C_R_SHIFT (back in INPUT [7..0]
               front bi OUT_CSHR [7..0]
               left  in SHIFTCODE [2..0]);

(* externally declared cell S_R_SHIFT *)
cell S_R_SHIFT (back in INPUT [7..0]
               front out OUT_SHR [7..0]
               left  in SHIFT_CODE [2..0]);
external;
```

```
(* declaration of the function SHIFTPARAM *)
func SHIFTPARAM (SHIFT_CODE [2..0]);
constant SHIFT_MAX [3..0] .= 8;
        Z .= 0;
node     SHIFTPARAM [2..0]; SUB [3..0];
begin
terminal SUB .= SHIFT_MAX - Z.SHIFTCODE;
        SHIFTPARAM .= SUB [2..0];
end;

constant ONE .= 1;
node     OUT_SHR, REF_SHL, OUT_L_SHIFT [7..0];
tribus   OUT_CSHR [7..0];

begin
(* instantiation to form a shift right operation *)
make S_R_SHIFT (back  in  INPUT
              front  out OUT_SHR
              left   in  SHIFT_CODE);

(* instantiation to form a shift left operation *)
make S_R_SHIFT (back  in  reflect(INPUT)
              front  out REF_SHL
              left   in  SHIFTPARAM(SHIFT_CODE) );

terminal OUT_L_SHIFT .= reflect(REF_SHL);

(* tristate connection of shifter outputs to the bus OUT_CSHR *)
bus  OUT_CSHR .= ONE enables OUT_SHR;
     OUT_CSHR .= ONE enables OUT_L_SHIFT;
```

Fig. 17: KARL-III description of the circular shift right cell

3.3 EXAMPLE OF A SHUFFLE NETWORK

As an example of a shuffle network we will illustrate the indirect binary n-cube network [29] [31] which is used to interconnect a set of processing elements. The network has N inputs and consists of n stages of N/2 two function interchange boxes (see Fig. 18) with individual box control. The topology of the binary n-cube network is given by the construction rule, that at stage i of the network the two addresses of the input lines to an interchange box only differ in the ith. bit position. In the network data passes through stage 0, then stage 1, and finally through stage n-1.

The block diagram of the case n=3 is shown in Fig. 19. Thereby CUBE is the composition cell with eight input and output ports, each 6 bit wide. In Fig. 20 you can find the corresponding network description in the hardware description language KARL-III. In [16] Klein and Sastry describe the same example using their model of parameterized modules and interconnections.

Considering Fig. 19 the cell CUBE is composed of the three subcells COLUMN0, COLUMN1, and

COLUMN2. In KARL-III this is indicated by their instantiations in the make-statements. This feature

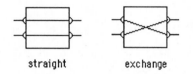

straight   exchange

Fig. 18: Block diagram of interchange boxes

Each instantiation of the cell COLUMN consists of four instantiations of the interchange box cell SWBOX. In KARL-III the vertical arrangement of the cells SWBOX is expressed by the layer abutment operator '@'. All computation is described in the SWBOX modules. The rest of the network consists of interconnections. Each input port of an interchange box consists of 3 data and 3 tag bits, which indicate whether it should be routed to the upper or lower output of the output port of the box. When data lines arrive an interchange box the tag bits of the two inputs are decoded. If both are destinated to the same output port an error condition is activated otherwise the inputs are connected in straight or exchange mode to the output ports.

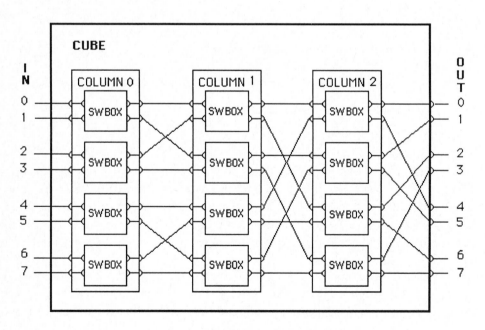

Fig. 19: Indirect binary n-cube network

```
cell CUBE ( left   in  IN_7_0[7..0;5..0]
            right out OUT_7_0[7..0;5..0] );

(* declaration of one stage of the the binary-n-cube network *)
cell COLUMN ( left   in  IN_7_0[7..0;5..0]
              right out OUT_7_0[7..0;5..0] );

(* declaration of the processing cell *)
cell SWBOX ( left   in  IN0, IN1 [5..0]
             right out OUT0, OUT1[5..0] );
switch ERROR [5..0];
node    OUT0, OUT1 [5..0];
begin
terminal OUT0 .= case IN0[5].IN1[5] of
                    0: ERROR
                    1: IN0[5..0] (* straight mode *)
                    2: IN1[5..0] (* exchange mode *)
                    3: ERROR
                endcase;
         OUT1 .= case IN0[5].IN1[5] of
                    0: ERROR
                    1: IN1[5..0] (* straight mode *)
                    2: IN0[5..0] (* exchange mode *)
                    3: ERROR
                endcase;
end;

node OUT_7_0 [7..0;5..0];
     OUT0, OUT1, OUT2, OUT3, OUT4, OUT5, OUT6, OUT7 [5..0];
begin
make (* instantiation of four SWITCH_BOXes to form a column *)
     (* of the binary-n-cube network                        *)
  SWBOX @ SWBOX @ SWBOX @ SWBOX
  ( left in  IN_7_0[7;5..0]; IN_7_0[6;5..0];
             IN_7_0[5;5..0]; IN_7_0[4;5..0];
             IN_7_0[3;5..0]; IN_7_0[2;5..0];
             IN_7_0[1;5..0]; IN_7_0[0;5..0]
    right out OUT0; OUT1; OUT2; OUT3; OUT4; OUT5; OUT6; OUT7 );

terminal OUT_7_0 .= OUT0|OUT1|OUT2|OUT3|OUT4|OUT5|OUT6|OUT7;
end;

node (* declaration of the cube output:        *)
     OUT_7_0 [7..0;5..0];
     (* declaration of internal cube nodes:    *)
     IN_COLUMN1_74, IN_COLUMN1_30,
     IN_COLUMN2_74, IN_COLUMN2_30 [3..0;5..0];
     OUT_COLUMN0, OUT_COLUMN1, OUT_COLUMN2 [7..0;5..0];

begin
(* instantiation of column 0 *)
make  COLUMN&0 ( left in IN_7_0 right out OUT_COLUMN0 );

(* instantiation of column 1 *)
terminal IN_COLUMN1_74 .= OUT_COLUMN0[7..4;5..0];
         IN_COLUMN1_30 .= OUT_COLUMN0[3..0;5..0];
```

```
make COLUMN&1 (left  In merge&2(IN_COLUMN1_74) |
```

```
terminal IN_COLUMN2_74 .= OUT_COLUMN1[7..4;5..0];
        IN_COLUMN2_30 .= OUT_COLUMN1[3..0;5..0];
make COLUMN&2 ( left  In merge&2(merge&2(IN_COLUMN2_74) |
                               merge&2(IN_COLUMN2_30))
             right out OUT_COLUMN2);

(* output generation *)
terminal OUT_7_0 .= merge&4(OUT_COLUMN2);

end.
```

Fig. 20: KARL-III description of indirect binary n-cube network

CONCLUSIONS

In parallel to concise data path description concepts in HDL's there is a necessity of a design algebra. This design algebra should include well-grounded rules to transform the structure of the described hardware without changing its behavior. The rules should support the design of key cells and of wiring cells to form a regular placement and routing on a chip. However the design algebra also may be the basis to further design systems and to board designs e.g. to connect several copies of a chip to a circuit with the same function but a multiple data path width.

REFERENCES:

HARDWARE SYNTHESIS:

[1]   Cleemput, W.M., Hierarchical Design for VLSI: Problems and Advantages, in: Seitz,C.L., Proceedings of Caltech Conference on Very Large Scale Integration (1979).

[2]   Eshraghian, K. Weste, N., Principles of CMOS VLSI Design: A Systems Perspective, Addison Wesley (1985).

[3]   Hartenstein, R.W., Fundamentals of Structured Hardware Design (North-Holland, 1977).

[4]   Mead, C. and Conway, L., Introduction to VLSI Systems (Addison-Wesley, 1980).

[5]   Synder, L., Supercomputers and VLSI: The Effect of Large-Scale Integration on Computer Architecture, in: Yovits, M.C., Advances in Computers vol. 23 (Academic Press, 1984).

[6]   Swartzlander, Earl E., Computer Arithmetic, Stroudsburg, Pa: Dowden, Hutchinson u. Ross (1980).

HARDWARE DESCRIPTION LANGUAGES

[7]   Barbacci, M.R., A comparison of register transfer languages for describing computers and digital systems, IEEE Transactions on Computers vol. c-24 (1975) 137-150.

[8]   Chu,Y., Introducing CDL, IEEE Computer vol. 7 no.12 (1974) 42-44.

[9]   Dietmeyer,D.L., Introducing DDL, IEEE Computer vol. 7 no. 12 (1974) 34-38.

[10]   Girardi, G., ABL data structure, CVT-report, CSELT Torino, Italy (March 1984).

[11]   Hartenstein, R.W., Lemmert, K., and Wodtko, A., KARL-III Reference Manual, Department of Computer Science, Kaiserslautern University (1986).

[12]   Hartenstein, R.W. and Borrmann, B., hyperKARL-III Language, Reference Manual, Department of Computer Science Kaiserslautern University (March 1985).

[13]   Hill, F.J., Introducing AHPL, IEEE Computer vol. 7 no. 12 (1974) 28-30.

[14]   Hill, F.J. and Peterson, G.R., Digital Systems: Hardware Organisation and Design (John Wiley and Sons, New York, 1978).

[15]   Kato, S. and Sasaki, T., FDL: A structural behavior description language, in: Uehara, T. and Barbacci, M. (eds), 6th. Sym. on Computer Hardware Description Languages an their Applications (North Holland , Amsterdam, 1983).

[16]   Klein, S. and Sastry, S., Parameterized modules an interconnections in unified hardware descritpions, in: Breuer M. and Hartenstein, R. (eds), Int. Conf. on Computer Hardware Description Languages and their Applications (North Holland, Amsterdam, 1981).

[17]   Shiva, S.G., Computer Hardware Description Languages - Tutorial, Proceedings of the IEEE vol. 67 no. 12 (1979) 1605-1615.

[18]   Siewiorek, D.P., Introducing ISP, IEEE Computer vol. 7 no. 12 (1974) 39-41.

[19]   Veiga, P.M.B. and Lanca, M.J.A., Harpa: A hierarchical multi-level hardware description language, IEEE 21th. Design Automation Conference (1984) 59-65.

MUTLISTAGE INTERCONNECTION NETWORKS:

[20]   Batcher, K.E., Sorting Networks and their Applications, Spring Joint Computer Conference (1968) 307-314.

[21]   Bhuyan, L.N. and Agrawal, D.P., Design and Performance of Generalized Interconnection Networks, IEEE Trans. o. Comp. vol. c-32 no. 12 (1983) 1081-1090.

[22]   Franklin, M.,Wann, D., et al., Pin Limitations and Partitioning of VLSI Interconnection Networks, IEEE Trans. o. Comp. vol. c-31 no. 11 (1982) 1109-1116.

[23]   Kruskal, C. and Snir, M., The Performance of Multistage Interconnections Networks for Multiprocessors, IEEE Trans. o. Comp. vol c-32 no. 12 (1983) 1091-1098.

[24]   Kung, H.T.,The Structure of Parallel Algorithms, in: Yovits, M.C. (ed), Advances in Computers vol. 19 (Academic Press 1980).

[25]   Lakshmivarhan, S., Dhall, S.K. and Miller, L.L., Parallel Sorting Algorithms, in: Yovits, M.C. (ed), Advances in Computers vol. 23 (Academic Press, 1984).

[26]   Lawrie, D.H., Acces Alignment of Data in an Array Processor IEEE Trans. o. Comp. vol. c-24 no 12 (1975) 1145-1155.

[27]   Lee, D. and Shen, J., Testing and Testability of (N,K) Shuffle/Exchange Networks, Research Report no. CMUCAD-83-19, Dept. of Electr. and Comp. Engineering, Carnegie

[28]   Padmanabhan, K. and Lawrie, D.C., A Class of Redundant Path Multistage Interconnection

[30]   Siegel, H.J., Analysis Techniques for SIMD Machine Interconnection Networks and the Effect of Processor Addres Mask, IEEE Trans. o. Comp. vol. c-26 no. 2 (1977) 153-161

[31]   Siegel, H.J., Interconnection Networks for SIMD Machines, IEEE Computer (1979) 57-65.

[32]   Stone, H.S., Parallel Processing with the Perfect Shuffle, IEEE Trans. o. Comp. vol. c-20 no. 2 (1971) 153-161.

SHIFT APPLICATION NETWORKS:

[33]   Hartenstein, R.W., KARL-III Primer, Computer Science Department, Kaisrslautern Univerity (1985).

[34]   Linford, J.R., Binary-to-BCD Conversion with Complex IC Functions, Computer Design (1973).

[35]   Schreiber, F.A. and Stefanelli, R., Two Methods for Fast Integer Binary-BCD Conversion, IEEE Transactions on Computer 1978 200-207.

SYSTOLIC ARRAYS:

[36]   Kung, H.T., Let's Design Algorithms for VLSI Systems, in: Seitz, C.L., Proceedings of Caltech Conference on Very Large Scale Integration (1979).

[37]   Kung, H.T. and Leiserson, C.E., Systolic Arrays for VLSI, in: Mead, C. and Conway, L., Introduction to VLSI Systems (Addison-Wesley, 1980) 263-332.

[38]   Lam, M. and Mostow, J., A Transformational Model of VLSI Systolic Design, in: Uehara, T. and Barbacci, M. (eds) Computer Hardware Description Languages and their Application (North-Holland, 1983).

[39]   Shareen, M., muFP, a Language for VLSI Design, ACM Symposium on LISP and Functional Programming (1984) 104-112.

[40]   Shareen, M., The Design and Verfication of Regular Synchronous Circuits, report, Computer Science Depart., Chalmers Tekniska Hogskola Goteborg.

[41]   Shareen, M., µFP - An Algebraic VLSI Design Language, Phil. d. Thesis, Programming Research Group, Oxford Universtiy Computing Laboratory (1983).

FUNCTIONAL PROGRAMMING:

[42]   Backus, J., Can Programming be liberated from the von Neumann Style? A Functional Style and Its Algebra of Programs, Communications of the ACM vol. 21 no. 8 (1978) 613-643.

HARDWARE DESCRIPTION LANGUAGES
R.W. Hartenstein (Editor)
© Elsevier Science Publishers B.V. (North-Holland), 1987

3.2. CONTROL PART DESCRIPTIONS AND LANGUAGES FOR MICROPROGRAMMING

Paolo PRINETTO

*Politecnico di Torino, Dipartimento di Automatica e Informatica
Corso Duca degli Abruzzi 24, I-10129 Torino TO, Italy*

In the paper, microprogramming languages are examined. Following the taxonomy proposed by Davidson, they are classified according to two criteria: the "level" of the language (high or low) and whether it is host machine "independent" or not.

1   INTRODUCTION

The use of microprogramming, first introduced by Wilkes in 1951 [Wilk_51], has become a widely accepted technique for both implementing control units and emulating instruction sets. Due to this growing interest, the process of microprogram writing has been given a peculiar attention and a lot of engineering efforts have been put in the development of tools for the construction of efficient and reliable firmware. (Good surveys of the activities related with firmware engineering can be found in [RaAd_80], [DaSh_80a], and [DaSh_85]).

Among the different branches of this new discipline, an important role is played by the so called microprogramming languages: languages proposed and adopted to be compiled into microcode. These languages can be classified according to many criteria, but one of the most suitable ones is the taxonomy proposed by Davidson in [Davi_83] and [Davi_86]. He classifies them according to two criteria: the "level" of the language (high or low) and whether it is host machine "independent" or not.

As a preliminary distinction, a high level microprogramming language usually provides:

- the possibility of defining "virtual" storage resources and some data structures, without caring about their actual hardware implementation;

- a set of control structures similar to those available in most high level programming languages, usually including while and for loop control and case test;

- a set of primitives that will be mapped by the compiler into a sequence of microinstructions of the target machine.

In compiling high level microprogramming languages it is necessary to perform an assignment of data and functional resources defined at the source level to the physical resources of the target architecture. For instance, a data resource, i.e., a variable of the high level microprogramming language, must be stored into a register in the target. As another example, a functional resource of the language, such as the "add" operator, must be implemented by an ALU, or an adder present in the target. Such a process, usually referred to as "binding", is automatically carried out by the compiler, even if the user is usually given the possibility of forcing the allocation of certain variables into some specific registers.

On the opposite side, a low level microprogramming language typically provides:

- the capability of specifying the actual hardware resources of the host machine;

- A very primitive set of control structures, usually limited to loop handling, goto facility, and, sometimes, multi-way branches similar to those available in microprogram sequencers.

- A set of primitives that can be directly mapped into one (and seldom more) microinstruction(s) of the host machine. These usually include simple arithmetic and logical operators and data transfers between the hardware resources of the target and namely registers and memory locations.

- As a consequence of this machine bounding, they very seldom provide the capability of defining data structures.

As far as machine dependence/independence is concerned, a machine dependent microprogramming language can be defined as a language in which all available operations and data elements have a direct mapping to a resource of the target machine. This means that the binding to data and functional resources of the target machine is performed at the language definition level. As a consequence, the language can be used for one target machine, only. Assembly languages are one class of machine dependent languages.

In a machine independent microprogramming language, the binding is done not at the language definition level, but it is performed by the compiler at a later stage of the compilation process. At the beginning, the compiler allocates storage for all the variables of the program, in a pure machine independent manner, and the control and data operations are not tied directly to the target microinstruction set, but usually transformed into an intermediate language. The final binding is performed later by the code generator, which takes into account the hardware organization of the target. Thus, it is possible to translate a microprogram written in a machine independent language into microcode for several target machines, given a code generator for each.

In the following, I shall examine some of the most important microprogramming languages. I do not claim to be exhaustive: I shall outline the characteristics of those microprogramming languages that, in my opinion, are particularly significant, either from a research or from a practical point of view. Only the main features of each language will be presented, without going into lexical and syntactical details. For such details please refer either to the original papers presenting the languages or to survey papers. Among these, a good one, although now somewhat out of date, is [Sint_80]. Davidson presents in [Davi_83] and [Davi_86] an excellent taxonomy of high level microprogramming languages and examines problems concerned with their usage. The role played by hardware description languages in microprogram systems is deeply examined by Dasgupta in [Dasg_85].

In the rest of the paper, section 2 is concerned with high level microprogramming languages, distinguishing between machine independent, and machine dependent languages. Section 3 deals with low level microprogramming languages and microprogram meta-assemblers.

2   HIGH LEVEL MICROPROGRAMMING LANGUAGES

As sketched in the introduction, a high level microprogramming language is characterized by the fact that one statement can be compiled into a sequence of microinstructions and that the language provides control constructs and data structures that may not be directly supported by the target machine.

The last feature is indeed common to most high level programming languages too, and, in fact, the reasons pushing toward the use of high level microprogramming languages are mostly the same pushing toward high level programming languages. They can be summarized as follows [Davi_86]:

- they reduce the number of code lines to be written and consequently they increase firmware productivity, reliability and maintainability;

- they make the development of large firmware projects more manageable;

- their machine independence allows microprograms sharing between different targets;

- they allow the adoption of most of tools and methodologies developed in software engineering.

In spite of all these advantages, even if high level microprogramming languages have been deeply investigated, no widely accepted high level language exists today, and some researchers are still arguing about the real need for them. On the converse, a lot of high level programming languages exist and nobody doubts about their necessity. The main reason for such a difference relies in the need for efficient microcode generation and in the high costs in implementing efficient high level microprogramming language compilers, with respect to the limited number of possible applications.

In spite of this skepticism, the more recent trends in firmware engineering are toward the implementation of integrated microprogramming systems for microcode development [DWhe_86]. These are CAD systems that assist microprogrammers in all phases of firmware design and implementation and, specifically, in microcode synthesis, verification, and simulation.

Integrated systems may be either machine independent (or "retargetable") i.e., usable for different classes of target machines, or not retargetable, i.e., tailored to a specific microarchitecture only. It is however a common opinion that, due to the high development costs, microprogramming systems will become an industrial reality only if the same system can be used and re-used in different microprogrammable architecture development projects.

These systems usually accept as input two kinds of information, provided through two different languages. The former is the source program, mostly expressed in a high level machine independent microprogramming language. The latter is a description of the target architecture in an "Architecture Description Language" (ADL). This can be either a usual Hardware Description Language at the register

transfer level, or an ad hoc language. A detailed analysis of ADLs is
out of the scope of this paper, but the requirements for an ADL are
clearly examined and discussed in [DWHe_76].

Before going through the great variety of proposed high level
microprogramming languages, it is worthwhile to examine in brief their
fields of application.

These languages are quite similar to high level languages, and
are useful in all those cases that do not resort to features peculiar
for a target microarchitecture and not available on other targets.
These applications usually include "application firmware", such as
parts of a high level language program to be migrated into microcode,
computation intensive applications that can be sped up by
microprogramming, or proprietary programs migrated into microcode to
protect them. Their development strictly resembles application
software development. In these cases microprograms are usually very
large, machine independent and resort to features available on most
architectures. As a consequence, they are the most suited to the use
of high level microprogramming languages.

With regard to machine dependence/independence, three classes of
high level microprogramming languages ought to be considered: machine
dependent, machine independent, and mixed.

2.1   High Level Machine Independent Languages.

In these languages, all storage allocation is done by the
compiler, although hints about which variables should be placed in
registers may sometimes be given by the microprogrammer.   Binding is
done later, during the actual microcode generation phase.   The
provided set of operators usually consists of those functions
available on most targets.

One of the great advantages of the high level machine independent
language is that if the language is designed to be compatible with a
high level language, programs written in that language can be easily
migrated into microcode. Adopted languages are, in this sense:

- the C language, as in the case of micro-C (see below);

- PL/1, as in the case of PL/MP approach at IBM [KiTa_79];

- ADA, as in the case of the systems developed at Intel and at the University of Southwestern Louisiana [LSDa_83].

In some cases, such as in the SIMULGEN system developed at AT&T Bell Labs [Eldr_84], the high level language itself (namely, in SIMULGEN the C language) is used to describe the circuit, with few facilities added for symbolic reference to simulator entities.

The main problem with this class of microprogramming languages is still the efficiency of the produced microcode, since the requirements for strong microcode optimization pose great difficulties in writing efficient microcode compilers. in fact, some language features that would be absolutely innocuous in a high level language environment can cause grave efficiency problems here.

A compiler for a machine independent language can be transferred to other machines by rewriting the code generator. However, these compilers must be very sophisticated to generate good code. Thus they are cost effective only when a large number of different target machines may be used, and this is a rather uncommon situation.

Among the great variety of proposed languages, we shall briefly examine: micro-C, VMPL, HLL, S*M, SUILVEN and MIMOLA.

micro-C

Within the languages proposed to guarantee a certain degree of compatibility with a high level programming language, C language plays a key role. In fact, numerous versions of C for microprogramming, called micro-C [Gurd_83], [HHAr_85], [DuMu_84], [DuMu_85] have been proposed. Micro-C is especially interesting, since it is well known, and compilers are readily available; micro-C programs can be recompiled and run as C programs, making microprogram debugging easier. In addition the features of C correspond well to those useful in microprogramming and C compilers tend to produce efficient code. One company has recently announced a microengine and compiler that

will allow acceleration of any C program by compilation into
microcode. We are likely to see large application libraries of
compute intensive code migrated into microcode in next years.

VMPL

Many characteristics of high level machine independent languages are
well summarized by the language VMPL, designed and implemented at
Oregon State University by Lewis and his colleagues [MaLe_78],
[MaLe_80], [MaLe_81]. A program written in this language is
essentially an abstract microprogram within which the declared
variables refer to objects in the target machine and all specified
executable statements result in transformations on the state of these
variables. In order to execute this abstract microprogram on some
given host, the program is first translated into a machine independent
"intermediate machine language" named IML and then compiled into host
machine microcode. The compilation requires three phases: the first
is supplied with a description of the host and produces a "machine
dependent intermediate language" (MDIL); the second translates this
into a vertical microprogram and the third produces the target
microcode performing complex scheduling, allocations, and
optimizations (a local microcode compaction algorithm is adopted).

An interesting feature of VMPL is the association of the
attributes Global/Local and Temporary/Permanent with variables
declared inside a block. The purpose of these attribute is to assist
the compiler in allocating or de-allocating the host's resources, by
explicitly assigning priorities to declared variables. Thus, when the
compiler has to bind a variable to a register and no free registers
are available, the priority of the variables determine which of them
is to be deallocated, so that the freed register can be reassigned.
Global variables have a higher priority than local (i.e., get
deallocated first) and temporary variables take precedence over
permanent ones.

The VMPL system has been implemented and used to generate
microcode for PDP 11/40E host machine. This work represents therefore
the first successful attempt at building a retargetable
microprogramming system.

HLL

A high level machine independent language named HLL (Higher Level
Language) has been developed at JRS Labs by Sherada and Gieser
[ShGi_81], [GiSh_82]. They implemented it to allow the translation of
computation intensive application programs, written in high level
languages, for distributed processing environments, into microcode.
Source programs are compiled first to a machine independent
intermediate language and then to a machine dependent form consisting
of elementary microoperations, with optimization performed during each
step. The microoperations are then compacted into executable
microinstructions for a specific target machine. The procedure has
been implemented for experimental purposes and used to compile several
types of application programs. Experimental results obtained by
comparing the microcode produced by HLL with hand coded microprograms
showed that the compiled version was from 5 to 10% longer than the
hand coded one [ShGi_83]. Anyway, programmers' productivity was
greatly increased by the use of HLL.

S*M

Dasgupta and his co-workers have developed S*M: a functional language
for describing the behaviour of clocked microarchitectures, i.e., of
microarchitectures whose operations and behaviours are under the
control of one, or more, synchronized clocks [DWHe_86], [Wils_85].
S*M was developed from a revised model for clocked microarchitectures
that was originally proposed by Dasgupta in [Dasg_84b]. In the design
and implementation of S*M the authors have been strongly influenced by
some of the ideas originating in the software domain on functional
specification, axiomatic (or assertional) proofs of program
correctness, and non procedural (or rule based) invocation of actions.
According to the intentions of the proposers, S*M could be used for:

- the description of clocked microarchitectures for automatic
  retargetable systems for microcode generation, optimization,
  compaction and verification;

- simulating the operations of machines at the
  microarchitecture level, for performance evaluation and
  microprogram debug;

- the formal specification of clocked architectures at
different levels of abstraction;

- the specification of architectural behaviour of a system to
feed an automated design synthesis system.

On the basis of these considerations, S*M has been implemented as
an axiomatic language, whose main feature is that the behaviour of a
system is expressed in terms of assertions in some axiomatic or
deductive system. The primary advantages of such organization arise
when the designer wishes either to postpone making operational design
decisions or when he wishes to express his ignorance of the internal
structure of some modules.

SUILVEN

The SUILVEN language, proposed by Sommerville [Somm_79], has been
developed from an earlier machine description language called SMDL.
It provides facilities for the programmer to specify exactly the size
and structure of the data elements. In addition, it allows programs
to be split into independent procedures and has control constructs
naturally leading to structured programming. The SUILVEN compiler
allocates addresses to each data area, thus removing the necessity for
a run time interface between a SUILVEN program and its associated
machine.

MIMOLA

Marwedel and his co-workers implemented MIMOLA (Machine Independent
MicroprOgramming LAnguage), which has been designed as a unique
language capable of describing both the high level microcode and the
target architecture [Marw_81], [Marw_84b], [Zimm_76], [Zimm_80a],
[Zimm_80b]. The hardware model underlying MIMOLA has been dictated by
its use in the hardware design system MSS2 [Marw_85] and is
characterized by "hardware modules" and "data paths" [Marw_84a]. For
some applications this structural description may be lower than
necessary. Data paths represent the resources, whereas the operators
may be either the standard ones or user-defined. This extension
mechanism allows adding target dependent operators to the language,

without changing the syntax of the compiler. By this facility, machine dependent programs can take advantage of special purpose hardware, which may be mapped into non standard operators. The meaning of non standard operators has to be declared, in a way that strongly resembles ADA function declaration.

MIMOLA has been extended to include most of PASCAL's high level language elements, without providing, however, a strong typing mechanism. The basic data type is thus the bit string of arbitrary length.

A MIMOLA program is a sequence of parallel instructions called esb's (elementary statement block). Each esb contains statements, which are assumed to be executed in parallel, if enough hardware is available.

Automatic and manual storage allocations are possible; in the latter case each variable must be associated with its address.

A total of about 20 different targets are reported to have been programmed in MIMOLA [Marw_84a].

2.2   High Level Languages With Machine Independent And Dependent
       Components.

These languages allow the inclusion of some machine dependence for efficiency reasons: they are mostly machine independent languages that may be properly adapted to a given target. There is no strict dividing line between machine dependent and independent languages. Most languages have machine dependent and independent features, and the proper mixture to obtain a proper degree of portability and efficiency is not yet known.

Several techniques have been developed to get the advantages of both machine dependence and independence [Davi_86]. The common characteristic of all these approaches is to reduce the amount of work required to port the language and compiler by attempting to maximize the common features of what can be viewed as a set of different machine dependent languages.

We shall now examine three of the proposed approaches.

2.2.1   First Approach

The first approach aims at designing a language that is only somewhat machine dependent, perhaps using a standard set of operators and the data resources of a particular target. This strategy makes the compiler much easier to build and mostly portable, and keeps the language easy to use.

2.2.2   Second Approach

A second strategy is to design a machine independent language framework, that can be altered in a well specified way to produce a machine dependent language. It is important to note that the use of a given language family does not necessarily depend on the availability of complex compilers and simulators, whose implementations may not be economically justified. As long as there exist well-defined syntax and semantics for the language(s), these can still support systematic design in an acceptable fashion. It has been reported [Davi_83], for instance, that microcode for the Intel iAPX 432 was developed along this line, using ADA both as an architecture description and a firmware design language. High level algorithms were first written in ADA without any binding of variables or operations to specific hardware resources. The ADA description was then manually transformed into a lower level representation using an ADA subset, with some hardware bindings, intended to facilitate the specification of machine specific microcode. Finally, further manual translation into assembly level microcode was done.

Certainly the most interesting problem that the "family of languages" approach poses is to demonstrate the correctness of inter-level transformations. The method developed at SRI

International [Role_77], [MSSc_82] uses a rigorous and formally verifiable inter-level mapping technique.

Three different approaches will now be outlined, and namely: S*, microAPL and OHNE.

S*

The language schema S*, proposed by Dasgupta [Dasg_78] [Dasg_80] is based on the consideration that, for efficiency reasons, the syntax and the semantics of some constructs, such as those corresponding to assignment statements and to arithmetic and logic operations, should be tailored to a specific host. On the contrary, control structures should be expressed in a machine independent manner, since they are very similar in most machines. In addition, the microprogrammer should be given the possibility of having very low-level control of hardware resources, so that microcode segments that are time critical or heavily used may be optimally specified, if necessary, by the user, rather than be generated by the compiler. Finally it is desirable to free the microprogrammer from having to be concerned with control unit architectures. Dasgupta deduced that a language satisfying these issues may not be directly applicable to all host microarchitectures, but it may provide a detailed language "schema", i.e., a partially specified language, containing constructs that are usable across a wide range of hosts. S* was proposed as such a schema and it denotes a family of languages: for a given microarchitecture M1, a particular language S*(M1) may be obtained by completing the incomplete syntax of S*, taking into account the specific characteristics of M1: S* is thus "instantiated" to a particular language S*(M1), with respect to M1.

The design of S* was largely influenced by PASCAL and by many important principles of programming methodology. The primitive data type in S* is the bit, consisting of the values {0,1}. All other data types are structures from this primitive and include the "sequence", "array", "stack", and "tuple" (similar to PASCAL's record) data types. Anyway, since a specific instantiation S*(M1) is a high level machine dependent language, its data objects must be specific and tailored to M1 architecture. Thus the microprogrammer may not arbitrary invent new data objects.

The machine independent constructs in S* include the usual
control statements (sequential composition, if-then-else, while and
repeat, parameterless procedure call, and goto). In addition, the
construct "cocycle" allows the microprogrammer to specify low-level
parallelism statements: all the statements enclosed between "cocycle"
and "coend" are executed "simultaneously" in that they all begin and
terminate within a single microcycle, whose duration and definition is
machine independent. The parallel statements in S* are rather general
and powerful, allowing the specification of statements that satisfy
monophase, polyphase, and multicycle timing behaviour. An additional
construct "region" has been added: it forces the compiler not to
optimize or compact all the statements enclosed between "region" and
"endreg".

The main strength of the S* philosophy is that it permits
instantiation of the schema at different levels of abstraction of a
host machine. Indeed, a particular instantiation may view the given
host machine at more than one level of abstraction, so that one may
mix "high level" and "low level" statements within the same program.

A preliminary instantiation of S* for the VARIAN 75 machine has
been done in 1980. Later on it has also been instantiated with
respect to Nanodata QM-1 [Dasg_84a], [KlDa_82].

A more elegant technique for instantiation has been proposed by
Damm and his colleagues [Damm_84a], [Damm_84b]. They have implemented
a syntax directed proof system for the verification of horizontal
architectures. The system is based on the axiomatic architecture
description language AADL [DDMS_86], which is sufficiently rich to
allow the specification of target architectures while providing a
concise model for clocked microarchitectures. For each description N
written in AADL they systematically construct a (Hoare-style)
axiomatic definition of an N-dependent high level microprogramming
language based on S*, allowing a complete axiomatic treatment of the
timing behaviour and dynamic conflicts of microprograms written in
S*(N) [Damm_85], [DaDo_85]. In fact, a microarchitecture's given
formal description N in AADL defines an N-dependent microprogramming
language S*(N) such that [DDMS_86]:

- variables of S*(N) correspond exactly to storage elements
    declared in N;

- assignments of S*(N) correspond exactly to transfer or function microoperations declared in N;

- test conditions of S*(N) correspond exactly to sequencing microoperations declared in N;

- control structures of S*(N) correspond exactly to those of language schema S*.

microAPL

The idea of instantiating a schema for a particular host machine has also been explored by Hopson, using an APL like language. The language, named microAPL [HHTh_81] is particularly suited for machines in which up to 4 or 5 microoperations may be explicitly referenced in a microinstruction. This number is restricted by the condition that each microinstruction must not exceed one line of microAPL code. In spite of this strong drawback, the language allows a complete firmware engineering environment to be set up with a common language.

OHNE

A particularly encouraging industrial demonstration of the efficiency of the schema principle is OHNE: a language developed at Burroughs Corporation as part of a suite of tools called the E-machine workbench [WaMa_83]. This has been used for the development of a Burroughs family of stack-based architectures. The schema approach was reported to be particularly effective since microcode was required to be developed for a number of similar though not identical processors. Like S*, OHNE actually defines a family of languages and each implementation is an instantiation reflecting the characteristics of a particular host machine.

The basic execution entities in the language are subroutines and operations. The latter ones are, in fact, special subroutines that are bound to particular host machine operators and are thus similar to the notion of bound resources observed in MARBLE [Davi_80].

Probably the most distinctive feature of OHNE is its concept of

"definitions", which provides effective support for stepwise
refinements. They are string of characters enclosed between "<<" and
">>" that can be inserted within a microprogram and signify
unspecified pieces of code. These can then later on be specified and
amplified by executable OHNE code and could, in turn, embed other
"definitions".

2.2.3   Third Approach

A third approach to the realization of machine
dependent/independent high level microprogramming languages is to
model the target machine resources that must be used by the source
microprogram in terms of machine independent language features. The
program remains machine independent, and thus portable, but the
modelled target resources can be bound directly to actual hardware
resources by the compiler.

Among the examples belonging to this class, we shall examine
MARBLE, EMPL and LUKKO.

MARBLE

MARBLE, proposed by Davidson [Davi_80], [DaSh_80b], is one example of
this approach. The goal in the design of MARBLE was to provide a
medium in which microprograms could be written either for some
abstract machines or for specific host architectures. MARBLE allows
the implementation of host machine data objects in terms of standard
PASCAL-like type constructors. Similarly, MARBLE functions can be
used to represent machine-specific functional resources. These
constructs can be recognized by the compiler and passed on to the code
generator. In addition, some facilities have been added to facilitate
resource binding specification at the source level.

Basically the MARBLE compiler first partitions the source program
into a machine independent part and a machine dependent one. The
former is translated into a machine independent intermediate form
which is then passed to the code generator, where its resources are

bound and merged with the machine dependent code.

EMPL

Another example of this third approach is represented by EMPL
(Extensible MicroProgramming Language): a language proposed by DeWitt
and using an approach based on the class concept of SIMULA [DeWi_76].
The language is designed to be extensible so that it may take
advantage of special microinstructions available on the target
machine. The extensibility mechanism works by allowing the user to
define types, consisting of data structures plus associated
operations. The specification of a "machine dependent microoperation"
in an operator body guarantees a certain degree of machine
independence, while retaining the chance to produce efficient code.
However, the provided control statements are not sophisticated enough
to allow optimal microinstruction sequencing (case and nested
if-then-else constructs are in fact absent).

LUKKO

Along the line of EMPL, Heinanen proposed a language named LUKKO
[KuHe_80], [Hein_83], characterized by having two abstraction
mechanisms, called "procedure declaration" and "type declaration", by
means of which higher control operators and abstract data types can be
built from the host machine level data types and microoperations.
LUKKO is thus very much in the spirit of microprogramming languages
based on data abstraction principles, such as Modula2. in fact, in
contrast to the traditional approach, data objects with
microprogrammed operations are typed objects in LUKKO, not untyped bit
configurations. However, their implementation is based on a
machine-oriented view of the microprogrammable machine, with a single
untyped primitive data structure called word. LUKKO is the natural
evolution of the low level machine dependent language MLUKKO (see
below).

2.3   High Level Machine Dependent Languages.

High level machine dependent languages usually provide high level
control resources, which are bound to target control resources by the
compiler. Data and functional resources are bound at language
definition time, although rather complex expressions are usually
available. The details of data transfers between memory and registers
are handled by the compiler, as is some of the detailed bookkeeping
required by the target (such as checking for interrupts every so many
cycles).

Applications for these languages include those of the low level
machine independent languages, as well as larger applications that
would benefit from the productivity improvement possible with higher
level languages. These applications include operating system assists,
console programs and other large microprograms suitable for only one
target.

Compilers for machine dependent languages are easier to write
because the binding process, which is at the heart of compilation, is
straightforward. The code that is produced by these compilers is
likely to be efficient, since the programmer has the responsibility
for making best use of target machine resources. However, these
compilers cannot be moved to other targets, and this forces people
using other targets to build up new compilers from scratch.

Since advances in compiler generators have reduced the amount of
work needed to implement parsers and lexical analyzers, thus allowing
more of the effort to be put in the code generation task, the ·number
of adopted high level machine dependent languages is expected to
increase.

We shall examine here: MPL, MPGL, STRUM and microTAL.

MPL

An early example of high level machine dependent language is MPL, a
PL/1 like language developed by Eckhouse [Eckh_71]. Data objects are
declared in MPL as one or two dimensional "arrays of bits" or as

"events".   For   instance,   registers   would   be   represented   as
one-dimensional integer arrays, main memory as a two-dimensional array
and testable flags as events. The main statements are the assignment,
if-then, if-then-else, and goto. MPL programs are composed of
procedures, and variables may be defined as being local or global with
respect to procedures. Although the MPL was originally intended to be
a   host   machine   independent   language,   it   seems   oriented   toward
vertically   organized   control   unit   architectures   and,   more
specifically, toward Interdata 3.

MPGL

A language comparable with MPL is MPGL, proposed by Baba [Baba_77],
[BaHa_81]; its structuring facilities are rather poor but it presents
the interesting feature that the complete specification of the target
is part of the source program and not embedded in the compiler. The
MPGL compiler uses this specification to generate the code and thus it
can be fed with a program for any machine having a microarchitecture
that can be described in the specification formalism of MPGL.   Note
that, however, this does not imply that the MPGL programs are machine
independent.

STRUM

Another language in this category is STRUM, an ALGOL like language
designed and implemented by Patterson [Patt_76], [Patt_77], [RFFi_72]
and oriented to the implementation of Burroughs D Machine.   The
philosophy underlying this effort was to construct a language and a
programming   system   permitting   well   structured,   verifiable
microprograms to be written for the D Machine, without having to
sacrifice object code efficiency. Because of this, Patterson paid a
great deal of attention to the structuring facilities in the language,
which, as a result, has four types of repetition statements, four
alternative selection statements, a macro statement, and the procedure
statement. A STRUM program can be organized into blocks thus enabling
the scope of a variable to be explicitly outlined.

STRUM represents an important contribution to firmware
engineering, since it has demonstrated the feasibility of both

efficient microcode generation from a high level microprogram and
formal verification using axiomatic semantic and mechanical
verification techniques

A modified version of STRUM, named STRUM2, has been used as input
language for the V-Compiler: a retargetable microprogram generation
system [PGPS_81], [Patt_81].

microTAL

A further example of high level machine dependent language is
microTAL, implemented by Bartlett [Bart_81]. It is a subset of an
existing system programming language named TAL. The purpose of the
development of microTAL was to allow the migration of system and
application functions written in TAL to microcode, without expensive
reprogramming. The language has been used to write microcode for the
Tandem Nonstop II processor. The microTAL compiler, written in LISP,
automatically generates code to handle interrupts and page faults
which may occur during the execution of a procedure.

3   LOW LEVEL MICROPROGRAMMING LANGUAGES

Low level microprogramming languages are characterized by
providing a set of primitives that can be directly mapped into one
(and seldom more) microinstruction(s) of the host machine. They find
their wider usage in the development of the so called "system
firmware", i.e., of those microprograms mainly related, for instance,
to the implementation of instruction sets, operating system assist
code and diagnostics. This kind of application is in general not well
suited for high level microprogramming languages, for many reasons.
Each new application is bound to a new architecture and is thus
strongly machine dependent. In some cases, such as in the
microinstruction set implementation, microprograms are rather small
and must be extremely efficient. Moreover, in most cases, this kind
of microcode must be available in the very early project phases, but,
as already mentioned, the design of an efficient compiler for a high
level machine independent microprogramming language is a cumbersome

task and its development may become the bottleneck of the global project.

Also for low level languages machine independent and machine dependent languages have been proposed.

3.1 Low Level Machine Independent Languages

These languages implement a set of low level control and functional resources available on a wide range of targets. The data resources are simple, often a generic register type and arrays. Binding is usually done by the compiler, even if most implementations provide some binding to specific target registers in the language definition. This makes programs in the language not really portable.

These languages are suited for applications in which machine independence is important, but efficiency is still required. These needs are somewhat contradictory.

We shall here outline the main characteristics of YALLL and MIDL.

YALLL

An example of low level machine independent language is YALLL (Yet Another Low Level Language) [PLTu_79]. Its basic assumption is that is possible to choose a set of low level operators that can serve as the basis for efficient microcode generation. Primitive operators correspond to typical primitive functions present in microarchitectures, such as arithmetic operators add, subtract, increment and decrement, logical and shift operators, memory load and store, and register to register transfer. The syntax of primitive YALLL statements resembles assembly language statements, whereas the concept of data type is absent. Variables are all considered to be general purpose registers, with the exception of the memory address register (MAR) and the memory buffer register (MBR). Variables may be declared in YALLL programs and bound to specific registers in the host machine.

YALLL contains conditional and unconditional branch statements, including a multi—way branch, statements for calling and returning from procedures and an exit statement.   There are no means for expressing parallelism in the source program and parallelism detection and compaction must be done by the compiler.

YALLL was implemented on two different host machines, the VAX 11/780 and the HP 3000. The philosophy and premises underlying YALLL are quite similar to those of RISC architectures and therefore it appears to be quite suitable as the assembly language for such architectures. On the other side, the very primitiveness of YALLL operators and control structures, the absence of data structuring capabilities and the fact that parallelism cannot be expressed at the source level, seem to suggest that the language can not be implemented efficiently for more complex architectures.

MIDL

MIDL (MicroInstruction Description Language), proposed   by   Sint [Sint_81] is a declarative language, in which the semantic of microoperations is defined in term of YALLL operators.   In addition, it defines triggering conditions for microoperations and their operands, and it defines how operations are timed.   The "during" construct, in fact, associates timing information with all operations within its scope. Inner "during" constructs take precedence over outer ones, thus allowing an easy modelling of local exceptions to a global timing scheme.

The specifics of YALLL do not affect the structure of MIDL, since the replacement of YALLL by some other languages would simply result in another dialect.

3.2   Low Level Machine Dependent Languages.

In these languages the binding to target resources is done at language   definition   time.   The control, data, and functional primitives in the language closely match those available on the target, with control loop constructs sometimes added. Registers may be often renamed to match registers used by the algorithm, although storage allocation is done exclusively by the microprogrammer.

These languages are the closest to the target machine, and are therefore most useful for applications in which efficiency of the resulting code is paramount. These applications include emulation and device control, which constitute a great part of the microprograms written. are also the easiest to design and implement. However, the benefits of using a low level machine dependent language over a microassembly language are slight, reducing the incentive to create them. Another disadvantage is that they are useful for only one target. A compiler generator for these languages would increase their use, but most work in this area is being applied to higher level languages.

We shall briefly examine GMPL and MLUKKO.

GMPL

An example of low level machine dependent language is GMPL, proposed by Guffin [Guff_82] and adopted for the microcode development for the Hewlett Packard HP 3000 processor. It is mainly characterized by relatively powerful constructs, such as partial field operations and by permitting appropriate sub-options to be coupled with each of the various constructs.

MLUKKO

MLUKKO is a low level machine oriented language, with some high level feature, developed by Kurki-Suonio for the Ukko computer [KuHe_80]. The language is typeless, the only data objects being machine registers, memory words and constants. Memory allocation is static and can be fully controlled by the programmer. Language LUKKO has been derived from it.

3.3   Microprogram Meta-assemblers

As a concluding remark on low level   microprogramming   languages,
some words must be spent on microprogram assemblers. One of the major
requirements is here to pursue a certain degree of flexibility,   i.e.,
the possibility of utilizing the same tool in the development of
different target machines.   Such a need arises particularly from
implementations   based   on bit-slice microprocessors. In these cases,
in fact, the development of hardware and firmware must in   general   be
carried on concurrently, owing to their strict interdependence; a
change in the architecture of the processor usually requires a
redefinition of microinstruction formats.   A   possible   method for
achieving the necessary flexibility is   the   implementation   of
microprogram assemblers which can assemble code for any user-defined
target machine, i.e., of the so called "microprogram meta-assemblers".
A good survey on the activities in this area has been done by
Skordalakis [Skor_83], and by Powers and Hernandez [PoHe_78].

Different strategies have been proposed for   the   meta-assembler
organization.   With the help of the assembler definition, the
meta-assembler can either generate the desired assembler or adapt
itself to operate as the desired assembler. The formers are often
referred to as "generative meta-assemblers" and, at my best knowledge,
no generative microprogram meta-assemblers exists. The latter ones
are called "adaptive meta-assemblers" and can be divided into two
sub-categories,   according to the nature of their meta-assembly
language. To the first sub-category belong those which have a
meta-assembly language which is a macro language. The DGCAS system
developed by Skordalakis is the most significant example in this   area
[Skor_79a],   [Skor_79b], [Skor_80]. The second sub-category is by far
the most populated one, including those meta-assembly languages which
are not macro languages.   To this sub-category belong, among the
others, the following microprogram meta-assemblers:

      - AMDASM, developed at Advanced Micro Devices   [Hern_77],
         [Amd_76], [Amd_77a], [Amd_77b];

      - CROMIS, developed at Intel [Crom_75];

      - DAPL, developed at Zeno Systems [Dapl_76];

- DEFASM, developed at Politecnico di Torino, Torino, Italy [MPRo_81], [GPTa_84];

- HP64000, developed at Hewlett Packard [Yack_80];

- MACE, developed at Motorola [BaBl_79], [Mace_78];

- MAG, developed by Watabane [Wata_77];

- MICRO 8, developed by Greenberg [Gree_81];

- MICROMON, developed by Habib [HaYa_81];

- MICRO-AID, developed at Monolithic Memories [Graz_77], [Micr_77];

- MSFC, developed at McDonnell Douglas [HoEd_75], [MFS_77];

- M29, developed at Advanced Micro Devices [Eage_83];

- QMXASM, developed at Nanodata [Berg_80];

- RAPID, developed at Scientific Micro System [Rapi_74];

- RAYASM, developed at Raytheon [Raya_76];

- SIGNETICS, developed at Signetics [Sign_77].

The main characteristics of all these microprogram meta-assemblers are well summarized in [PoHe_78] and [Skor_83]. In most of them the required flexibility is achieved by splitting the processing of the source microprogram into two phases [PoHe_78]: microinstruction definition phase and assembly phase. The former sets up the microprogram word structure and mnemonic assignment, and is made once and for all for a given target machine. The latter translates each statement of the microprogram source (expressed in the mnemonic terms established during the microinstruction definition phase) into actual microcode, i.e., processor compatible bit patterns.

In some of these tools, (for instance, in DEFASM [MPRo_81]) the user is given the possibility of specifying some assertions about microcode organization, in terms of semantic correlations and logical relationships among microinstruction fields, sub-fields or mnemonic codes, that must always hold. This allows the meta-assembler to perform some automatic control on the code appearing in a source microprogram statement, resorting to the user provided semantic rules.

In the area of microprogram assemblers, it is worthwhile mentioning the work done at NEC to translate microprograms written for a target A into microprograms suitable for target B. To maintain the universality of a conversion tool, knowledge base techniques have been used. The tool should work for both horizontal and vertical microcode, even if troubles have been reported in converting horizontal microprograms [TTBS_84]. By using such a tool, a 50,000 step microprogram has been converted with just 40% of the man power required by the ordinary method. 2,500 step conversion rules have been inserted in the knowledge base for gaining such a result.

4   CONCLUSIONS

In the paper a survey on microprogramming languages is presented. From the analysis of current and future trends, it is evident that many research and development efforts are put in the implementation of integrated multi-purposes microprogramming systems for microcode description, generation, verification, and simulation. One of the main issues in this area is retargetability, i.e., the possibility of generating object microcode for different target machines from the same high level language source, by simply changing the description of the target architecture. This creates a problem area in which both firmware engineers and designers of hardware description languages must strongly cooperate. Some of the issues in this domain have recently been addressed in [Dasg_84b], [Damm_84a] and [Damm_84b].

5   REFERENCES

[Amd_76]   "CSC AMDASM Reference Manual," Computer Sciences Corp., El Segundo, CA, USA, 1976

[Amd_77a] "AMDASM/29 Reference Manual," Advanced Micro Devices, Sunnyvale, CA, USA, 1977

[Amd_77b] "AMDASM/80 Manual," Advanced Micro Devices, Sunnyvale, CA, USA, 1977

[Baba_77] T.Baba: "A microprogram generating system, MPG," Information Processing 1977 North Holland Publ., Amsterdam, 1977, pp. 739-744

[BaBl_79] T.Balph, W.Blood: "Assembler Streamlines Microprogramming," Computer Design, December 1979, pp. 79-89

[BaHa_81] T.Baba, H.Hagiwara: "The MPG System: A Machine Independent Efficient Microprogram Generator," IEEE Transactions on Computers, Vol.C-30, No.6, June 1981, pp. 373-395

[Bart_81] J.F.Bartlett: " MicroTAL - A Machine-Dependent High-Level Microprogramming Language," MICRO14: 14th Annual Microprogramming Workshop, Chatham, MA, USA, October 1981, pp. 109-114

[Berg_80] G.R.Berglass: "A Meta-assembler for High-Parallel Microprogrammable Systems," MICRO13: 13th Annual Microprogramming Workshop, Colorado Springs, CO, USA November 1980, pp. 181-189

[Crom_75] "CROMIS (series 3000 Cross Microprogramming System) Reference Manual," Intel Corp., Santa Clara, CA, USA, 1975

[DaDo_85] W.Damm, G.Dohmen: "Verification of Microprogrammed Computer Architectures in the S* - System: A Case Study," MICRO18: 18th Annual Microprogramming Workshop, October 1985

[Damm_84a] W.Damm: "Automatic Generation of Simulation Tools: A Case
          Study in the Design of a Retargetable Firmware Development
          System," Proc. Euromicro-84, B.Myrhaug (Ed), North-Holland,
          Amsterdam, 1984, pp. 165-176

[Damm_84b] W.Damm: "A Microprogramming Logic," Bericht   Nr.94,
          Technische Hochschule Aachen, Schriften Zur Informatik Und
          Angewandten Mathematik, Aachen, West Germany, May 1984

[Damm_85] W.Damm: "Design and specification of microprogrammed
          computer   architectures,"   MICRO18:   18th   Annual
          Microprogramming Workshop, October 1985

[Dapl_76] "DAPL Microprigramming Language   User's   Manual,"   Zeno
          Systems, Santa Monica, CA, 1976

[Dasg_78] S.Dasgupta: "Toward a microprogramming language schema,"
          MICRO11: 11th Annual Microprogramming Workshop, October
          1978, pp. 144-153

[Dasg_80] S.Dasgupta: "Some Aspects of High Level Microprogramming,"
          ACM   Computing   Surveys, vol.12, No.3, September 1980,
          pp. 295-324

[Dasg_84a] S.Dasgupta: "The Design and Desription of Computer
          Architectures," John Wiley & Sons, New York, 1984

[Dasg_84b] S.Dasgupta: "A model for clocked microarchitecrtures for
          firmware engineering and design automation applications,"
          MICRO17: 17th Annual Microprogramming Workshop, New
          Orleans, LU, USA, October 1984, pp. 298-308

[Dasg_85] S.Dasgupta: "Hardware   Description   Languages   in
          Microprogramming Systems," IEEE Computer, 18, 2, February
          1985, pp. 67-96

[DaSh_80a] S.Davidson, B.D.Shriver: "Firmware   Engineering: An
          Extensive   Update,"   IFIP   TC-10   Conference   on
          Microprogramming, Firmware and Restructurable Hardware,
          (G.Chroust and J.Mulbacher eds), North Holland, Amsterdam,
          1980, pp. 1-36

[DaSh_80b] S.Davidson, B.D.Shriver: "MARBLE: a high level machine independent language," IFIP TC-10 Conference on Microprogramming, Firmware and Restructurable Hardware, (G.Chroust and J.Mulbacher eds), North Holland, Amsterdam, 1980, pp. 1-36

[DaSh_85] S.Dasgupta, B.D.Shriver: "Developments in Firmware Engineering," Advances in Computers, Vol.24, Academic Press, New York, 1985, pp. 101-176

[Davi_80] S.Davidson: "Design and Construction of a Virtual Machine Resource Binding Language," Ph.D. Thesis, Department of Computer Science, University of Southwestern Louisiana, Lafayette, December 1980

[Davi_83] S.Davidson: "High Level Microprogramming: current usage, future prospetcs," MICRO16: 16th Annual Microprogramming Workshop, Downingtown, PA, USA, October 1983, pp. 193-200

[Davi_86] S.Davidson: "Progress in High-Level Microprogramming," IEEE Software, July 1986, pp. 18-26

[DDMS_86] W.Damm, G.Dohemen, K.Merkel, M.Sichelschmit: "The AADL/S* approach to firmware design verification," IEEE Software, July 1986, pp. 27-37

[DeWi_76] D.J.DeWitt: "A machine independent approach to the production of horizontal microcode," Ph.D. Thesis, University of Michigan, Ann Arbor, 1976

[DuMu_84] M.R.Duda, R.A.Mueller: "Micro-C Microprogramming Language specification," Technical Report CS 84-11, Dept. of Computer Science, Colorado State University, Fort Collins, CO, USA, October 1984

[DuMu_85] M.R.Duda, R.A.Mueller: "Micro-C Microprogramming Language (Version 3.1) Reference Manual," Technical Report CS 85-11, Dept. of Computer Science, Colorado State University, Fort Collins, CO, USA

[DWHe_86] S.Dasgupta,   J.Henanen,   P.A.Wilsey: "Axiomatic
specifications   in   firmware   development   systems,"
microarchitectures," IEEE Software, July 1986, pp. 49-58

[Eage_83] M.J.Eager: "M29 - An advanced retargetable microcode
assembler," MICRO16: 16th Annual Microprogramming Workshop,
Downingtown, PA, USA, October 1983, pp. 92-100

[Eckh_71] H.Eckhouse: "A high level microprogram language (MPL),"
Ph.D.   Thesis, Dept. of Computer Science, State University
of New York, Buffalo, USA, 1971

[Eldr_84] J.Eldridge: "A metasimulator for microcoded processors,"
MICRO17:   17th   Annual   Microprogramming Workshop, New
Orleans, LU, USA, October 1984, pp. 129-137

[GiSh_82] J.L.Gieser, R.J.Sheraga: "Microarchitecture Desription
Techniques,"   MICRO15:   15th   Annual   Microprogramming
Workshop, Paolo Alto, CA, USA, December 1982, pp. 23-31

[GPTa_84] S.Gai, P.Prinetto, R.Tasso: "DEFASM: User   manual,"
Internal report Dipartimento di Automatica e Informatica -
Politecnico di Torino - I.R. DAI/CAD, 13/84, Torino, Italy,
1984

[Graz_77] S.Graziano: "MICRO-AIDM A Microcode Assembler,"   Proc.
MICRON/77, November 1977

[Gree_81] K.F.Greenberg: "The Micro8 MIcrocode Assembler," MICRO14:
14th Annual Microprogramming Workshop, Chatham, MA, USA,
October 1981, pp. 78-82

[Guff_82] R.M.Guffin: "A   Microprogramming   Language   Directed
Microarchitecture," MICRO15:   15th Annual Microprogramming
Workshop, Paolo Alto, CA, USA, December 1982, pp. 42-49

[Gurd_83] R.P.Gurd: "Experience Developing Microcode Using a High
Level Language," MICRO16:   16th Annual Microprogramming
Workshop, Downingtown, PA, USA, October 1983, pp. 179-184

[HaYa_81] S.Habib, X.L.Yang: "The Use of a Meta-assembler to Design an M-code Interpreter on ADM2900 Chips," ACM Sigmicro Newsletter, Vol.12, No.4, 1981, pp. 38-50

[Hein_83] J.Heinanen: "A data and control abstraction approach to microprogramming," Publ. No.18, Tampere University of Technology, Tampere, Finland, 1983

[Hern_77] J.H.Hernandez: "AMDASM, A Microprogram Assembler for Bit-Slice Microprocessors," Proc. MICRON/77, November 1977

[HHAr_85] W.Hopkins, M.Horton, C.Arnold: "Micro-C, Experiments in High Level Firmware Generation Techniques," MICRO18: 18th Annual Microprogramming Workshop, October 1985

[HHTh_81] R.F.Hobson, P.Hannon, J.Thornburg: "High-Level Microprogramming with APL Syntax," MICRO14: 14th Annual Microprogramming Workshop, Chatham, MA, USA, October 1981, pp. 131-139

[HoEd_75] B.C.Hodges, A.J.Edwards: "Support Software for Microprogram Development," ACM Sigmicro Newsletter, January 1975, pp. 17-24

[KiTa_79] J.Kim, C.J.Tan: "Register assignement algorithms for optimizing microcode compilers, Part.I," Report RC7639, IBM T.J.Watson Research Center, Yorktown Heights, NY, USA, March 1979

[KlDa_82] A.Klassen, S.Dasgupta: "S*(QM-1): an instantiation of a high level microprogramming language schema S* for the Nanodata QM-1," MICRO15: 15th Annual Microprogramming Workshop, Paolo Alto, CA, USA, December 1982, pp. 126-130

[KuHe_80] R.Kurki-Suonio, J.Heinanen: "A data abstraction language based on microprogramming," MICRO13: 13th Annual Microprogramming Workshop, Colorado Springs, CO, USA, November 1980, pp. 154-161

[LSDa_83] J.Linn, B.Shriver, S.Dasgupta: "Component identification for a portable retargetable firmware development system," IEEE Int. Workshop on Computer System Organization, 1983, pp. 164-170

[Mace_78] "Motorola MACE 29/800 Development System User's Guide," Motorola Semiconductor Products, Phoenix, AZ, 1978

[MaLe_78] K.Malik, T.Lewis: "Design objectives for high level microprogramming languages," MICRO11: 11th Annual Microprogramming Workshop, October 1978, pp. 154-160

[MaLe_80] P.R.Ma, T.Lewis: "Design of a machine independent optimizing system for emulator development," ACM Trans on Progr. Lang. Syst. 2(2), 1980, pp. 239-262

[MaLe_81] P.R.Ma, T.Lewis: "On the design of a microcode compiler for a machine indipendent level language" Trans on Software Eng., vol SE-7, No.3, May 1981, pp. 261-274

[Marw_81] P.Marwedel: "A retergetable microcode generator system for a high level microprogramming Language," MICRO14: 14th Annual Microprogramming Workshop, Chatham, MA, USA, October 1981, pp. 115-123

[Marw_84a] P.Marwedel: "A Retargetable Compiler for a High Level Microprogramming Language," MICRO17: 17th Annual Microprogramming Workshop, New Orleans, LU, USA, October 1984, pp. 267-274

[Marw_84b] P.Marwedel: "The MIMOLA Design System: Tools for the Design of Digital Processors," ACM IEEE 21th Design Automation Conference, Albuquerque, (USA), July 1984.

[Marw_85] P.Marwedel: "The MIMOLA Design System: A Design System which Spans Several Levels," in: W.Giloi (ed.): Methodology for Computer System Design, North Holland, 1985

[MavM _80] S.D.Marcus, D.van-Mierop: "The ISI Microcode Development System," IFIP TC-10 Conference on Microprogramming, Firmware and Restructurable Hardware, North-Holland, Amsterdam, 1980, pp. 243-249

[MFS_77] "Meta-assembler User's Manual, Cosmic Program," Report No. MFS-23449, Computer Center, University of Georgia, Athens, GA, USA

[Micr_77] "MICRO-AID Micro Assembler User's Manual," Monolithic Memories, Sunnyvale, CA, USA, 1977

[Patt_76] D.A.Patterson: "STRUM: a structured microprogram development syustem for correct firmware," IEEE Transactions on Computers, Vol.C-25, No.10, 1976, pp. 974-985

[Patt_77] D.A.Patterson: "Verification of microprograms," Ph.D. Thesis, Dept. of Computer Science, University of California, Los Angeles, 1977

[Patt_81] D.A.Patterson: "An experiment in high level language microprogramming and verification," Communications ACM, 24(10), 1981, pp. 699-709

[PBBD_84] R.Piloty, M.Barbacci, D.Borrione, D.Dietmeyer, F.Hill, P.Skelly: "The CONLAN report," Lecture Notes in Computer Science No. 151, Springer Verlag, 1984

[PiBo_85] R.Piloty, D.Borrione: "The CONLAN Project: concepts, implementations and applications," IEEE Computer, vol. 18, no. 2, February 1985, pp. 81-92

[PLTu_79] D.A.Patterson, K.Lew, R.Tuck: "Towards an Efficient, Machine-Independent Language for Microprogramming," MICRO12: 12th Annual Microprogramming Workshop, Hershey. PA, USA, November 1979, pp. 22-35

[PoHe_78] V.M.Powers, J.H.Hernandez: "Microprogram Assemblers for Bit-Slice Microprocessors," IEEE Computer, Vol.11, No. 7, July 1978, pp. 108-120

[PGPS_81] D.A.Patterson, R.Goodell, D.M.Poe, S.C.Steely: "V-Compiler: a next generation tool for microprogramming," National Computer Conference, May 1981, pp. 103-109

[RaAd_80] T.G.Rauscher, P.M. Adams: "Microprogramming: A Tutorial
          and Survey of Recent Developments," IEEE Transactions on
          Computers, vol.C-20, No.1, January 1980, pp. 2-20

[Rapi_74] "RAPID (A Configuration-Independent Symbolic Assembler for
          Read-Only Memory) Reference Manual," Scientific Micro
          Systems, Mountain View, CA, USA, 1974

[Raya_76] "RAYASM (General Purpose Microcode Assembler)," Raytheon
          Semiconductor, Mountain View, CA, USA, 1976

[ShGi_81] R.I.Sheraga, J.L.Gieser: "Automatic Microcode Generation
          for Horizontally Microprogrammed Processors," MICRO14: 14th
          Annual Microprogramming Workshop, Chatham, MA, USA, October
          1981, pp. 154-168

[ShGi_83] R.J.Sheraga, J.L.Gieser: "Experiments in Automatic
          Microcode Generation," IEEE Transactions on Computers,
          vol.C-32, No.6, June 1983, pp. 557-568

[Sign_77] "Signetic Micro Assembler Reerence Manual," Signetics Corp.,
          Sunnyvale, CA, USA, 1977

[Sint_80] M.Sint: "A Survey of High Level Microprogramming
          Languages," MICRO13: 13th Annual Microprogramming Workshop,
          Colorado Springs, CO, USA, November 1980, pp. 141-153

[Sint_81] M.Sint: "MIDL - A Microinstruction Description Language,"
          MICRO14: 14th Annual Microprogramming Workshop, Chatham,
          MA, USA, October 1981, pp. 95-106

[Skor_79a] E.Skordalakis: "A Generalized Cross-Assembly System for
           Microcomputers," Euromicro J., Vol.5, No.2, 1979, pp. 82-88

[Skor_79b] E.Skordalakis: "Democritos Generalized Cross-assembly
           System," Proc. Informatica 79, Bled, Yugoslavia, 1979

[Skor_80] E.Skordalakis: "DGCAS as a Microprogram Development Tool,"
          ACM Sigmicro Newsletter, Vol. 11, No.1, 1980, pp. 10-16

[Skor_83] E.Skordalakis: "Meta-assemblers," IEEE Micro, April 1983, pp. 6-16

[Somm_79] J.F.Sommerville: "Towards Machine Independent Microprogramming," Euromicro Journal, 5,(1979), pp. 219-224

[TTBS_84] K.Takahashi, E.Takahashi, T.Bitoh, T.Sugimoto: "A new universal microprogram converter," MICRO17: 17th Annual Microprogramming Workshop, New Orleans, LU, USA, October 1984

[WaMa_83] G.Wagnon, D.J.W.Maine: "An E-Machine Workbench," MICRO16: 16th Annual Microprogramming Workshop, Downingtown, PA, USA October 1983, pp. 101-111

[Wata_77] M.Watanabe: "The Macro-assembler Generator for Microcomputers," Euromicro Newsletter, Vol.3, No.4, 1977, pp. 83-85

[Weid_80] T.G.Weidner: "CHAMIL, A Case Study in Microprogramming Language Design," ACM SIGPLAN Notices, 15 (1980), 1, pp. 156-166

[Wilk_51 M.V.Wilkes: "The best way to design an Automatic calculating machine," Electrical Engineering Department,. Manchester University, Englang, July 1951

[Wils_85] P.A.Wilsey: "S*M: An Axiomatic, Non-Procedural Hardware Description Langauge for Clocked Architectures," M.S. Thesis, University of Southern Lousiana, Lafayetter, Luisiana, USA, November 1985

[Yack_80] B.E.Yackle: "An Assembler for All Microprocessors," Hewlett-Packard Journal, October 1980, pp. 28-30

[Zimm_76] G.Zimmermann: "Eine Methode zum Entwurf von Digitalrechnern mit der Programmiersprache MIMOLA," Informatick Fachberichte 5, GI-6. Jahrestagung, Springer Verlag, Berlin, 1976, pp. 465-478

[Zimm_80a] G.Zimmermann: "Microprogram structures for high level
        language elements,"   MICRO10: 10th Annual Microprogramming
        Workshop, Niagara Falls, NY, USA, October 1977, pp. 47-54

[Zimm_80b] G.Zimmermann: "MDS The MIMOLA Design Method," Journal of
        Digital Systems, vol 4 3(1980)

3.3. INTERFACE AND I/O PROTOCOL DESCRIPTIONS

Alice C. PARKER and Nohbyung PARK

Department of Electrical Engineering-Systems
University of Southern California
Los Angeles, California, U.S.A.

3.3.1. Introduction

Recent multiprocessing research and development has spawned an increased interest in input/output and interconnections. It has become important to document, simulate, and formally verify entire systems, including their interconnections.

Naturally, the more detailed an interconnection description becomes, the more accurate the simulation can be and the more information the description can contain. With conventional hardware-descriptive language techniques, interconnections and their interfaces can be described accurately at both the gate and circuit levels.

This low level of description is not adequate for all applications, however. Sheer size and execution time of simulation programs have precluded simulations of large interconnected systems at the gate level, even when hardware accelerators are used. Verification of system behavior is difficult at this level since abstract function must be derived by the verifier from the description; also, the unstructured nature of low-level hardware descriptions may make some aspects of the verification indeterminate. Finally, low-level interface and interconnection descriptions contain details which the reader does not need and which tend to obscure his or her understanding of the behavior of the hardware. For these reasons, gate-level descriptive languages do not provide the kind of hardware description needed for the above tasks; a description at a higher level is needed. Such a description is often referred to as *behavioral*.

Behavioral descriptions differ from structural descriptions because they describe only the functions intended by the designer and not the hardware structure. Storage locations and register-transfers which exist in the hardware may be absent from the behavioral description; control hardware is implicit rather than explicit.

Behavioral descriptions can convey to the reader the overall operation of interfaces better than structural descriptions, since much of the unnecessary detail is eliminated.

Simulations proceed more rapidly, and can encompass larger systems. Verification is possible since the behavior is explicit, the implementation is hidden, and the description can be structured.

The Nature of I/O and Interface Operation

Consider the following example system configuration (Figure 3-7) which illustrates some basic aspects of interface and I/O operation. A single bus connects a device controller, CPU, and memory. The device controller is reading data in from an I/O device and writing it to memory; at the same time, the CPU is executing a program, and therefore accessing memory for instructions and data. At any time either the device controller or the CPU might be transferring information across the bus. At the same time, either or both might be requesting the bus for future transactions. Between the two devices there are four separate sequences of control, two for bus requests and two for bus transactions. At any time a maximum of three can be executing (only one bus transaction can occur at a time). This illustrates an inherent property of interface and I/O operation - complex control flow. More precisely,

- there can be multiple sequences of events executing concurrently and independently of each other,

- an event in one sequence can alter the execution order of events in another concurrent sequence,

- the time steps between events can be different for different sequences, and

- the onset of execution of one sequence can initiate or terminate another sequence.

Each sequence of events in reality represents an independent control environment or a finite state machine; we shall refer to each of these sequences hereafter as a *process*. The general meaning of process is an independent executing environment residing in a piece of hardware such as a device controller or a bus arbiter.

In light of the above example, we now describe what we believe to be the fundamental interconnection behavior which an I/O descriptive language must model:

- concurrent execution of multiple communicating processes,

- contention for shared resources between processes,

- critical sections of execution within processes,

- simultaneity of execution both within and across processes,

- processes which behave in a subordinate fashion with respect to other processes, and

- interconnections which exhibit behavior.

All of the above, except the last, are analogous to the fundamental behavior of multiple software processes executing on a multiprocessor (which is not surprising).

Concurrent execution and *shared resource contention* imply the more specific mechanisms of *mutual exclusion, synchronization, arbitration, priority allocation, and preemption of resources.* The shared resource in the above example is the bus; the four control flows represent concurrent processes; mutual exclusion results when one of the contending processes is actually given the bus resource for a data transaction and the others are excluded. Synchronization must occur when two concurrently executing processes communicate. Arbitration and priority allocation occur because the processes contend for the shared resource. Preemption occurs, for example, when a specific device process fails to release the bus after a specified time period. The bus is preempted, and the errant process may be terminated or suspended.

Critical sections of execution refer to sequences of events which cannot be suspended or terminated, but are within a process which could otherwise be interrupted. For example, a device transacting over the bus may be preempted by a faster device needing bus access, but only at the completion of a transfer protocol sequence.

Simultaneity means that actions may occur at the same instant of time. It also means that any side effects from simultaneous actions may conflict. For example, the assertion of a control line while synchronizing with a process may not produce the expected effect if another input to the process is simultaneously asserted.

A *subordinate process* (subprocess) is an independent execution environment which has certain dependencies on a more global process. The global process retains the ability to allow the subordinate process to start, and execution of the subordinate process is meaningless when the global process is not executing. An example is given later in this chapter.

In addition, interconnections exhibit *behavior.* The use of a tri-state bus implies certain semantics about the behavior of the bus; the delay through a crosspoint switch is part of the behavior; the gating in other interconnection strategies is also part of the interconnection behavior. For example, writing a "high" signal to a *wired-or* line does not mean that the line will contain a "high" value; it depends on whether other devices are asserting "low" on the same wire. In general, this means that interconnections have implied storage, time delays, and Boolean functions associated with them, just as storage is associated with variables in a conventional hardware descriptive language.

There is another consideration which should be mentioned here; it is an aspect of interconnection *description*. This is the notion of viewing interconnection structures as data structures. This view, allows us to treat declaration and replication of buses, crossbar switches, and daisy chains with the same ease we treat lists, arrays, and bit vectors in many programming languages.

A language designed to model the aspects of interconnection behavior described above must possess powerful and unconventional control constructs. In particular, semantics should exist to allow the following:

- priority orderings between processes,

- initiation, termination, and suspension of processes,

- interprocess synchronization primitives such as *signal* and *wait*,

- communication between processes, and

- description of timing dependencies, time-outs, and data I/O at fixed bit rates.

These semantics must include some notion of nonprocedurality to reflect the situation where many processes do not execute *until some condition becomes true*. Also required is the ability to express actions in parallel as well as sequentially.

In addition to these, language primitives should support description of operations common to I/O such as protocols, bit manipulation, code conversion, FIFO buffering, parity and error checking, manipulation of synchronous I/O data, electrical characteristics, and modeling of combinational logic.

It is interesting to note that two excellent comparisons of hardware descriptive languages (HDLs) which were published some time ago [Barbacci 75], [Figueroa 73] revealed some desirable properties of general-purpose HDLs. Barbacci cited the following:

- readability,

- familiarity with naming and usage conventions,

- use across several levels of detail,

- simplicity, small number of language primitives,

- extensibility,

- fidelity to the system organization,

- timing and concurrency,

- syntactic simplicity, and

- ability to separate data and control.

These were chosen presumably because of his emphasis on the *description* of digital systems. Barbacci defined *timing and concurrency* as "parallel actions," and explained "At the RT level, concurrent activities are described by allowing them to be activated simultaneously (i.e., under the same condition)." Figueroa, focusing on design automation, also found some of the above properties important; and in addition discussed the following:

- modularity,

- block structuring, including the existence of global and local variables,

- facilities for functions, subroutines, and macros,

- facilities for specification of parallel processes, and

- facilities for determining and controlling process interaction explicitly.

We now present a brief survey of languages and descriptive techniques which gives a historical perspective to the I/O description problem. Formal specification techniques for description of interface and interconnection operation have been used in the past. We subdivide these into the following:

- the state diagram approach,

- the flowchart approach, and

- the formal language approach.

Flowcharts, timing, and state diagrams have been used to describe a number of buses including the IEEE-488 bus (see [Knoblock 75]). Figs. 3-1 and 3-2 illustrate the use of these techniques. However, languages useful for bus I/O and port descriptions have been adopted more slowly.

Bell and Newell [Bell 71] laid the groundwork for the semantics of interface languages with their port semantics. Curtis, working with the Purdue Workshop on Industrial Computer Systems, Data Transmissions, and Interface Committee, proposed IDS, an Interface Description System [Curtis 75]. The system involved the use of the PMS (Processor-Memory-Switch) notation (from Bell and Newell) at the top level, and the use of a new language for description at the programming and register-transfer levels. In addition, the system was to cover other levels of description as well, but these were not defined at the time. The new language Curtis proposed was a version of the ISP (Instruction Set Processor) language with features of AHPL and with necessary timing constructs added. (Most of these have later been added to the ISPL version of ISP [Barbacci 76]).

At the same time Vissers had developed a language based on APL with timing constructs added which could be used to produce a formal description of the state diagrams describing interfacing functions. SDLC (Synchronous Data Link Controller) and the IEEE-488 interface bus had been described using this approach [Vissers 76], [Knoblock 75]. Vissers had intentionally restricted the coverage of the language to the gate and register-transfer levels, with the added timing. An example of the language is shown in Fig. 3-3, which illustrates the states and transitions of an example taken from the UNIBUS. In this example the granting process for nonprocessor requests on the UNIBUS is described, just as in Figs. 3-2 and 3-1.

The automation, or process, is named npr.grant. There are five major states in this process: $W1$-$W5$. State $W1$ is executed when there is a nonprocessor request and busgrants are enabled (nprbg.enable). The next state is $W2$, where there is a delay until the selection acknowledge line (SACK) is raised by the device receiving the nonprocessor grant (npg), or until 5000 time units elapse, whichever comes first. Then the nonprocessor grant line is released, and state $W4$ is reached. State $W4$ is a delay state until the selection acknowledge is released. Then state $W5$ is entered to reenable grants. The next state is again $W1$.

Marino proposed a hierarchical descriptive system for computer interfaces, MPLID [Marino 78]. This language system was modular. However, an overall control structure which reflected the aspects of interconnection behavior we discussed earlier was lacking. Another publication in this area, [Sorensen 78], described a system modeling language based on BCPL [Richards 69]. This language had three interesting features: the ability to declare processes (concurrency), mailboxes for interprocess communication (synchronization), and uninterruptible code sequences (critical sections). The language was designed to model systems at a high function level.

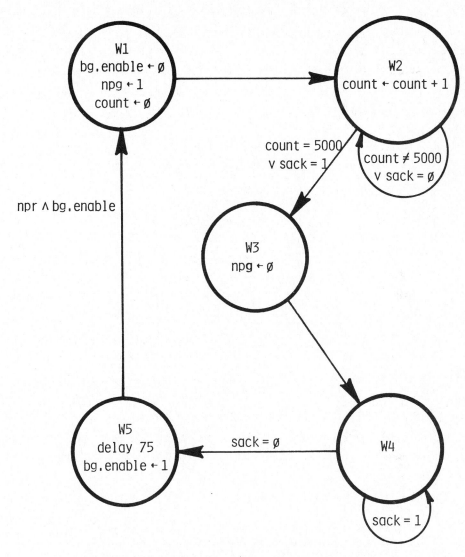

Figure 3-1: The UNIBUS[tm] nonprocessor grant process, described in state diagram form.

Along similar lines, SOL (simulation oriented language) [Knuth 64], based on ALGOL-60, did not describe hardware structures explicitly. However, it did allow concurrent processes, timing, subordinate processes, and shared-resource contention to be dealt with explicitly. One other system-level language, ASPOL [MacDougall 73], also reflected these important aspects of interconnection behavior. Concurrent processes could be declared, and processes could have priorities of execution. *Set* and *Wait* primitives were provided for synchronization. While these languages provided synchronization and process-level constructs, they did not provide the level of detail possessed by register-transfer languages.

AHPL III [Hill 79] and DDL [Dietmeyer 78] could also be used to accurately describe and simulate interconnection behavior. However, the main reason these two languages were not entirely suitable were

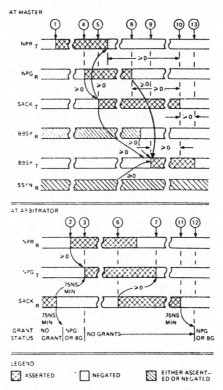

Figure 3-2: Timing of the UNIBUS[tm] nonprocessor grant process:
Copyright 1979, Digital Equipment Corporation. All rights reserved.
Reproduced with the permission of Digital Equipment Corporation.

- Descriptions represented hardware structure rather than behavior and became large quickly.

- Fundamental aspects of interconnection behavior were implicit in the descriptions. They had to be constructed by the user with available primitives.

- An overall notion of interconnection behavior in terms of the model discussed previously was lacking in the languages.

However, there were advantages to these languages. AHPL III [Hill 79] addressed

```
          ∇ npr.grant
[1] w1:→(~nprv~bg.enable)/w1
[2]        bg.enable←0
[3]        npg←0
[4]        npg←1
[5]        count←0
[6]        →w2
[7] w2: →(count ≠ 5000 ∧ sack = 0,count=5000 ∨ sack=1)/w2,w3
[8]        count←count+1
[9]        →w2
[10]w3: npg←0
[11]w4: →(sack=0,sack=1)/w5,w4
[12]w5: delay 75
[13]       bg.enable←1
[14]       →w1
```

Figure 3-3:   The UNIBUS[tm] nonprocessor grant
process, described using Vissers' language.

the "structuring of interconnections" issue. Both languages provided some basic primitive operations on data which were general and powerful. Thus, past languages were either at a very high level, or synchronization mechanisms and processes had to be implicit and were lost in the details of the descriptions.

The remainder of this chapter is divided into sections dealing with timing issues (Section 3.3.2), concurrency (Section 3.3.3), protocols (Section 3.3.4), exception handling (Section 3.3.5), and electrical characteristics (Section 3.3.6). Each of these sections presents the descriptive task in general terms, followed by specific examples of selected hardware descriptive languages.

3.3.2. Timing Issues

The major descriptive issues involved with timing include

- pulse widths,

- clocks,

- delays, and

- propagation delays.

Pulse widths are sometimes important in controlling/sensing mechanical devices, such as disk controllers. Most hardware descriptive languages describe pulses by changing the value on a port, delaying a fixed period of time (to be described below), and then changing the value again. For example,[1] in SLIDE [Parker 81], the description of the generation of a 34

---

[1]For our examples, we will adhere to the convention of reserved words in capital letters, and user labels in lower case whenever possible.

nanosecond pulse on a connection called **sync** would be

    sync ← 1 NEXT
  DELAY 34 NEXT
    sync ← 0

NEXT implies sequencing. The DELAY statement will be explained in more detail later.

Clocks are provided in many hardware description languages used for I/O. Usually, however, these clocks are single phase, used primarily to synchronize sequences of data coming from the outside world. In some languages with clocks, delays and other timing information are specified in clock *ticks* rather than in seconds. In SLIDE, clocks can be declared by the statement

    CLOCK 23;

which specifies a single-phase clock with 23 nsec cycle time. In the SAKURA language [Suzuki 82], the clock is implicit, and time is measured in ticks.

Delays are often inserted into hardware descriptions. In many cases, the delay is merely a command to a simulator to advance the current time; in such cases the language is more like a simulation language than a descriptive one. In some cases, however, the designer might wish to specify a delay between two operations which occur in sequential fashion. Such a delay is shown graphically in Figure 3-4. This figure shows a *control and timing graph*. The graph contains an arc, representing a time range, labeled *delay*. Two other arcs are connected to this range through points, which have no time duration. These arcs represent the operations before and after the delay. Point 1 represents the termination of operations prior to the delay. Point 2 represents the start of operations after the delay. The graphical representation explicitly shows that the earlier operations have all terminated prior to the delay, and the later operations will not begin until the delay is completed. Such a delay facility is found in SLIDE, with the DELAY statement.

Propagation delays through functional units or interconnections are modeled in a similar fashion. Usually occurring after an operation, a delay statement models the functional delay. Interconnect delays are often modeled by delay statements before writing to an outside world interconnect, or after reading from one. In some representations, like timed Petri nets, delays can be attached to transitions. Graphically, function propagation delays and wire propagation delays are shown in Figure 3-5. This figure illustrates a function, $f1$, an input value $a1$, and an output value $a2$. The time range during which value $a2$ exists begins some delay δf_1 after the point during which the value $a1$ begins to exist.

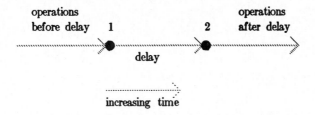

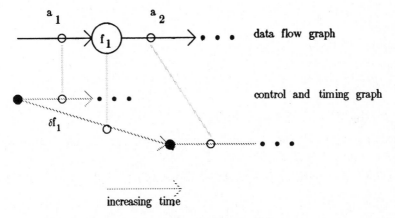

Figure 3-4: Delay illustrated with a
control and timing graph.

Figure 3-5: Modeling a propagation delay.

The SARA (System Architects' Design Apprentice) system [Razouk 80] represents
systems in three modeling domains, a control graph, a data graph, and an interpretation. The
control graph contains tokens, arcs which direct the flow of tokens, and nodes which represent
events. The data graph is a structural view of the system, and the interpretation describes the
behavior of each module in the data graph, using an extended version of PL/1 which includes
timing.

In SARA the interpretation contains the statement

@DELAY=3 @msec;

which models the propagation delay of the **process** description in which the delay statement
is found.

3.3.3. Concurrency

Concurrency is the dominating issue in modeling I/O and interfaces. Descriptions of concurrency involve

- asynchronous signals (e.g. reset),

- synchronization,

- communicating processes,

- subprocesses, and

- competing processes.

Asynchronous signals are signals (perhaps values or wires) which trigger immediate actions in the hardware. The classic example of an asynchronous signal is a reset line, which is not synchronized with a clock and which causes an immediate "escape" to a new state. SLIDE uses such a signal in a limited capacity. Processes can be initiated asynchronously (nonprocedurally) whenever the INIT expression becomes true. For example,

INIT transfers WHEN $(a = 3) \land (b < 5)$

states that the **transfers** process should be started at the precise moment that the expression $(a = 3) \land (b < 5)$ becomes true. The action of initiating a process can terminate other processes, allowing SLIDE to model "escapes".

FLEX [Comer 85], achieves a similar effect with *triggers* and *input wire specifications*. These statements have the form

[trigger-name|input-wire-expression] $\rightarrow$ statement;

Execution of these statements is delayed until the input wire expression or expression associated with the trigger-names becomes true. The trigger becomes inactive once handled. The value of the input-wire-expression does not change as a result of being handled.

Synchronization of multiple processes can be handled in a number of ways. Processes communicate in HDL descriptions by reading/writing shared (global) variables, wires or registers, or by messages. If communication is performed through shared variables, then *semaphores* or *signal-wait* commands allow access to the variables to be controlled. *Busy-waiting* is used to delay execution, or execution is specifically delayed through a *delay* statement. In SLIDE, a process can suspend itself with a statement of the form

[DELAY UNTIL|WHILE] $<$ condition $>$

which waits for a condition (on a shared register or on external wires) to become true (UNTIL) or false (WHILE). This DELAY suspends its own process execution, while the INIT statement specifies the conditions under which another process can begin. Statements following the DELAY are suspended; statements following the INIT are processed as usual. [Parker 81] also proposed two synchronization primitives, *signal* and *receive*.

SAKURA has guarded statements which suspend execution until the guard becomes true, similar to the SLIDE DELAY statement. The syntax is

WHEN <u>Event</u> → Statement

Events are associated with changing Boolean identifiers, and can be designated as UP, DOWN or CHANGE. For example,
   WHEN init UP → initialized ← TRUE
specifies that when **init** is raised, **initialized** gets set to TRUE. A statement with a set of guarded commands implies that the first event which occurs selects only the associated commands to be executed.

Path Expressions ([Lauer 80], [Campbell 74], [Anantharaman 84]) express a coarser grain of synchronization. Though normally used to describe software processes, they are also applicable to hardware descriptions, as shown in [Anantharaman 84]. This paper gives a simple example with two reader processes R_1 and R_2 and a writer process W:

path R_1 + W end,
path R_2 + W end.

This states that either one or both of the reads could occur, or the write could occur, but read and write are mutually exclusive. The temporal part of Flex allows specification of path expressions.

S_A^* [Dasgupta 82] provides an abstraction called the *synchronizer* with operations AWAIT and SIG. Variables are declared of type BIT as follows: (taken from [Dasgupta 82]).
   SYNC ibuffer_full:BIT(0)

   and used as follows:

AWAIT ibuffer_full

      .
      .
      .

SIG ibuffer_empty

   where the process is waiting for the **ibuffer_full** signal and then signaling
   ibuffer_empty.

[Piatkowski 82] describes a state architecture notation (SAN) which provides for communicating finite state machines (FSMs). FSMs change state upon the arrival of pulsed inputs (inputs defined only at discrete times). As the FSMs change state, the outputs are pulsed. This notation is shown in Figure 3-6, taken from [Piatkowski 82]. **cs** refers to current state, **nts** to next state, the **x**'s are inputs and the **ntzs**'s and **ntzp**'s are outputs.

FNS : list *next state information*

    cs / xp.1, xp.2,... / xs.1, xs.2,... ⇒ nts ;
    cs / xp.1, xp.2,... / xs.1, xs.2,... ⇒ nts ;
    .
    .
    .

END;

FOUTP : list *pulsed output information*
    cs / xp.1, xp.2,... / xs.1, xs.2,... ⇒ ntzp.1, ntzp.2,... ;
    cs / xp.1, xp.2,... / xs.1, xs.2,... ⇒ ntzp.1, ntzp.2,... ;
    .
    .
    .

END;

FOUTS : list *static output information*
    cs ⇒ ntzs.1, ntzs.2,... ;
    cs ⇒ ntzs.1, ntzs.2,... ;
    .
    .
    .

END;

Figure 3-6:   Finite-state machine specification format.

In SAKURA, processes communicate by reading and writing a shared storage element called a *node*, through their own bidirectional ports. Two parallel processes, reading and writing, are described in this fragment of a SAKURA description, taken from [Suzuki 82].

```
PAR{
   DO -- Reading process
     WHEN Read Req UP:
       Data Av  ← FALSE;
     WHEN Read Req DOWN:
       Data Av  ← ~rp
   ENDLOOP
    //
   DO --Writing process
     WHEN Write Req UP:
       Space Av  ←  FALSE;
     WHEN Write Req DOWN:
       Space Av  ← ~(wp MOD size) + 1
   ENDLOOP}
```

Execution is held up until the WHEN event occurs, which allows parallel processes to be started or resumed.

SARA describes process communication through **send** and **receive** signals which describe state transitions. Each process is described with a separate control graph and a signal causes state transitions by controlling the movement of tokens through each control graph. FLEX describes process communication by allowing multiple processes, each with its own flow of control, to execute in parallel. The process can share global variables, or communicate through the *trigger* and *input-wire-expressions* described earlier. Mutual exclusion in shared variable access is provided via uninterruptible atomic actions, which can be any normal action (including a procedure).

An example of FLEX [Comer 85] is the segment

```
PAR
   DO alive  → reset_idle(alive.up_signals)OD;
   DO timer  → inc_idle()OD;
RAP;
```

Where **alive** and **timer** are triggers, **reset_idle** and **inc_idle** are procedure calls, and the two DO statements represent parallel processes, indicated by the PAR...RAP construct[2]. **reset_idle** is invoked when the **alive** trigger must be handled, and **inc_idle** invoked when the the **timer** trigger must be handled. Actual trigger expressions would have been given earlier in the trigger declarations.

Concurrent Prolog has been used for hardware description and simulation [Suzuki 85]. Concurrent processes are subgoals to be repeated, and communicate through shared parameters. The direction of data flow is indicated with a "?" after the variable which is to receive the value. For example (taken from [Suzuki 85]),

G:-A(x),b(x?)

describes two concurrent processes **A** and **B**, and a shared variable **x**. A value is assigned to **x** by **A** and is used by **B**. A *commit* operator does not allow variables to be accessible by other processes until all subgoals which might affect the value of the variable succeed within a process. Concurrent Prolog deals with time as a shared variable, like any other.

---

[2]In FLEX, parallel processes are actually intended to be executed sequentially, since the target machine being described is a uniprocessor. This is a limitation of the target machine rather than the language itself.

Behavior Expressions (BE's) [McFarland 83] describes the behavior of a single *process* by describing all possible sequences of events. Events can be reads or writes to the external world. Predicates are used to describe the conditions under which events occur. Thus, process communication can be described implicitly by specifying the read and write events. The Behavior Expression

$$[R(x_{in}):true \cdot W(x_{out}):(x_{out})=1 \wedge x_{in}(j_{in})=1)]*l.$$

repeatedly reads x_{in} and writes a 1 to x_{out} only if the value it has just read is a 1. If the last value read is not a 1, it does not write anything, but begins the next repetition immediately. l is a loop counter.

Subprocesses are the required subordinate processes we describe in the Introduction. Subprocesses are processes started (initiated) by outer processes. A SLIDE description consists of one *main process* which syntactically encompasses all other subprocesses (much like an ALGOL program consists of one main program which encompasses all subroutines). Variables global to the entire description are declared within the main process. Variables which are local to a subprocess are declared within that subprocess. Processes nested at the same level model either concurrent process execution or processes competing for shared resources. Since each process describes the operation of a piece of hardware, each one is an asynchronous executing environment. Processes which need to communicate with each other can do so by using global variables (e.g., by asserting a shared line) or by using *signals*.

Processes are started (called *initiation*) nonprocedurally. When each process (except for the main process) is defined, the conditions under which is to be initiated are given. A *priority mechanism* exists which can be used to allow some processes to terminate execution or mutually exclude execution of others. Subprocesses terminate when the outer processes in which they are declared are terminated.

By the same token, subprocesses can only start when their outer processes are executing. An example of this is shown below. Process **flag.detect** detects a flag of six continuous one bits with preceding and succeeding zero bits. Once a flag is detected, it allows process **serial.to.parallel** to start. However, flag detect continues to execute, and when the end flag is detected, terminates process **serial.to.parallel**.

PROCESS flag.detect;
    INIT serial.to.parallel:0 WHEN flag;

    BEGIN
    (end flag detection is performed here)
    END;

PROCESS serial.to.parallel;
    BEGIN

    END;

    The zero in the INIT statement indicates the priority of the **serial.to.parallel** process.

    Competing processes exist when processes compete for a shared resource, such as a hardware module or bus. Path expressions allow description of mutually exclusive processes as in the following:

    PATH reset + normal END;

which states that either the **reset** or **normal** process is to execute but not both concurrently. Predicate Path Expressions (PPEs) [Andler 79] allow process execution only when the predicate associated with a process is true. For example,

    PATH <reset [init] | normal [$\sim$ init]>END

describes the same **reset** and **normal** processes, but the conditions under which each could execute are now specified. The | specifies exclusive selection.

    In SLIDE, each process has an explicit priority. Informally, a process starts executing when: 1) the process of which it is a subprocess is executing, 2) its initiation conditions are true, and 3) no process at the same subprocess level with a higher priority is executing. When a process starts executing, all sibling processes which are at the same subprocess level, have a lower priority, and are executing are terminated (i.e. killed).

    Priorities can be used to time-order the execution of processes. For example, assume we have three processes, A, B, and C, no two of which can execute concurrently. A is to always execute as soon as its initiation conditions become true; B is to execute when its initiation conditions become true, but only if A is idle; and C can execute only of A and B are idle. This can be done by giving A priority 0, B priority 1, and C priority 2. Then as soon as A's initiation conditions become true, it will start executing, terminating B or C if they were executing. When *A* finishes executing, it will restart if its conditions are still true. If not, *B* may start if its conditions are true. If not, *C* may start.

In our example, if terminating a process once it has started executing is undesirable, a 1-bit variable can be used which prevents other processes from starting while any process is executing. This simple type of synchronization allows for mutual exclusion, and critical sections of processes can be protected from preemption (termination).

A detailed example follows. The example system configuration is shown in Fig. 3-7. Assume we are writing a SLIDE description for a disk controller interfaced to a UNIBUS. The controller is to do a transfer operation to a memory location over the UNIBUS whenever the internal **dataready** line rises from logical 0 to logical 1. The controller is to reset (i.e., stop any on-going transfer and reset itself) whenever the **initline** rises from logical 0 to logical 1. Part of a SLIDE description for the controller is in Fig. 3-8. Lines 1 and 2 specify the conditions under which the **Dreset** and **Dtransfer** processes start executing. **Dreset** is given a higher priority than **Dtransfer** since a reset should terminate any on-going transfer operation.

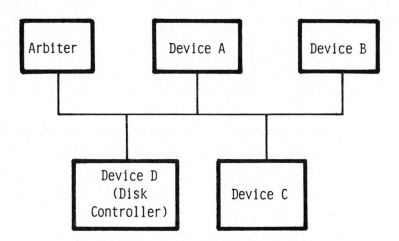

Figure 3-7:   The example system configuration.

The expression **initline EQL/** and **dataready EQL/** are true at the moment the line rises from 0 to 1; not before or afterwards. These have different semantics than the expressions **initline EQL 1** and **dataready EQL 1,** which are true whenever the lines contain logical 1.

If many controllers, each with its own **transfer** and **reset** processes, are to be connected to a UNIBUS, the overall structure of the resulting SLIDE description is shown in Fig. 3-8. Here we have an arbiter, a set of devices, and their interconnections. Notice that the arbiter can always be *initiated* as long as it is not being initialized via the **arbiter** PROCESS, since the **arbiter** PROCESS *initiation* condition is always true.

```
MAIN PROCESS UNIBUS;
    global bus declarations
    INIT Iarbiter:0 WHEN. initline EQL /;
    INIT arbiter:1 WHEN TRUE;
    INIT Dreset:0 WHEN initline EQL /;
    INIT Dtransfer:1 WHEN dataready EQL /;
    INIT Areset:0 WHEN initline EQL /;
    INIT Atransfer:1 WHEN adataready EQL /;
      .
    other device processes are initiated similarly
      .
    PROCESS arbiter;
        .
        BEGIN
        arbitrate the bus
        END
        .
    PROCESS Iarbiter;
        .
        BEGIN
        initialize devices on the bus
        END
        .
    PROCESS Dreset;
        .
        BEGIN
        reset the disk controller
        END
        .
    PROCESS Dtransfer;
        .
        BEGIN
        transfer data on the unibus
        from disk to memory
        END
        .
    other processes are similar

    BEGIN          !Main process execution!
        DELAY WHILE TRUE          !idle forever!
    END;
```

Figure 3-8: The overall structure of an example SLIDE description.

3.3.4. Protocols

The study of protocols includes sequential synchronous data, handshaking, latching and buffering. The focus in this chapter is on description of *hardware* protocols, although some concepts are common to both hardware and software protocols.

Sequential, synchronous data is data arriving or departing to the outside world sequentially, at a fixed rate. Description of synchronous I/O (i.e., I/O transfer which occurs at a fixed rate) is difficult because of the different implementations of hardware which perform synchronization. Ada has been used to describe such situations as a history of values [Barbacci 84]. *Step signals* correspond to transient values, and *time signals* are sequences of step signals, one step signal per unit time. Time signals can be examined at any point in the past, or at the current time. This is done by keeping a record of step signals and time signals as part of the pin manager package. Thus, all values which appear on pins are recorded according to the specifications found in a "pin manager" package.

Piatkowski [Piatkowski 82] also describes this situation. The State Architecture Notation supports streams of input and output data. Only the most recent data is accessed, however. Stream input and output data has reserved names. For example, **XS.1** is static input 1 (A static input is one where values only change immediately following each sample time), and **ZS.1** is static output 1. Pulsed variables (e.g. **XP.1**) are defined only at the variable's sample times.

SLIDE allows a limited description of synchronous I/O with the SYNC declaration. For example,

$$\text{SYNC @ 100000 LINE Tape } 1{<}8{:}0{>}$$

declares the synchronous transfer of 9 bits in parallel from/to a line named **tape 1** at a rate of one transfer per 100000 clock pulses. The use of a variable declared as synchronous carries with it an implicit wait for the data to synchronize.

We can test for a *time-ordered* sequence of values on synchronous line(s). for example if s is declared as

$$\text{SYNC @ 50000 LINE } s{<}{>}$$

then the statement below delays execution until s takes on the values of 5 ones followed by a zero

DELAY UNTIL s EQL |1|1|1|1|1|1|0|.

This reads *delay until s equals the time-ordered sequence of values 1,1,1,1,1,1, then 0.*

Handshaking is not usually dealt with as a primitive. In most HDLs which describe I/O, handshaking consists of setting and resetting shared variables or interconnections. SARA supports description of handshaking in this manner, as does Piatkowski's SAN.

SLIDE supports handshaking by the values in shared registers or on shared wires, in the same way that synchronization is performed. An example handshaking sequence is shown in figure 3-9.

```
[1]     npr ← / NEXT        ! grant the bus !
[2]     DELAY 5000 UNTIL sack EQL /NEXT!Clock is a 2 nanosecond!
                                !clock!
[3]     npg ← \;     !lower the grant line!
```

Figure 3-9:   An example handshaking sequence.

The left arrow indicates wire **npg** is being written to. The / and \ symbols will be explained later.

Latching and buffering are also usually implicit in reads or writes to and from external variables, registers or wires. In FLEX, however, such operations are modeled in a more complex fashion. Output wires can be latched. Normally, wires output values for only one clock cycle, but latched wires must be explicitly changed. Such wires are declared as follows (taken from [Comer 85]):

d:OUTPUT TO LATCH;

where **d** is the name of the wire. Wires can be buffered, also. Buffering here implies FIFO storage, and in Flex up to three outputs can be stored at a time and later accessed. For example (also from [Comer 85]):

g:ARRAY [0..7] OF INPUT FROM BUFFER (3);
h:ARRAY [0..7] OF OUTPUT TO BUFFER (3);

This is similar to the SLIDE BUFFER declaration. An assignment to a named buffer in SLIDE puts the item at the end of the buffer, and assignment from a buffer removes the first item.

Many HDLs are able to model external signals and protocols more accurately by describing signal transitions or edges. Such transitions are modeled in SAKURA with the identifiers **UP** and **DOWN**, in temporal logic with up and down arrows, in FLEX with **change**, and in SLIDE by **/** and ****.

3.3.5. Exception Handling Unique to I/O

Three major types of exception handling are important to I/O:

- data errors,

- missing data or missing signals, and

- asynchronous signals which signal exceptions or invoke exception-handling procedures.

SLIDE has operators that allow lookup tables to be accessed, so that data can be coded. For example,

```
TABLE gray  <2:0><2:0>
    '000 ⇒ '000,
    '001 ⇒ '001,
    '010 ⇒ '011,
    '011 ⇒ '010,
    '100 ⇒ '110,
    '101 ⇒ '111,
    '110 ⇒ '101,
    '111 ⇒ '100;
```

is a gray code conversion table specified in binary[3].

SLIDE also has a parity bit generation and checking operation.

Time-out capability is very important. For example, assume a bus arbiter grants the bus to a process by raising a **busgrant** line. Within 5 μs it expects the process to acknowledge by raising the **sack** line. If this does not occur, the arbiter will time-out, then lower the grant line to reset the bus. A SLIDE description of this is in Fig. 3-9. (A NEXT used as a statement delimiter forces sequential execution.) A DELAY-UNTIL-ELSE construct which is more complex allows for exception handling. (The ELSE statement is executed when a time-out occurs.)

Timeouts are supported in SAN by declaring clocks. The **start** input starts the timer. If the clock runs without being reset until it times out, it pulses out a **timeout** signal to all inputs given in the clock specification.

---

[3] In SLIDE, a single quote (') indicates a binary number; a pound sign (#) indicates an octal number.

3.3.6. Electrical Characteristics

SLIDE supports some technology information. For example,

OCAL LINE d<15:0>, a <17:0>

declares two 16-and 18-bit wide TTL *Open Collector Active Low* bus segments, one named **d**, and one named **a**. These correspond to the data and address segments of the UNIBUS, respectively. Because the lines are *typed*, writing to a line implies certain behavior. Writing to an open collector line, for example, implies an attempt to set or reset the line, but may not result in a change in state of the line unless other devices wired to the line cooperate. For this reason, hardware declarations, which may seem low level, are needed in order to model abstract behavior.

References

[Anantharaman 84]
> Anantharaman, T. S., et. al.
> *Compiling Path Expressions Into VLSI Circuits.*
> Technical Report CMU-CS-85-102, Carnegie-Mellon University, August,
> 1984.

[Andler 79]
> Andler, S.
> *Predicate Path Expressions: A High-Level Synchronization Mechanism.*
> PhD thesis, Department of Computer Science, Carnegie-Mellon University,
> August, 1979.

[Barbacci 75]
> Barbacci, M.
> A Comparison of Register Transfer Languages for Describing Computers and
> Digital Systems.
> *IEEE Transactions on Computers* C-24(2):137-150, February, 1975.

[Barbacci 76]
> Barbacci,M.
> *The Symbolic Manipulation of Computer Descriptions: ISPL Compiler and
> Simulator.*
> Technical Report, Dept. of Computer Science, Carnegie-Mellon University,
> Pittsburgh, Pa., April, 1976.

[Barbacci 84]
> Barbacci, M., Grout, S., Lindstrom, G., Maloney, M., Organick, E., and
> Rudisill, D.
> *Ada as a Hardware Description Language: An Initial Report.*
> Technical Report CMU-CS-85-104, Dept. of Computer Science, Carnegie-
> Mellon University, December, 1984.

[Bell 71]
> Bell, C., Newell, A.
> *Computer Structures: Readings and Examples.*
> McGraw-Hill Book Co., New York, 1971.

[Campbell 74]
> Campbell, R.H., and Habermann, A.N.
> The Specification of Process Synchronization by Path Expressions.
> *Lecture Notes in Computer Science.* Volume 16.*Operating Systems*.
> Springer Verlag, 1974, pages 89-102.

[Comer 85]
> Comer, D.E., and Gehani, N.H.
> Flex: A High-Level Language for Specifying Customized Microprocessors.
> *IEEE Transactions on Software Engineering* SE-11(4):387-396, April, 1985.

[Curtis 75]
> Curtis,D.
> IDS, An Interface Description System.
> November, 1975.
> Unpublished note, ALCOA.

[Dasgupta 82]
> Dasgupta, S.
> Computer Design and Description Languages.
> *Advances in Computers.*
> Academic Press, 1982, pages 91-154.

[Dietmeyer 78]   Dietmeyer, D.
                     Logic Design of Digital Systems.
                     Allyn and Bacon, 1978.

[Figueroa 73]   Figueroa, M.A.
                     Analyses of Languages for the Design of Digital Computers.
                     Technical Report, Coordinated Science Laboratory, Univ. of Illinois, Urbana,
                         IL., May, 1973.

[Hill 79]   Hill, D. and vanCleemput, W.
                     SABLE: A Tool for Generating Structured, Multi-level Simulations.
                     In *Proceedings of the 1979 Design Automation Conference.* IEEE and
                         ACM, 1979.

[Knoblock 75]   Knoblock, D., Loughry, D., and Vissers,C.
                     Insight Into Interfacing.
                     IEEE Spectrum 12(5):50-57, May, 1975.

[Knuth 64]   Knuth, D. and McNeley, J.
                     A Formal Definition of SOL.
                     IEEE Transactions on Computers C-13:409-414, August, 1964.

[Lauer 80]   Lauer, P. E.
                     Project on the Design and Analysis of Highly Parallel Distributed Systems.
                     Technical Report, the University of Newcastle upon Tyne, December, 1980.

[MacDougall 73]   MacDougall, M. and J. McAlpine.
                     Computer System Simulation with ASPOL.
                     In *Proceedings of the Symposium on the Simulation of Computer Systems.*
                         September, 1973.

[Marino 78]   Marino, E.
                     Computer Interface Description.
                     In *Proceedings of the 17th Annual Technical symposium.* ACM and
                         National Bureau of Standards, June, 1978.

[McFarland 83]   McFarland, M. and Parker, A.
                     An Abstract Model of Behavior for Hardware Description.
                     IEEE Transactions on Computers C-32(7):621-637, July, 1983.

[Parker 81]   Parker, A. and Wallace, J.
                     SLIDE: An I/O Hardware Description Language.
                     IEEE Transactions On Computers C-30(6), June, 1981.

[Piatkowski 82]   Piatkowski, T. F., Ip L., and He, D.
                     State Architecture Notation and Simulation: A Formal Technique for the
                         Specification and testing of Protocol Systems.
                     Computer Networks 1982(6):397-417, 1982.

[Razouk 80]   Razouk, R., and Estrin, G.
                     Modeling and Verification of Communication Protocols in SARA: The X.21
                         Interface.
                     IEEE Transactions on Computers C-29(12):1038-1052, December, 1980.

[Richards 69]   Richards, M.
                   BCPL: A Tool for Compiler Writers and Systems Programming.
                   In *Spring Joint Computer Conference*. 1969.

[Sorensen 78]   Sorensen, Ib Holm.
                   System Modeling.
                   Master's thesis, Computer Science Department, University of Aarhus,
                       Denmark, March, 1978.

[Suzuki 82]   Suzuki, N., and Burstall, R.
                   Sakura: A VLSI Modelling Language.
                   In Artech House (editor), *Proceedings of the Conference on Advanced
                       Research in VLSI*. Dedham, Mass., 1982.

[Suzuki 85]   Suzuki, N.
                   Concurrent Prolog as an Efficient VLSI Design Language.
                   Computer 18(2):33-40, February, 1985.

[Vissers 76]   Vissers, C.
                   Interface, A Dispersed Architecture.
                   In *Proceedings of the Third Annual Symposium on Computer Architecture*,
                       pages 98-104. ACM SIGARCH and IEEE Computer Society, 1976.

3.4. GRAPHIC HARDWARE DESCRIPTION LANGUAGES

Udo WELTERS

Department of Computer Science, University of Kaiserslautern
Postfach 3049, D-6750 Kaiserslautern, F.R.G.

The growing complexity of circuit design at layout level resulted in new design methods and a stepwise improvement of design aids. Some milestones are drawing masks by hand, using a computer for drawing and then drawing systems with online consistency checkers. At other design levels we can find the same trend. The first hardware description languages (HDL) only needed a text editor for capturing the description of a circuit. With growing complexity designers at first made schematic diagrams by hand and then translated these into the HDL. Later they used schematic entry systems and translated the diagram by themselves or automaticllay by the computer. In the last step there will be schematic entry systems with online checks of HDL rules.
This work gives the definition of the term graphic hardware description language (GHDL). It shows the difference between a GHDL programming system and a common schematic entry system, and gives an example for a GHDL.

1. INTRODUCTION

The term of a graphic hardware description language (GHDL) is composed of the terms hardware description language (HDL) and graphic language (GL). A HDL is a textual description language. Defining a GHDL we want to combine the advantages of the two languages. In the field of hardware design, the most important GHDL application is to provide an alternative graphic companion notation to the HDL, comparable to logic diagrams as companions to Boolean equations. But what are the advantages? Y. Chu [CHU74] lists the following advantages of a HDL:

- they serve as a means of communication among computer engineers
- they permit a precise yet concise description
- they provide convenient documentation
- they are amenable to simulation on a computer
- they aid greatly in an integrated and total design automation ranging
  from the computer structures to VLSI circuits

In addition to this the first most obvious advantage of a GL over a textual HDL is illustrated by the statement 'A picture says more than thousand words'. Said with other words: the very best perceptive capabilities of human beings are achieved by using suitable visual presentations. Another advantage is the fact, that software and hardware of modern interactive graphic user interfaces using a mouse, hierarchies of pull-down menues and window techniques are self-explanatory and offer a highly efficient user guidance. That's why a GHDL is a much better user interface than a textual HDL [FAR86, BAL86].

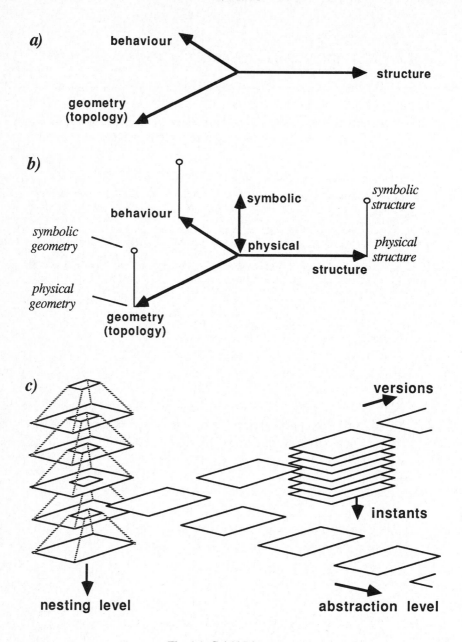

Fig. 1.1: Gajski Diagramm

Information to be handled by interactive graphic editors may be classified into three kinds: structural, behavioural, and geometric information (Gajski diagram, see fig. 1.1a). However, we have to distinguish between application information (e.g. physical information) and presentational information (symbolic information, also see fig. 1.1b). Physical geometry shows the real geometry of, for instance, an NMOS transistor, whereas symbolic geometry is used to draw the transistor symbol such as used in circuit diagrams. For more details see [HAR77] within this volume. HDLs are used to represent behavioural information. Structural information cannot be avoided. But it is only an aid to describe complex behavioural models in terms of more simple models, picked from a repertory or predefined by the user. It is a partitioning aid needed in any kind of software engineering.

For VLSI design we may distinguish three classes of graphics editors: structural editors, geometry editors, and syntax-oriented editors. There are structural editors, which mainly have been implemented to capture wiring lists (e.g. [NNV86]). Geometry editors [OAL84, OUS81, DA85, MEM84] are mainly needed for layout design. Syntax-directed graphics editors are subject of this paper. In case of HDL applications these editors are primarily used to represent behavioural information. So these editors are mainly viewing aids and interactive graphics entry devices for non-geometric concepts, mostly having been defined by a textual language and its syntax. In these editors structural and geometrical information is used as an aid to represent application-specific conceptual information. The geometry of drawing primarily is not used to present geometric information. Lines drawn graphically just illustrate structural information. For instance, a line connecting two objects in a block diagram just illustrates graphically, that the circuits being represented by these objects are interconnected. This structural information is not used to show physical structure. It is just an aid to represent conceptual information of even higher level: both circuits communicate with each other. The combination of structural presentation and text yields the conceptual information. This means, that syntax-oriented editors have two underlying grammars: a graphical grammar showing the composition rules of graphical objects, as well as a textual grammar showing the syntax of text and its association with graphical objects.

1.1 Terminology and Syntax Relations / Geometry

Before looking for more advantages of the combination of the two languages, we define a GL and a *GHDL diagram*. (Although in textual HDLs we often say *HDL program* for a hardware description, we prefer to use *GHDL diagram* for GL descriptions.) A GL diagram is a picture using graphic primitives. These graphic primitives can be composed to more complex graphic entities, defining a picture hierarchy. A 'GHDL diagram' is a GL program using the primitives and the grammar of the underlying HDL. A GHDL diagram can be generated by drawing a picture of the hardware we have to describe under consideration of the HDL grammar rules. The generation is computer-aided in such a way that the user needs not to know exactly the grammar rules of the HDL. This is comparable to a programming system which interactively accepts a flowchart and generates the textual form of the program.The picture hierarchy in a GHDL corresponds to the block structure of the HDL. Fig 1.2 shows a HDL program example using the textual language KARL-3 [HAL86]. Fig.1.3 shows the corresponding GHDL diagram using the graphic language ABL (A Block diagram Language) created with the editor ABLED [GIR85].

<u>cell</u> FA (<u>front in</u> A; B <u>right in</u> CI <u>back out</u> COUT <u>left out</u> CO) ;
<u>node</u> CO; COUT;
<u>begin</u> (* of FA *)
    <u>terminal</u> CO .= ((A <u>and</u> B) <u>or</u> (A <u>and</u> CI) <u>or</u> (B <u>and</u> CI)) ;
    <u>terminal</u> COUT .= (A <u>exor</u> (B <u>exor</u> CI)) ;
<u>end</u> (* of cell FA *) ;

Fig. 1.2: KARL-III Description of Cell named Fulladder

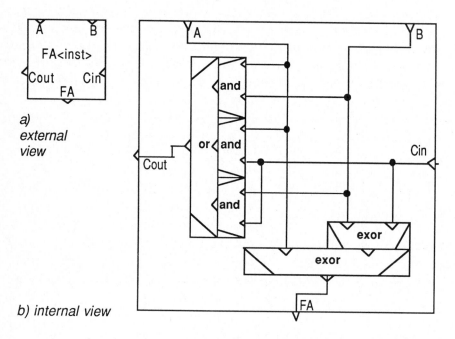

a)
external
view

b) internal view

Fig. 1.3 : GHDL or GDL diagram of Cell named Fulladder
a) external view b) internal view

To store and process a GL or GHDL diagram on a computer we need an application-oriented data structure in addition to the GL code driving the display. For this case we have an intermediate form of the picture in the main storage. We call it working data structure (WDS). For a permanent storage and manipulation of the picture we need a data base form. This normally is a textual representation of the picture information. In case of a GHDL this information follows the grammar rules of the underlying HDL. From this data base or the intermediate WDS version of a picture an interpreter or translator produces the executable code for a simulator (e.g. RTcode for the KARL-III system [RTC85]), or, for the display processor, respectively. The display processor generates the corresponding GL drawing command sequence to produce the drawings of the GHDL diagram on the screen. A translator may also generate an HDL program (see fig 1.4).

Such a system is called *schematics entry system*. But we will distinguish between such schematics entry systems, which are only *GL systems*, from those ones which are *GHDL systems*. In chapter 4 a classification scheme for existing schematics entry systems will be introduced.

In the next chapter we describe a GL programming system and give an example for the textual description of the circuit in fig. 1.2. Then we give the characteristics of a description with an HDL, analyzing several HDLs and synthesize GL and HDL into a GHDL. Then we show an example of a GHDL system. In all these chapters we only consider the register transfer level. In the last chapter we consider the term multi level GHDL, multi level concerning the representation levels or abstraction levels of an hardware description. We look for existing GHDLs and extend the example GHDL to a multi level GHDL.

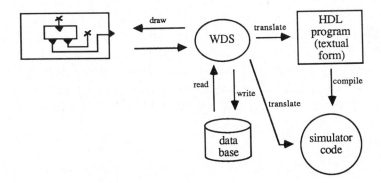

Fig. 1.4: Example of a GHDL Programming System

2. GRAPHIC LANGUAGES (GLs)

This section gives a brief introduction on graphic languages (GLs) and graphic programming languages (GPLs). The goal of this section is, to clarify the differences between ordinary interactive drawing programs and schematics editors on one side and the implementation of graphic hardware description languages on the other side. The first class is interesting as far as a GL implementation is needed as a subsystem of a GHDL implementation. A survey on existing GHDL implementations is given in section 4.2.

2.1 Picture Editors

How can we generate a GL diagram? One way is to draw it by hand. Using a digitizer we can get a digitized version of the picture stored in a bit map format. However this approach is very time consuming, especially if the picture is modified very often. Another point against hand drawing picture is that also nongraphic information may be connected with the picture, which cannot be captured from its bit map representation. A more powerful approach uses a computer to generate the bit map of a drawing. By the way: techniques of generation, representation, manipulation, processing, or evaluation of graphic objects by a computer as well as the association of graphic objects with nongraphic data structures are subject of the discipline of computer graphics [GIL78].

A system is called an interactive picture editor (PE) if it can generate and modify pictures under control of human interaction. A PE is based directly or indirectly on a graphic programming language (GPL). The purpose of a GPL is it to tell the computer how to draw elementary patterns into its display file (i.e. onto its screen). Various GPLs have been developed and implemented [KUL68, SHA69, WAR76]. A GPL implementation may be a graphics subroutine package, an extension to another language [PFI76] or its own language[KUL68]. Each approach has its advantages and disadvantages [GIL78].

Fig. 1.3 shows a picture having been generated by the GL subsystem within the ABLED [GIR85, GIR84A, GIR84B, GIR84C], a portable picture editor which is the implementation of the ABL language. Since ABL is the companion notation of KARL-3 (a textual HDL), ABLED is a GHDL implementation. With this editor we can interactively create and modify ABL block diagram drawings. It is used for the design of integrated circuits at register transfer level based on KARL-3 structures [GIR85, HAR84, HAR77, HAR79, NN86]. ABLED is subject of section 4.2.

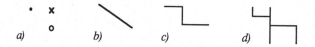

Fig. 2.1 : Drawing primitives: a) point, b) vector, c) polyline, d) net

2.2 Picture Hierarchies

A picture may be described by a set of graphic primitives. [GIL78] defines a graphic primitive as a graphic element which can be drawn by a special hardware unit of the display processor. Examples of possible primitives are: dots, straight-line segments (vectors), alphanumeric characters, special symbols, special curve segments (e.g. arcs), or even "surface patches". Fig. 2.1 shows a few examples of drawing primitives, which use only dots and straight lines. Together with text these few drawing primitives are sufficient for a GHDL implementation. This is illustrated by the ABL drawings in fig. 1.3, 3.1 and 3.2. E.g. a black and white picture can be described by an enumeration of black dots. Using only the primitive generation commands the display processor language (DPL) program can be compared with an assembler program. With such a picture description we cannot reflect the structure of our picture. We also have to consider the relationship between the elements of a picture. E.g. a box is not recognized as a box but as four different lines. The picture editor must have the possibility for capturing the picture hierarchy. The editor must provide commands not only for the generation of primitives but also for grouping these primitives into more complex entities. We define the following entities [GIL78].

Primitive: A primitive is either a graphic primitive or a character. A graphic primitive is specified by a point (for dot or vector) or by an ordered set of points (for circle or surface patch). A character is specified by a code word.

Item: An item is an ordered set of primitives of the same type.

Segment: A segment is a set of disjoint items. We should see, that the set of items is a subset of the set segments etc. (see fig. 2.2).

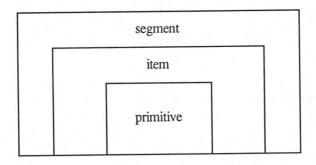

Fig. 2.2: Nested Hierarchy of Graphic Entities

With such a hierarchy we have a first definition of the general structure of a picture. This picture hierarchy can be achieved with most of the graphic packages. E.g. the standardized GKS [END83] language as an interface to the application program provides output primitives of polyline, text and others. Polyline and text are items in the hierarchy defined above. The output primitives may be grouped into sets which can be addressed and manipulated as a whole. These picture parts are called segments. There is no further hierarchy, however the nesting of segments is supported by GKS.

Often graphic entities are related to non-graphic entities. We may have different hierarchies within non-graphic objects, hierarchies which do not meet the three graphical levels having been introduced above. Such hierarchies are implemented by nested subroutines and more complex data structures of the picture editor. The program interfaces, the corresponding primitives and ways of hierarchy structuring are illustrated in fig. 2.2. The following sections illustrate as an example the picture hierarchy of the ABLED editor. Its application picture hierarchy is implemented by a corresponding graphic subroutine hierarchy.

2.3 An Example : Picture Hierarchy within ABLED
Different versions of the GHDL editor ABLED have been implemented using different GPL interfaces: GKS (running under VMS on VAX 11/7xx and VAXstation), CORE (running under

UNIX 4.2 BSD on SUN workstations) and the Apollo graphic hardware interface (under AEGIS on all models of Apollo workstations). The graphic subroutine package we use in the example is that of the Graphical Kernel System (GKS) [END83]. This subroutine package contains basic graphical procedures (GPL procedures) to manipulate, generate and process a picture. The data structure hierarchy is corresponding to the GL hierarchy described above. The primitives are used in the device interface (driver-to-GKS interface). The user may generate items, texts, polylines and filled areas and he can group segments of items. Above the graphic package subroutine level we find the application program levels (see fig. 2.3).

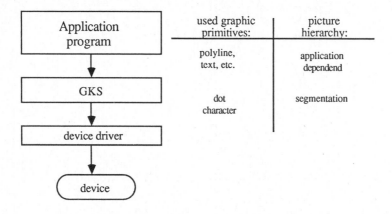

Fig. 2.3: Programm Levels

The application program uses more complex graphic primitives than the graphic subroutine package. The graphic primitives at the user level are the items defined above. [BER76] defines the picture primitives at user level in the following way.

> A feature of a graphics programming system is a "picture primitive" if:
> 1.   it is invoked by a single user level command
> 2.   it directly generates potentially displayable information, and
> 3.   the displayed information may only be identified, modified, and transformed as a single entity.

Within an application the items are composed to form higher level graphic elements. The objects of our example application are cells. A cell is composed of an external view and an internal view. The external view is composed of a boundary box, text objects and pins. The internal view is composed of components, ports, nets, text objects and a boundary box. A component is the instantiation of a cell or of a primitive. Ports and pins consist of a text and a symbol. Symbols and boundary boxes consist of polylines. Since the application uses the items of text and polyline, and, since the graphic package identifies the graphic entities by segment numbers, all items are created as segments. Fig.

2.4 shows the data structure hierarchy of the application program. This structure, may be implemented by using pointers for each connection line and the type declarations of each entity. Variables by are created interactively by creating a cell or by calling a reader which retrieves the cell descriptions from a data base .

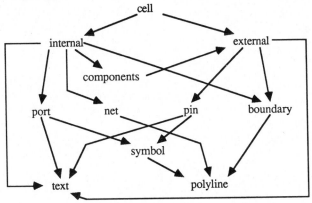

Fig. 2.4: Graphic Data Hierarchy of the Application

2.4 Procedure Hierarchies within ABLED

Corresponding to the picture and data structure hierarchy the ABLED editor has an application procedure hierarchy. For example, we have the procedure DRAWCELL, which draws a complete cell. Its input parameter is the cellname. Fig. 2.5 shows the structure of procedure DRAWCELLINT.

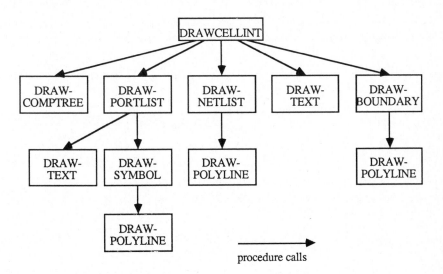

Fig. 2.5: Example of Procedure Hierarchy

Fig. 2.6 illustrates some user primitive creation procedures of GKS as well as some higher level graphic procedures of ABLED. Fig. 2.7 shows some examples of PASCAL graphical data type declarations within the WDS of ABLED.

Primitive Description Procedures

GPL (N,PX,Py)	(*generates a polyline*)
GTX (PX,PY,CHARS)	(*generates a text object*)

Higher Level Procedures

DRAWSYMBOL	(* draw one symbol *)
DRAWPORT	
DRAWPIN	
DRAWNET	
DRAWBOUNDARY	
DRAWCOMP	(* draw component *)
DRAWCELLEXT	(* cell external view *)
DRAWCELLINT	(* cell internal view *)
DRAWCELL	(* draw the entire cell *)

Fig. 2.6: List of GKS and ABLED Procedures

Graphic Data Type Definitions:

```
texttype = record
          posx: real;
          posy: real;
          string: packed array [1..n] of char;
          .....              (*display attributes*)
      end;


polyline = record
          points: array [1..n] of point;
          npoint: integer;
          ......             (*display attributes*)
      end;


cell = record
          name:   <cell name>
          cellext: record
          cellint: record .
```

Fig. 2.7: Parts of WDS Type Definitions

3. HARDWARE DESCRIPTION LANGUAGES (HDLs)

Graphic hardware description languages (GHDLs) are derived from textual hardware description languages (HDLs). To be able to talk about peculiarities of GHDLs we have to be aware of some essentials of HDLs. Some constructs of a HDL cannot be represented efficiently by graphical symbols. These are e.g. data path width specifications of primitives (constant, terminal,..). Such constructs can be represented by a textual description within a GHDL diagram. To maintain the compatibility with the underlying HDL in the GHDL diagram these texts follow the same syntax rules as in the HDL. Already the naming conventions for cell names, port names etc. follow the syntax of the HDL. There are also a lot of semantic rules a GHDL diagram has to follow in order to be consistent with its mother HDL. For example, it is not allowed to instantiate a non-primitive twice using the same name. To solve this problem an instant number has to be appended to the name. There is also no recursiveness concerning connections, an input pin of a cell cannot be connected with the output pin of the same cell, and concerning the calling hierarchy of subcells.

Till now a lot of hardware description languages have been developed [CLE77, HAP79, MPT82, NUR82]. HDLs describe partly the behaviour and partly the structure of a piece of hardware at one particular level of abstraction (e.g. gate level) or at several different abstraction levels (such as e.g. various mixtures of procedural level, register transfer level, logical level and switching level). In this chapter we define the restrictions and constraints imposed on a GL program by the underlying HDL. Only such GL diagrams are called GHDL diagrams which follow such conditions.

3.1 HDL Primitives

Many HDLs describe hardware at register transfer (RT) level. But there are also HDLs which cover other abstraction levels. For example, KARL-3 covers three different levels of abstraction: RT level, gate level, and switching level. This is illustrated by the example in fig. 3.1 where a simple multiplexer is shown at RT level (fig. a), at gate level (fig. b) and at switching level (fig. c). (Figures 3.1 d and e show different versions of it at circuit level.) Each of these abstraction levels has its own primitives. A primitive of a particular abstraction level is an object (the symbol for a particular kind of hardware element) which cannot be further decomposed at this level. I.e. a primitive is an atomic element of a particular level of abstraction. At RT level, e.g., we have the multiplexer or relational operators as primitives. These cannot be decomposed at RT level, since their logic equations use primitives of another abstraction level (of the gate level).

For each primitive of a particular level a HDL provides either a data type or a functor (operator or standard function). Data types are found in the declaration part of a description and functors in the statement part. This is different from HDL to HDL. KARL, for example, has the declaration types of *register*, *constant*, *terminal*, etc. For instance, it defines a multiplexer by a case statement. In a GHDL we define a drawing symbol for each primitive of its textual companion HDL. We call such a symbol a GHDL primitive. For each GHDL primitive having been instantiated its symbol is drawn onto the screen. So we may describe a hardware graphically by instantiating all GHDL primitives needed. Fig 3.2 shows the symbol of a multiplexer and a demultiplexer.

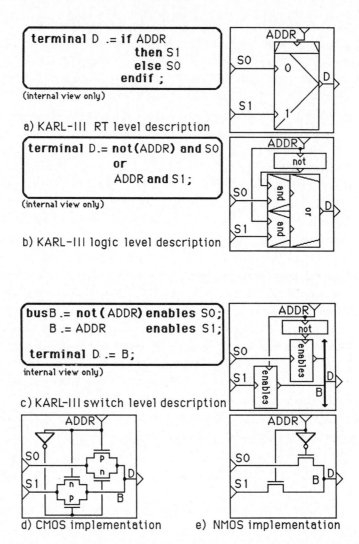

terminal D .= if ADDR
            then S1
            else S0
            endif ;

(internal view only)

a) KARL-III RT level description

terminal D.= not(ADDR) and S0
            or
            ADDR and S1;

(internal view only)

b) KARL-III logic level description

busB.= not(ADDR) enables S0;
    B.= ADDR   enables S1;

terminal D .= B;

internal view only)

c) KARL-III switch level description

d) CMOS implementation   e) NMOS implementation

Fig. 3.1 : The hierarchy of abstraction levels: textual (KARL, left side) and graphic (ABL, right side) representation of a simple mujltiplexer: at RT level (a), at gate level (b), at switching level (c), and of two versions at circuit level (d, e).

For a GHDL we have to follow that the primitives of the GL are the primitives of the underlying HDL. Instantiation of a primitive in a HDL corresponds to drawing and connecting the symbol of this primitive in a GHDL diagram.

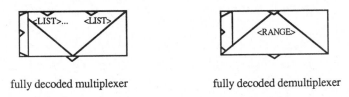

fully decoded multiplexer   fully decoded demultiplexer

Fig. 3.2 : Symbol for a Multiplexer and a Demultiplexer

3.2 Cell hierarchies

To cope with complexity normally hardware is partitioned into cells or other kinds of modules. A cell may have one or more daughter cells. Each daughter cell may again have daughters. The hierarchy of cells within a VLSI design may have up to a dozen or more nesting levels. This cell hierarchy is mostly of symbolic character: a chip having been fabricated from a design with hundreds of cells uses only a single silicon crystal. This illustrates, that the structural information which defines the cell hierarchy is symbolic information, but no physical information (compare fig. 1.1 b). The graphical representation of a cell hierarchy is illustrated by figures 1.3 and 3.3. Figure 3.3 a) shows the external view of a cell PLUS4. Fig. 3.3 b) shows its internal view, where the following kinds of cell components are visible in addition to the cell frame (external view): nets (wiring), primitive components (dot operator for data path concatenation, curly brackets for data path splicing) as well as four instances FA&3 thru FA&0 of the daughter cell FA (full adder). Fig. 1.3 a) again shows the external view of this cell FA. Fig. 1.3 b) shows its internal view. Cell FA contains primitive objects only (wires, as well as the logical KARL primitives *and, or* and *exor*). These objects are the leaves of the cell hierarchy of PLUS4.

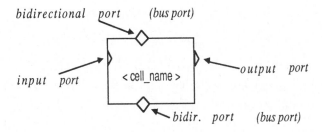

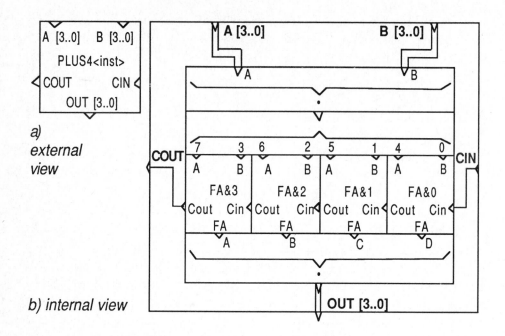

Fig. 3.3 : Cell Hierarchy Example : Cell PLUS4

3.3 Connectivity

An important application of an HDL description is to compile it in an object code serving as the
input of a simulator. HDL primitives are translated io corresponding simulator code elements. Also
the connections between these elements are translated. In an HDL program these connections may
be defined in two ways: explicitly (e.g. in SDL [CLE77]) or implicitly by special operators and
expressions (e.g. in KARL-3). In SDL all connections are defined in terms of nets. The format is
as follows:

<net name> = <pin list>; or
<net name> = FROM (<pin list>) TO (<pin list>);

e.g.: NET1 = FROM (.ENABLE) TO (G1.IN1);

In KARL-3 "electrically" two kinds of explicit connection definitions are distinguished: directed
connections (using the **terminal** statement: only a single data source allowed) and bidirectional
interconnections (using the **bus** statement: multiple data sources allowed). Sources to terminals and
buses are assigned by means of expressions. The following example shows a terminal, whose
source is the output of an adder :

TERMINAL <identifier> .= <expression>;

e.g.: TERMINAL HELP2 .= ZERO.IN + ZERO.HELP1;

This notation is much more concise, than the one available within SDL :

NET1 = FROM (ZERO.0) TO (CASCADE1.1);
NET2 = FROM (IN.0) TO (CASCADE1.2);
NET3 = FROM (ZERO.0) TO (CASCADE2.1);
NET4 = FROM (HELP1.0) TO (CASCADE2.2);
NET5 = FROM (CASCADE1.0) TO (PLUS.1);
NET6 = FROM (CASCADE2.0) TO (PLUS.2);
NET7 = FROM (PLUS.0) TO (HELP2.1);

In a GL diagram for drawing nets we only have polylines available. But in a GHDL diagram a net is represented by a composition of one or more polylines and text. Topologically KARL-3 also features another kind of interconnection definition: it uses *wiring by abutment,* where no polyline is used, since no wire is needed. In this case a net is represented only by referencing the ports to be connected. A good HDL compiler checks for matching path width and signal flow direction of the

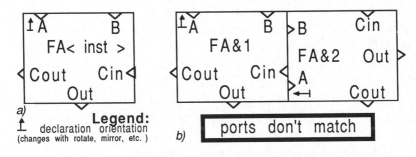

Fig. 3.4 : ABLED's Syntax Check before Abutments

ports to be connected. The ABLED editor, which is an implementation example of a GHDL, has an abutment primitive with such on-line check features. Fig. 3.4 illustrates the ABLED graphic on-line syntax check. The full adder (a) of fig. 1-3 is instantiated twice, where the second instantiation has been rotated left by error. This way the two slices cannot be abutted. Fig. 3.4 b) illustrates this along with ABLED's diagnostic message : *'ports don't match '.* ABLED refuses to execute the abutment command.

4. PECULIARITIES OF GHDLs

A GHDL combines the advantages of GL and HDL. In this chapter we define a GHDL, look for existing implementations and give an example for it. The difference between a GL and a GHDL are the grammar rules. GHDL primitives are more complex than the ones of a GL. Using the block structure features of a GL we can generate cells like the primitives of the GHDL, but these cells are not restricted to primitives allowed by the underlying HDL. We call all these checks to be applied to a GHDL diagram to guarantee the compatibility with the underlying HDL, 'HDL checks'.

We already mentioned, that we digitize the picture during programming it, and compile the digitized version either directly into simulator code or into the HDL language. The HDL checks applied to the diagram can be done in batch or on-line mode. Using the batch mode we translate the picture into a HDL program and use the HDL compiler for the HDL checks or we use directly an HDL checker to prove the correctness of the GHDL diagram. The errors can be reported to the user by messages or by marking the error in the diagram. Such systems have the disadvantage, that errors are not detected immediatelly to avoid subsequent errors. Another disadvantage is that there may be structures in the schematic diagram that cannot be translated into a corresponding HDL structure. Those systems are often called schematic entry systems and are comparable with a GL system.

Systems with on-line checks generate correct GHDL diagrams. Such diagrams we can translate directly without further checks into simulator code.

4.1 Schematic Entry Systems

This section gives a summary on graphic hardware design aids, which are not really GHDL implementations. In the literature we find some systems belonging to the field of schematics entry systems and GHDL programming systems.

A typical schematic entry system is **CASS** (Computer Aided Schematic System) [BAY77]. With CASS it is possible to define new components in a library. Any two dimensional figure may be a CASS-symbol. CASS includes an interactive graphics editor offering all facilities for editing and modifying drawings. Connections to other systems are possible via an integrated data base and by postprocessing CASS-DUMP. CASS can be used in connection with other systems, e.g. for a batch HDL checker, but an online check of HDL rules is not possible.

A system using CASS as schematic entry system and which processes the data captured with CASS is presented by [BBK79]. With this system it is possible to capture the behavioral description of hardware by events and conditions. The graphical elements are similar to a traditional flowchart. For example, CASS may be used for design data entry in logic simulation [BAY78].

Another schematic entry system is **CLOS** [KAW82]. Schematic data is stored in a data base and can be accessed by different translators and CAD tools. Detected errors are reported either on line printer or hard copy drawings. Components can be defined for four HDLs (LDS, ALS-4, MASCAD, SPICE). None of these languages cover the register transfer level.

The **SCALD** system [WWI78] uses as schematic entry system the Stanford University Drawing System (SUDS) Graphics Editor. SUDS supports the generation of logic diagram data further processed by SCALD.

[OST84] presents **SFDL** (a Symbolic Functional Description Language) which supports the capture of design data at the functional or procedural level, i.e. levels above the register transfer level. SFDL is translated into PASCAL. The system is comparable with the system described in [BBK79]. But it has one peculiarity, to recognize hand drawn diagrams. SFDL itself is part of a system which also allows drawing of block diagrams to describe the structural data.

Concerning layout editors with a GHDL or a schematic entry system we have the format of the textual description of the layout, a layout description language (LDL), e.g. CIF instead of a HDL. This LDL is input to a other processors, e.g. mask generator. The checks comparable to HDL checks are design rule check, connectivity check and others. We find systems with an on-line check e.g. MAGIC [OAL84], VLDL [BER83], SIDS [CKS80], IGS [IAL78], and systems without on-line check e.g. CAESAR [OUS81], KIC [KEB83]. Some of them, IGS, VLDL and SIDS covering symbolic layout and layout level.

4.2 GHDL Implementations

A GHDL implementation is an interactive CAD system for the synthesis of hardware designs. It uses a specialized GL with its interactive features as a subsystem. This GL provides a set of application specific drawing symbols like special boxes for multiplexers, registers, decoders and many other elements of hardware structures.

The first GHDL having been published is **FLOWWARE** [CHT77]. It is an extended graphic version of a subset of Chu's hardware description language CDL [CHU65]. The extensions include a graphic flow chart language, so that FLOWWARE is a behavioural multi-level hardware design language including functional and procedural levels. The package is no more available. It has been running on an IBM/360 model 50 using several NOVA 800 minicomputers as graphic front ends.

Before going on we should first specify the term multi level. In hardware design several methodological abstraction levels are needed. For terminology and classification schemes see [CLE79, HHA87, DUS83].

Another GHDL has been proposed and specified in 1977, the language **ABL** (A Block diagram Language) [HAR77]. This language is the graphic companion language of the textual KARL language. ABL feature *DOMINO notation* which is the graphic representation of *wiring by abutment,* where no lines have to be drawn to show interconnection (see also fig. 3.4).

ABL has been proposed together with a specification of the textual hardware description language KARL-1 (also see [HAR77]). The first textual HDL implementation of the DOMINO feature is part of KARL-3 ([KLE84, AWO84, HL84]), a successor of KARL-2, which did not have this abutment ([PH77, HE80, HLSW, ES81, BW81]). The first graphic implementation of ABL is the interactive graphic editor **ABLED** [GIR85, BBG86, GIR86], which is running under VMS on VAX11/7xx, microVAX and VAXstation, under UNIX on SUN and APOLLO workstations, as well as under AEGIS on APOLLO. ABLED is connected to the KARL system by an ABL-to-KARL translator [WEL84], such that it may be used as an interactive graphic front end of the KARL3 simulator system. The KARL-3 system is running on all of the above mentioned computers (additionally it is installed on IBM PC-AT and ATARI ST).

FLOWWARE and ABLED are the only GHDL implementations which are not just specialized schematics editors. FLOWWARE had also an I/O-facility of simulation data integrated into the interactive graphic editor and its menu control. Simulation results could be inserted at proper locations into the GL diagram display. ABLED has powerful graphic on-line syntax check features, which cover almost all the application oriented diagnostics of the textual KARL compiler [UWE87]. Compared to the use of textual KARL-3 only, ABLED has improved the functional/logic design turn-around time, so that the designer productivity has increased by about a factor of five [BAL86, FAR86]. The KARL simulator is linked to ABLED via the translator. This link is currently a batch process, so there is a bottle neck between editing and simulation, which are both interactive (see [MAV87] for interactive simulation).

GHDL implementations are strongly syntax-directed graphic editors, which substantially improve design turn-around time. This frees time to experiment with alternative architectures [BAL86, FAR86]. With respect to ABLED this will be further improved by a new version which directly works on RTCode [NN86], the executable form of KARL-3 descriptions used by the simulator. So the ABL-to-KARL translator and the KARL-3 compiler would be needed anymore and interactive simulation could be an integrated part of experimenting with alternative architectures at the same session in front of the ABLED screen. Since SCIL-3, the new KARL-3 interactive simulation and testing language [MAV87, SHE81] supports the integration of interactive simulation and interactive testing (prototype testing as well as testpattern development), design for testability and functional testpattern development should also be integrated [HAR86, HCC87].

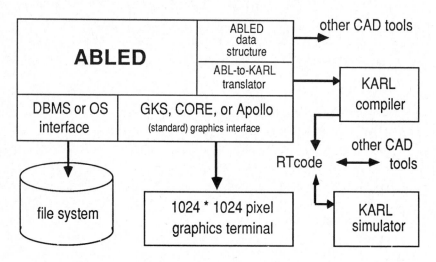

Fig. 4.1 : Interfaces of the ABLED editor used as tool integrator

Quite a number of CAD tools use the RTCode as an interchange format to the ABLED including the KARL system.Most of these tools have been developed within the CVT project under partial support by the commission of the European Communities:

Tool Name	Application	Implementer
VERENA	logic verification	University of Hamburg, F.R.G. University of Passau, F.R.G.
PRIMITIVE	microprogram verification	University of Passau, F.R.G.
MIAKA	microprogram assembling	University of Bremen, F.R.G.
FERT	layout fault extraction	CSELT, Torino, Italy
OFSKA	functional fault simulation	OLIVETTI, Ivrea, Italy
OFGKA	fault model generation	CSELT, Torino, Italy
OTAKA	testability analysis	OLIVETTI, Ivrea, Italy
TIGER	test patter generation	CSELT, Torino, Italy
IRENE	RTlanguage for micro- computer descriptions	INPG, Grenoble, France
KARENE	simulator for micro- computer structures	INPG, Grenoble, France
ARIANNA	chip planning	University of Genua, Italy
ASMA	sequential circuit generation	CSELT, Torino, Italy
SCOD	sequential circuit minimization	CSELT, Torino, Italy
KMIN	logic minimization	University of Karlsruhe, F.R.G.
LOGE	sequential circuit and logic generation	University of Karlsruhe, F.R.G.
FOLD	PLA folding	University of Genua, Italy
FLAP	layout generation	University of Genua, Italy
KARATE *)	testpattern generation	University of Kaiserslautern, F.R.G.
REX	register-transfer extraction	University of Kaiserslautern, F.R.G.

*) not yet implemented

Table 4.1 : CAD tools around ABLED / KARL-3

With the new ABLED version mentioned above, these tools will have an even more efficient front end.

4.2.1 ABLED - An Example of a GHDL Implementation

As an example for a GHDL we present ABLED (A Block diagram Language EDitor). ABLED is not just another schematic entry system but an architectural level design tool providing strong guidance and on-line diagnostic features to its users [HGW85]. The notation of the ABLED primitives is derived from the close connection of a schematic entry system with a HDL. The HDL companion is KARL-3. On the following pages we will list some primitives of ABLED, show some HDL rules the designer must not break.

Fig. 4.2 shows some ABL symbols. The shape of the symbols can be modified by the user. The shown symbols are the models for some KARL-3 logic primitives.

Primitive Name   GHDL Primitive

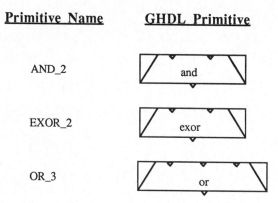

AND_2

EXOR_2

OR_3

Fig. 4.2 : Primitives Used in GHDL diagram of Fig. 1.3

When a primitive is instantiated, the name of a rule category is written on the scree. This editor query has to be answered by the designer. An example for this instantiation mechanism is shown in fig.4.2. The designer instantiates a clock and has to enter the name and the correct clock declaration.

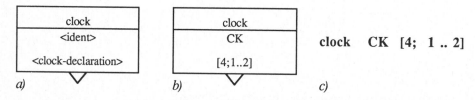

Fig. 4.3: Example of Syntactical Rule Query and Reply: a) after picking the symbol; b) after entering the textual information; c) KARL textual representation (clock declaration)

When a designer creates an ABL diagram of a cell, first he has to create the external view of the cell. This corresponds to the header description in a KARL-3 program. Then he can create the internal view of the cell corresponding to the body description of KARL-3. If he wants to instantiate a cell not yet declared, the editor does not allow this. If all subcells are defined, he can instantiate them one or more times. The instantiated subcells and primitives can be connected with nets. The editor checksthe compatibility of the connected pins and ports (pin is the notation of connection points of the external view of a cell, port of the internal view).

4.2.2 A Multi-Level GHDL

KARL and ABL are multi-level languages covering the switching level, the gate level and the non-procedural RTlevel. The question is: why not have a GHDL covering much more levels ? We try to answer this question by the current implementation efforts on MLED (Multi-Level EDitor) [HWE87, HWA87]. This is also conceptually a real multi-level editor, whereas ABLED merges the above thee abstraction levels down to a single-level implementation.

MLED, however, features simultaneous representations of the same object at several levels of abstraction (see fig. 1.1 c) for illustration). MLED has a modern window-oriented user interface (see fig. 5.1).

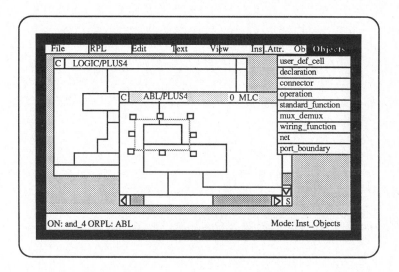

Fig. 5.1: An Example of the MLED's User Interface

The user may select at any time a view to be edited. The number of levels within MLED is extensible by support of a menu design editor and a drawing function editor.

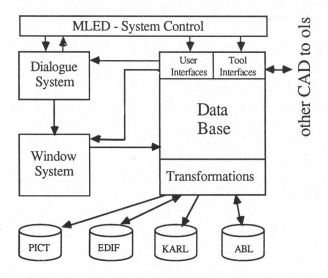

Fig. 5.2: Interfaces of the MLED

MLED provides version handling (fig. 3.1) and consistency restoring features for interconnection geometry needed for top-down design in cell hierarchies [HWA87]. So it may also serve as a design database management system or subsystem and as a tool integrator (fig. 5.2). It also features desk top publishing support by a PICT format generator [PICT, PIC87].

Conclusions

A brief survey on graphic hardware description languages has been given, as well as a little insight in its computer graphics foundations. It has been illustrated, that graphic hardware description editors as user interfaces substantially enhance the power of HDLs as an effective means of design tool integration.

Acknowledgements

I wish to give many thanks to Prof. Hartenstein for his encouraging and fruitful discussions on the subject of this paper. Without these intensive talks it would have never been possible. I owe many thanks to my colleagues, especially J. Blödel, R. Hauck, K. Lemmert, M. Ryba, M. Weber and A. Wodtko for reediting and reading the proofs of this article.

6. REFERENCES

[ABA86]   N. N.: ABAKUS (ABL and KARL User Group) Workshop, Passau, June 17 - 20, Collection of viewgraphs; Univ. Kaiserslautern / Univ. Passau 19861

[AWO84]   A. Wodtko: Semantics Processing within the KARL-3 Compiler; Diploma Thesis; Universität Kaiserslautern; 1984

[BAA78]   H. M. Bayegan, E. Aas: An Integrated System for Interactive Editing of Schematics, Logic Simulation and PCB Layout; 15th Design Auto- mation Conference, 1978

[BAL86]   G. P. Balboni, V. Vercellone: Experiences in Using KARL-3 in Designing a CMOS Circuit for Packet-switched Networks; ABAKUS Workshop, Passau, 1986 (see [ABA86])

[BAY77]   H. M. Bayegan: CASS: Computer Aided Schematic System; 14th Design Automation Conf., 1977

[BBG86]   A. M. Biraghi, A. Bonomo, G. Girardi: ABLEDitor: User Manual; CVT report; CSELT, Torino, 1986

[BBK79]   H. M. Bayegan, O. Baadsvik, O. Kirkaune: An Interactive Graphic High Level Language for Hardware Design; in (ed.: W. M. van Cleemput): Proc. CHDL'79 - Symp. on Computer Hardware Description Languages and their Applications, IEEE, New York,1979

[BER76]   R. D. Bergeron: Picture Primitives in Device- Independent Graphics Systems, Department of Mathematics, University of New Hampshire, Durham, New Hampshire 03824; Proc. ACM Symposium on Graphic Languages, ACM, New York, New York10036

[BER83]   K. van Berkel: VLSI-Layout Design Requires a New Language; report, Philips Research Labaratories, Eindhoven, 1983

[BW81]   B. Weber: PASCAL - Implementierung eines Simulators auf KARL-2 Basis; Diploma Thesis; Universität Kaiserslautern; 1981

[CHDL81]   M. Breuer, R. Hartenstein: Computer Hardware Description Languages and their Applications; North Holland Publ. Co, Amsterdam/NewYork, 1981

[CHT77]   Shigfat S. Ching, James H. Tracy: An Interactive Computer Graphics Language for the Design and Simulation of Digital Systems; University of Missouri-Rolla; Computer, Vol10, pp.35-41, June 1977

[CHU65]   Y. Chu: "An ALGOL-Like Computer Design Language", Communications of ACM, Vol.8, Oct. 1965, pp. 606-615

[CHU74]   Y. Chu: Why Do We Need Computer Hardware Description Languages ?; Computer, Vol. 7, Dec. 1974

[CKS80]   D. Clary, R. Kirk, S. Sapiro: A Symbolic Interactive Design System; 17th Design Automation Conference, 1980

[CLE77]   W. M. van Cleemput: A Hierarchical Language for the Structural Description of Digital Systems; 14th Design Automation Conf., 1977

[CLE79]   W. M. van Cleemput: Computer Hardware Description Languages and their Applications; 16th Design Automation Conference, 1979

[DA85]   Daisy Systems Corporation, CHIPMASTER: Reference Manual, 1985, Mountain View, CA 94039,

[DUS83]   S. Dudani, E. Stabler: Types of Hardware Description; in (ed.: M. Barbacci, T. Uehara): Proc. IFIP CHDL'83 - Symp. on Computer Hardware Description Languages and their Applications; North Holland Publishing Company, New York [Amsterdam, 1983

[ENC86]   J. Encarnaçao: CAD-Schnittstellen und Datentransferformate im Elektronik-Bereich; Springer-Verlag, Berlin/Heidelber/New York, 1986

[END83]   G. Enderle, K. Kansay, G. Pfaff: GKS - The Graphical Standard; Springer-Verlag, Heidelberg

[ES81]   E. Schaaf: PASCAL - Implementierung eines KARL Compilers; Diploma Thesis; Universität Kaiserslautern, 1981

[FAR86]   D. Farelly: The practical Application of ABLED and KARL in Designing a Speech Synthesizer; ABAKUS Workshop, Passau, 1986 (see [ABA86])

[GIL78]   W. K. Giloi: Interactive Computer Graphics: Data Structures, Algorithms, Languages; Prentice-Hall, Englewood Cliffs, N.J., 1978

[GIR85]   G. Girardi et al.: ABLED - a RT Level Schematics Editor and Simulator Interface; Proc. EUROMICRO Symposium, Brussels, Belgium, 1985; North Holland Publ., 1985

[GIR86]   G. Girardi et al.: KARL (textual) and ABL (graphic) : A User/Designer Interface in Microelectronics; in [ENC86]

[HAL86]   R. Hartenstein, K. Lemmert: KARL-3 Manual; CVT report, Univ. Kaiserslautern, Kaiserslautern, 1986

[HAR77]   R. Hartenstein: Fundamentals of Structured Hardware Design; North Holland Publ. Co. /American Elsevier, Amsterdam/New York, 1977

[HAR79]   R. Hartenstein, E. von Puttkamer; KARL - A Hardware Description Language as Part of a CAD Tool for LSI; in (ed.: W. van Cleemput): Proc. CHDL'79 - Symp. on Computer Hardware Description Languages and their Applications; IEEE, New York, 1979

[HAR86]   R. Hartenstein; High-Level Simulation and VLSI; Proc. Summer School on VLSI Design, Beatenberg, Switzerland, 1986 (eds. W. Fichtner, M. Morf); Kluwer Academic Publishers, 1986

[HAR87]   R. Hartenstein: The Classification of Hardware Description Languages; within this volume

[HCC87]   R. Hartenstein: Simulation von VLSI-Schaltungen und -Systemen; CADCAM Report, Februar 1987

[HE80]   U. Hedengran: A Parser for KARL-2; internal report; Dept. of Appliede Electronics, Royal Inst. of
            Technology; Stockholm, 1980

[HGW85]   R. Hartenstein, G. Girardi, U. Welters: An RT Level Schematics Editor and Simulator User Interface;
            in (ed.: K. Waldschmidt): Proc. EUROMICRO Symp. 1985, North Holland, Amsterdam 1985

[HL84]   R. Hartenstein, K. Lemmert: KARL-3 Reference Manual, CVT Report; Kaiserslautern, 1984

[HLSW81]   R. Hartenstein, P. Liell, E. Schaaf, B. Weber: KARL User Manual and Introduction; Report;
            Berkeley/Kaiserslautern, 1981

[HWE87]   R. Hartenstein, U. Welters: MLED: A Multiple Abstraction Level Graphics Editor; EUROMICRO'87,
            Portsmouth, UK, Sept 1987

[HWL87]   R. Hartenstein, U. Welters: Mehrebenen-Graphik-Editor als DBMS für VLSI-Simulation; Proc.
            ASIM '87; Zürich, 1987

[IAL78]   B. Infante, D. Bracken, B. McCalla, S. Yamakoshi, E. Cohen: An Inter- active Graphics System for the
            Design of Integrated Circuits; 15th Design Automation Conference, 1978

[KAW82]   E. Kawamoto: Schematic Entry System with a Query Feature; ICCC, Port Chester, 1982

[KEB83]   Ken Kell, Giles Billingsley: KIC2 - Integrated Circuit Layout Program; University of California at
            Berkeley, Berkely, CA, April 1983; Documentation Decus No. VAX-44

[KLE84]   K. Lemmert: A LALR(1) Parser for the RT Language KARL-3; Diploma Thesis; Universität
            Kaiserslautern, 1984

[KUL68]   H. E. Kulsrud: A General Purpose Graphic Language; Comm. ACM, Volume 11, No. 4, April 1968

[MAV87]   A. Mavridis: Languages for Simulator Activation and Tester Operation; (within this volume)

[MPT82]   J. D. Morison, N.E. Peeling, T.L. Thorp: ELLA: A Hardware Description Language; Int'l. Conf.on
            Circuits and Computers,1982

[NN86]   N. N.: KARL Manual; CVT report, Universität Kaiserslautern, 1986

[NNV86]   N. N.: IC Design Software; VTI, San Jose, CA 95131, USA, 1986

[NUR82]   G. M. Nurie: LOGAL+ - A Hardware Description Language for Hierarchical Design and Multilevel
            Simulation; Int'l. Conf. on Circuits and Computers, 1982

[OAL84]   J. K. Ousterhout, G. T. Hamachi, R. N. Mayo, W. S. Scott, G. S. Taylor: MAGIC: A VLSI Layout
            System; 21st Design Automation Conf., 1984

[OST84]   G. Odawara, J. Sato, M. Tomita: A Symbolic Functional Description Language; 21st Design
            Automation Conf., 1984

[OUS81]   John K. Ousterhout: Caesar: An Interactive Editor for VLSI Layouts; University of California,
            Berkeley; VLSI Design Fourth Quarter 1981

[PIC87]   D. Finkler: PICT-Generator für das MLED-Graphicsystem; Projektarbeit; Universität Kaiserslautern,
            1987

[PICT86]   MacIntosh Technical Notes; Number 21,27; September 1986

[PFI76]   Gregory F. Pfister: A High Level Language Extension for Creating and Controlling Dynamic Pictures;
            Proc. ACM Symposium on Graphic Languages, ACM, New York, 1976

[PH77]   P. Haselmeier: The KARL language compiler; Report no. 77-14; Instituto di Elettronica, Politecnico di
            Milano, 1977

[SHA69]   Alan C. Shaw: A Formal Picture Description Scheme as a Basis for Picture Processing Systems; Information and Control 14, 9-52 (1969)

[SHE81]   W. Sherwood: An Interactive Simulation Debugging Interface; in [CHDL81]

[UWE87]   U. Welters: ABL-to-KARL Relations; CVT Report; Kaiserslautern, 1987

[WAR76]   Caroline Wardle: ARTSPEK: A Graphics Language for Artists; Proc. ACM Symposium on Graphic Languages, ACM, New York, 1976

[WEL84]   Udo Welters: ABLTOKARL translator: Algorithm Description; CVT report, University of Kauserslautern, 1984

[WWI78]   T. M. McWilliams, L. C. Widdoes, Jr.: SCALD: Structured Computer Aided Logic Design; 14th Design Automation Conference, 1978

Appendix

ABL	A Block diagram Language
ABLED	A Block diagram Language EDitor
CG	Computer Graphics
DPL	Display Processor Language
FA	Full Adder
GL	Graphic Language
GHDL	Graphic Hardware Description Language
GKS	Graphical Kernel System
HDL	Hardware Description Language
KARL	KAiserslautern Register Transfer level Language
LDL	Layout Description Language
PE	Picture Editor
RT	Register Transfer
WDS	Working Data Structure

3.5. LANGUAGES FOR SIMULATOR ACTIVATION AND TESTER OPERATION

Angelos MAVRIDIS

*Department of Computer Science, University of Kaiserslautern
Postfach 3049, D-6750 Kaiserslautern, F.R.G.*

This paper tries to give a classification of activation languages and dataformats for simulators and for testers. It tries to show, what both kinds of languages have in common, and, how combinations of such languages could help to support synergisms between simulation, testing and design for testability.

PREFACE

For lack of sufficient space within this paper the presentation of product data and similar statements or descriptions about test equipment and its support software has been shortened drastically. Some of the sources having been available to us may already be incomplete and abbreviated substantially, or, fail to meet highlights properly. So I deny to guarantee the correctness, completeness, and/or adequacy of presentation of any features of products being mentioned in this paper. For more details and a lot of references see [MAV87].

For simplicity this paper always speaks about *languages,* also when it could be questioned, whether the term *'data format'* or other terms would not be better. That's since both may be specified by a formal grammar and analyzed by a parser. Why not talk about data languages?

1. INTRODUCTION

Future trends in CAD for VLSI also focus on developing a methodology of design problem capture and conceptual design at very early phases of the design process [HA87, HALE86, HAR87, HAWE86, LENE87], definitely before the costly logic design procedure has been started. The goal is to avoid expensive redesigns needed to correct conceptual design errors, such as e.g. missing the requirements, bad topology and bad structure of the circuit, too much area consumption or bad (VLSI-) architecture and others.

Design tools based on high level hardware design languages could be used for design problem capture, since a concise notation is needed be used to set up a *design specification* which formally expresses the design problem. To be sure to capture the design problem correctly the specification has to be checked against the requirements. Often several possible solutions have to be considered and analyzed, so that experimenting with alternative architectures is needed. Such experiments should be carried out at the highest possible level of abstraction to avoid incomprehensibility of descriptions and high labour cost because of high complexity. In using a suitable RT simulator, first the specification could be simulated (also see [GHW85]), as well as later on different conceptual design versions.

Also *Design for testability* should be a part of conceptual design efforts, rather than of logic design [HAWE86, LEM86], since this is a good strategy to reduce the cost of test pattern development and improve the quality of test patterns. Not only testability per se, but also the length of the test is a very important objective in design for testability. The best solution is to carry out test pattern development in very early phases of the design process, at least before logic design has been started. The most desirable time would be, after the specification is ready. So, instead of being part of the logical design, and thus being highly expensive, testability would be a subject of early design planning and partitioning definition. The designer could fully concentrate on the testability architecture of a circuit and could experiment with alternative architectures.

However, this would require, that the test patterns should be available at such an early time, so that testability data and test length data of different version architectures could be compared. Although functional test patterns are primarily developed for production testing of integrated circuits, it is useful to have such test patterns available already along with the design specification: ready for specification verification. Since simulation still is the most important means of high level verification, and, since for large circuits an exhaustive simulation is not possible anyway, subsets of the test patterns have to be selected in a clever way. So the designer has to think about activation patterns. This is an opportunity to integrate conceptual design, design for testability, and functional test development in a synergistic way. Such a synergism would be supported by integrating the environments of conceptual design and simulation as well as test pattern generation and evaluation (e.g. see [SM85, SMP86]). Such an integration could be supported by the development of suitable activation languages and interfaces. That's why such languages are subject of this paper.

2. SIMULATION VERSUS TESTING

In electronic design automation (EDA) environments we may identify *activation languages (AL)*, in contrast to *(hardware) description languages (HDLs or DLs)*. Activation languages may be used to describe stimuli / response time patterns to activate an existing physical object or its model having been predefined (e.g. by using a HDL or similar notation [HA87]). By *activation* the object accepts a sequence of stimuli, so that its sequence of internal and external responses may be observed. Such a sequence is called a *test* . A *description,* however, statically describes the structure, geometry and behavioural properties of an object, i.e. without activating it [HA87].

Looking at testing and simulation we may distinguish two kinds of activation languages. Testing languages are used to define tests to be applied to a *circuit under test (CUT)*. In this section we discuss the structure of a variety of languages having been implemented for testing and simulator activation. To achieve a more structured presentation of the area the languages will be subdivided into sublanguages. This could also help to find good candidates to form combined languages in order to support tool integration between both areas.

At a first glance simulator activation and testing look very much different. For simulation we need both, a description and an activation. But for testing the description has been replaced by a physical object, such as an integrated circuit or another hardware subsystem. The consequence is, that both, testing and simulation need an activation language. However, there is an important difference between these two kinds of activation. In simulation an activation may directly reference the DL description of the simulation models. In testing the CUT can be reached only via the interface of the testing device using TEL *(TEL* stands for *Test Equipment-oriented Language)*. Because of these different goals ALs may be subdivided into:

SAL   Simulator AL   TL   Testing Language

In this section it will be tried to illustrate, that despite of these differences testing and simulation have a lot in common. Most languages being subject of this paper (except TELs) are more or less source languages for compilers or interpreters finally generating TEL object code. If we identify sublanguages within languages used for both areas, we will be able to find such languages which show substantial similarities between testing and simulation.

2.1 Languages in Testing

Let us start with testing languages (TLs). Later on we will discuss languages used for simulator activation. Within testing languages we may distinguish at least different sublanguages for equipment-oriented set-ups and adjustments (TELs) and for the logic specification of the test (test specification languages: TSLs). TLs may be further subdivided into sub-sublanguages. So TLs may be subdivided into the following classes and subclasses:

TSL   Test Specification (Description) Languages:
   TVL   Test Vector Languages
   TPL   Test Programming Languages

TEL   Test Equipment-oriented Languages:
   TAL   Tester ALs
   ESL   Equipment Set-up Languages

In the following section we will talk about TELs. In the section following behind we will discuss TSLs. Such TSLs are mostly sublanguages within a larger system, which specify the logical aspects of the test, sometimes in a more equipment-independent form *(high level TSLs)* or in using more equipment-dependent notations *(Test 'Assembly' Languages)*.

2.1.1 Test Equipment-oriented Languages (TELs)

TELs are used to express equipment-dependant details. Let us call this sublanguage an *equipment set-up language (ESL)*. An ESL is needed to describe, how the *ATE (automatic test equipment)* has to be operated to drive and sense the physical object mounted into a socket and how it has to be set-up properly. An ESL allows a procedural form of indirect ATE description, since the parameter ranges of ESL instructions reflect the ATE specifications. Fig. 2.1 illustrates this equipment environment. To achieve an extremely fast application of testvectors to the CUT an ultra-fast vector burst memory of limited size is used.This memory has to be preloaded in using special tester commands.

Normally ESLs are sublanguages having been integrated into a TEL, together with a *TAL (tester activation language)*. TALs are used to describe the entire test procedures including monitoring the test equipment, handling messages, housekeeping, partitioning the test into bursts and the downloading of these bursts. So in this scheme ESLs and TALs are sublanguages of TELs.

2.1.2 Test Specification Languages (TSLs)

The activation patterns to be used in a TAL are derived from test descriptions having been formulated in using a language called a *TSL (test specification language)*. Such languages describe tests, preferably in a device-independent manner. A *test* is a sequence of *test vectors* (mostly bit vectors).

Such TSLs may also be subdivided into *test(er) programming languages (TPLs)* and*test vector languages (TVLs)*. TPLs are capable to describe tests algorithmically, whereas TVLs describe tests vector by vector, ready for ATE application. Also TVLs could be tester-independent. However, most TVLs are either tester-specific languages, or, languages being specific to a tester family of a particular manufacturer.

TPLs are important for test development. A test may be rather long so that it would be advisable not to store all its vectors. A dumb method of testing a 32 bit counter, for example, would need about 2^{32} test vectors. A more concise and more comprehensible activation description would be a *test program,* which is an *algorithmic test description,* in contrast to the*vectorized test description* having been used above. For our binary counter example it would be a simple program loop, which may be expressed in a programming language like Pascal. This illustrates, why special *test programming languages (TPLs)* have been developed, which are extended programming languages.

2.2 Languages in Simulation

Simulator activation languages (SALs) are used to control the simulator and to define tests to be applied to a *circuit under simulation (CUS)*. SALs are mostly simulator-specific. Also a SAL may be subdivided into sublanguages for test description and for controlling the simulator and other support features (also see table 2.1). Not many and no successful efforts have been made towards the implementation of a standardized SAL. Mostly the abstraction level in programming with a SAL is low. However, the number of high-level, but still simulator-specific languages, such as e.g. that of LASAR Version 6 [LASA85] is increasing.

Simulation and test languages (SATLs) are or should be, respectively, high-level, simulator- and tester-independent languages like TSLs. Such languages try to provide a common activation description notation for simulation, test development and testing which supports the integration of these three areas and which tries to be as tester- and simulator-independent as possible. SATLs are not yet widely spread. Examples will be dealt with later in this paper.

2.3 Tasks and Procedures

Simulation is rather similar to testing. If instead of a CUT the its simulation model is tested, this test normally is called simulation, although the same test patterns may be used. However, there are also differences. This section will try to give more insight into the different environments and tasks of testing and simulation and the processing needed for them. We may distinguish the following tasks and problem areas:

- testing strategies, algorithms and procedures
- tracing, diagnosis and debugging procedures
- ATE set-up and adjustments proecedures
- filing, I/O and logging management

We may also distinguish different subareas in the area of testing and simulation. Each of these areas has its own little subenvironment, which depends on the particular goals to be reached by their tasks. We may distinguish the following types of tasks:

- simulation for verification
- simulation using PME (physical model extension)
- simulation for test development
- test (program) debugging
- prototype testing (for design debugging)
- production testing (go / no go check only)

In a modern environment the testing equipment (or the PME equipment which is quite similar to testing equipment) is interfaced to a host computer. In most of these tasks (except simulation without PME and partly test debugging) equipment-dependant aspects are important. Some of these tasks may be carried out by software running on a host computer, some by an ATE device, and some by tracing-oriented testing devices, called *logical analyzers (LAs)*. Table 2.1 gives a survey on such environments and their particular processing requirements. Two major classes of testing may be distinguished, depending on the kind of specifications to be checked:

- functional testing (test by checking logical correctness)

- parametric testing (checking timing performance and electrical performance)

In parametrical testing AC, DC, timing, maximum operation speed and similar non-digital data are checked. Functional tests in an IC production environment are finished with a GO/NO GO-result, because IC's cannot be repaired and are discarded. But functional testing or simulation for debugging a prototype or a design at an engineers workstation requires much more detailed data, since it is needed for diagnosis to detect the fault location inside the CUT or CUS.

	task					
processing needed	simulation f. verification	simul. f. test development	simulation with PME	test program debugging	prototype testing	production testing
ATE-independent test definition	useful	useful	useful	useful	no	no
I/O, logging and diagnostics	software	software	software	software	logic analyzer	no
statistics capture	no	no	no	no	no	yes
tracing and debugging	software	software	software	software	by logic analyzer	no
ATE-specific set-up procedure	no	no	yes	partly	yes	yes
ATE-specific test definition	no	no	generated automatically	partly	yes	yes

Table 2.1: Processing requirements of different tasks in the area of testing and simulation

Timing parameters usually measured during AC testing are address access time, address and data set up time, address and data hold time, chip select access time, chip deselect to high impedance time, and minimum pulse width. Although AC tests are relative simple, they often require special test capabilities and high degree of accuracy. That's why normally parametric testing is carried out separately, whereas functional testing is done at a lower speed. This is similar to the situation in simulation, where normally timing simulation is done separatley using simulators different from those used for functional simulation.

2.4 Special Equipment in Testing and Simulation

The differences between simulation-based and tester-dependent functional tests are due to timing problems. Simulators are slow, whereas testers have to prove, that the CUT works correct also at highest speed specified. This causes differences in the basic models of stimuli generation and expected output comparison. In parametric testing for test vectors a very fast comparator and a very fast burst memory is used, which is expensive and thus of limited size (see fig. 2.1). Tests have to be partitioned for downloading. Parameters to define driving current levels, voltage levels, and timing parameters have to be provided for the device drivers and receivers.

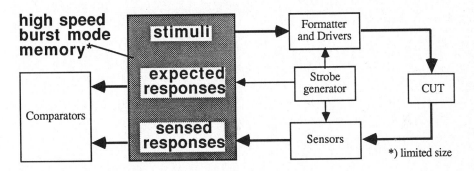

Fig. 2.1: Tester structure: illustration of burst memory use.

Physical Model Extensions and Hardware Modelling Systems. Hardware modelling systems like the Dynamic Prototype Test from HBB Systems [HADA85] or the Hardware Modelling Library from Mentor Graphics [EVAN86] are developed to use an actual hardware device as a functional and timing model during simulation to speed up simulation time. By converting this approach to the opposite, these two hardware modelling systems are announced to be able to be used as prototype verification equipment for limited functional testing. After fabrication, a new chip is plugged into the hardware modelling system, and its response is compared with the specific simulation model originally used in the design phase. In case differences between the software and the hardware model occur, the deviations can be seen as failures in the prototype. State of the art of hardware modelling systems provide poor timing resolution and little fine adjustable control over critical edge placement compared to instrumentation clusters, because their main intention is only to deliver the functional response of the reference device on them.

Production Testers. Large ATE such as production testers of Sentry, GenRad, or Teradyne are very powerful with respect to resolution, timing control and speed. Individual tester electronics per pin support low channel skew, up to subnanosecond resolution, and, DC and AC parametric measurement for precise device characterization. Diagnosis is not supported since not it is needed for go / no go tests.

Prototyping Testers. Verification testers are used for prototype testing, where the goal is the location and analysis of faults for design debugging. This means, that a good support of tracing, breakpointing and other debugging aids is essential. Speed is not so important, since most of the debugging is done interactively. An example of a verification tester is the Logic Master of Integrated Measurement Systems (IMS) [LIMI86], which offers channels with individual format generators for channel groups, a deep memory, and, since recently, a Parametric Measurement Unit (PMU) for DC characterization, which unlike the production testers contains only one DC measurement channel. By a switching matrix, this channel sequentially measures the parameters of the device pins one after the other, at the cost of measurement time. The Logic Master has the same screen user interface to program the functional tests and the parametric measurements, and it offers an Auto-Measure mode, which searches the functional test patterns and takes DC parametric measurements on the first low and the first high that occurs on each pin of the device. Other examples of verification testers are Daisy's Personal LOGICIAN CAE [DAIS85] and the Tektronix DAS9100 digital analysis system [TEK83].

3. LANGUAGES FOR SIMULATOR ACTIVATION

Simulator activation languages are front ends to simulators. In the last years, many advances happened in the user interfaces of simulators. Simulator activation languages are used to control the simulator and describe the test used to stimulate the simulation model. The languages vary in the programming abstraction level and in the functions they provide. This section deals with SALs, and gives a brief survey on the features of the user interfaces of a few SALs. The user interface is a critical factor in the acceptability and usability of the language, and his design must be considered as integral part of language design. The major goals of user interfaces are:

1. Shortening program preparation time

2. Making the system usable by humans of various skill levels (design engineer, test engineer, casual user, incomming inspection and production personnel)

3. Assistance in training new users

4. Availability of the full power of the simulator or the tester when the user is an expert

5. Protection from operator's errors

6. Easy modification or extension of the language

It is possible to satisfy the above criteria by implementing "multi-mode" command languages [DOWN83] . The modes which can be simulanteously available vary according to the simulator or the tester used:

      1. command mode (textual, may be, with a syntax-directed screen editor)

2. dialogue mode (smart queries sollicit the right data from the user).

3. menu mode (picking a choice from a menue of commands currently acceptable)

4. window mode (interactive graphic input in multiple schematics displays)

A combination of command mode and window mode, e.g. command mode and menu mode in a multiple window environment, is used in modern workstations to enter commands of SAL. The workstations have not only a comfortable input, but they also provide graphical output of the simulation data in logic analyzer manner.

In design verification and prototyping interactive simulation is at least as important as batch runs. Interactive simulation allows the designer to study and experiment with design alternatives during initial, creative phases of the system development. Modification of system descriptions, and evaluation of proposed design behavior can be accomplished rapidly using interactive simulation at register transfer level [HOLL83,THOM86,HALE86].

Simulation allows to study the design more in depth than testing, because every signal in the design, and not only the signals at the pins can be watched by moving 'virtual probes' over the design. These virtual probes are usually implemented by print or plot commands or via an interactive graphics display interface showing schematics of the design [GUSC84]. The signals in a simulation model can be locked for patching the simulation model or for fault simulation.

3.1 Some Simulator Activation Languages Examples

Exhaustive testing becomes economically unfeasible at a certain point in time [SEGE82], and the designer with his experience has to experiment with different pattern generation algorithms and testing methods (structural tests, functional tests, self-tests, designs for testability), and to decide, which method is most promising. The following section gives a brief survey on some examples of simulator activation languages. Completeness of this survey is not guaranteed. We may have missed some important languages, and we excuse for this. For a comparison of command user interfaces see [WERO83].

SCIL-II. SCIL-II (Simulator Command Input Language II) is the old version simulator activation language [SCIL85, HLW86] of the multi-level KARL-III simulation system [HAVP79, HAR77, HLW86]. SCIL-II has already been used for the KARL-II system [WEB81], a forerunner of KARL-III, which is no more supported. SCIL-II is a fully interactive simulator activation language with additional batch mode feature. It supports addressing within cell hierarchies with multiple instantiations using the same cell name. It has a restart feature as well as a memory (model) dump feature. Appendix B shows synopses of SCIL-II commands.

It is interesting that not only the KARL simulator is using SCIL. Also the output of the KARATE test pattern generator (currently being implemented in Kaiserslautern [HAWO85,HALE86]) and the output of the RT level test pattern generator TIGER [SM85,SMP86] are are using SCIL-II. This language also features diagnostic messages conveyed over from the simulator. Disadvantages of SCIL-II are that no parameters can be used during macro calling, that the values to be assigned cannot be calculated during simulation, and the lack of a modern user interface. A completely redesigned language version not having these disadvantages is SCIL-III [BAS87, MAV87, SW87]. Also see section 6 and appendix F.

THEMIS. The THEMIS (The Hierarchical Multi-level Interactive Simulator) restartable logic simulator of Prime Computer [DOSS84,SCHI84] performs simulation at the behavioral, RT, logic and switch level. THEMIS strongly supports breakpointing not only on values but also on signal edges. Controlled by the breakpoint mechanism also a logic analyzer emulation may be triggered. Any number of 'virtual logic analyzers' can be connected to the circuit. THEMIS has no way to declare soft variables.

SIMPLE. The SIMPLE Interactive Language [HOLL83] interacts with the SIMPLE (SImulator of Machine oPerations and Logic) RT simulator of National Semiconductor. The SIMPLE Interactive Language is easy to learn, and it has a simple syntax. With an "undo" command it is possible to step backwards. "Dump" and "restore" commands, together with "undo", allow flexible experimenting by stepping forward and backward and restarting from earlier points in the history of a simulation run. It also has features for breakpointing and to limit the number of cycles needed to reach a goal (e.g. the value of a particular variable). Disadvantages of SIMPLE is that no program variables or procedures can be declared, and the lack of arithmetic expressions.

ESIM. The ESIM functional level interactive simulator (used in a C-based environment [FREY84, BYLI84], in the AT&T engineering design system EDS), has only a few commands. The ESIM activation language is a low level language. In appendix C synopses of ESIM commands are listed.

ISPS. The command language [BOOT81] of the ISPS RT level simulator [BARB80,BARB81] also features a BREAK which breakpoints each time a new value is assigned to a specified variable, and TRACE which logs changes of specified values. A timing diagram generator (TDG) generates plots of SPICE output.

VTIsim. VTIsim is an interactive mixed-mode event-driven simulator within the VTItools window environment [VTIS85, VTIT85, VTI86]. It simulates transistor logic with MOS device timing. Waveforms may also be displayed graphically during the simulation, or plotted from a file. The VTIsim simulator commands are summarized by synopses in appendix D.

SCALD. The Logic Simulator of the SCALD system [VALID86] performs a dual window simulation and uses the schematics of the circuit in one window as a virtual breadboard for the simulation. The simulator itsself, running in the other window, becomes a virtual logic analyzer, where the user may open signals for viewing simply by pointing at them. The selected signals are displayed as waveform, bus mode or tabular outputs. Furthermore, breakpoints may be set on individual or on multiple signals or on complex logical equations of them, and conditional triggering is recognized.

DLS. Daisy's Virtual Logic Anlyzer of Daisy's Logic Simulator DLS [DAIS85, DENBE86] allows to trigger on various conditions and to display any combination of events desired, to acquire and to trace the values on nodes without the typical hardware limitations of real logic analyzers like limitations on the number of pads, number of channels, acquisition depth or sample rate.

3.1.1 Compact Stimuli Formats

A few simulators' command languages (e.g. DSL and WDL see below) feature a so called Compact Simulator Stimulus Descriptions (*CSSD*, this is not a product name). The goal of these languages is to provide a user a compact, more or less high-level pattern description possibility. CSSD's are compiled, and the resulting pattern files feed the simulator in a batch mode run, just like the Test Description Languages (TSL), discussed in a later chapter. But there are major differences between a CSSD and a TSL. While patterns described in a TSL are used to drive different simulators, CSSD's are strictly simulator-dependent. While TSL pattern descriptions include additional information like test pin assignments and timing, CSSD pattern files are not suitable for test equipment driving. In the following subchapters, various compact simulator stimulus description languages will be viewed. A general disadvantage of these languages is, that they don't support algorithmic pattern description really. The so-called algorithmic description of the vendors ends in the possibility to declare in the main test program callable test procedures or macros.

DSL. The Digital Stimulus Language (DSL) [GLAD85] is suitable for all CADAT simulators and CADAT derivate implementations including the FSIM fault simulator. To emulate timing features in production ATE's up to 16 clock phases can be used. Through the use of user defined macros, stimulus description can be structured. From a DSL source the test vectors are generated by expanding the macros and substituting arguments. Appendix E shows the reserved words of DSL for pattern description.

Waveform Description. HILO's-2 [HILO84] or -3 compact simulator stimulus description is called Waveform Description Language (WDL) [LEIM84, VALID86], and it is supported from all HILO vendors. WDL allows definition of both the input stimulus and the expected response using four values (0,1,X,Z) in different data formats (binary, octal, hex and decimal). Simple waveform clauses can be used to specify logic changes at relative or absolute time, and loops can be nested to describe more complex waveforms. Concurrent and sequential block structure allows description of parallel as well as repetitive stimulation activities.

The Graphics Waveform Editor [GUSC84] used in the CDX-series workstations of Cadnetix Corp. even allows an interactive graphic input of waveforms. It is an extension of the company's graphic waveform output display, which allows to interactively manipulate graphic displays of waveforms. Output waveforms of earlier simulations can be reedited for being reused as inputs. Along with this editor also the DSL language available for the CADAT simulator, which can be used to reedit patterns in textual or quasi-algorithmic from simulation patterns.

3.2 Physical Modelling

Simulation of highly complex systems suffers from two major problems: simulation is by orders of magnitude slower than real hardware, and if standard parts (e.g. microprocessors) are used within a design, normally their simulation models are not available. The writing of simulation models would cost man months or even more. In both cases it would help to combine simulation with the operation of real hardware: simulation models would communicate with the *physical models* of real hardware. This technique is called Physical Modelling. One possible way to implement such physical modelling would be to run the real hardware on an ATE device which communicates with the host computer of the simulator. In Kaiserslautern, for instance, the physical modelling feature of the KARL system uses a Tektronix 9100 tester interfaced to a VAX [TEK83, MAV86]. The advantages of physical models are:

- if a new version of a standard part is available, the PME may be adapted rapidly

- simulation is accelerated substantially.

The KARL physical model feature is currently used for research and teaching only. However, up to now we have information about the following physical model products being marketed by industry:

- RealChip Modeling System for the SCALD system workstations and for the Validation Designer VS on VAXStations II from Valid Logic Systems [WIHA84, VALID84]

- Physical Model Extension PMX for Daisy Personal Logician [DAIS85]

- HICHIP for the HILO-3 Logic Simulator from GenRad [SERT86]

- Dynamic Hardware Modeler from HHB Systems ([EVAN86])

- Hardware Modeling Library (HML) from Mentor Graphics ([EVAN86])

- DataSource for the LASAR Version 6 Logic Simulator from Teradyne [TERA86]

DataSource, for instance, offers features like dynamic pattern memory allocation, multi-user access to the same device, multiple devices in the same simulation run and features to add any custom circuitry to the system .

4. TESTER ACTIVATION LANGUAGES

Test programming languages and their implementations are problem-oriented tools to program ATE systems. Languages provided by ATE vendors are much more complex than simulator activation or pattern description languages. Considerable expertise on the ATE used is needed. Some are low level languages, others are high-level languages. The repertory of statements may distinguish:

1. ordinary programming statements

2. test(er)-specific statements (for tester control, set up and execution)

The cost and difficulty of generating effective test programs for VLSI circuits is increasing with circuit complexity. Some tests include megabytes of test patterns. Good languages are an important aid to the test design expert, but also to achieve the validation of design for testability considerations at very early phases of the design process. Structured test descriptions are needed to make tests more comprehensible [STEV82].

An important strategy is it, to use techniques of software engineering for test design. Techniques similar to those of structured programming help to make test description more comprehensible. Test programming languages add features of programming languages, so that tests could be described algorithmically, but also have statements and structures supporting the special features of the tester as hardware configuration, adjustable time delays and clocks, programmable power supplies, high precision measurement units allowing to execute effective tests. This helps to define tests more concisely, with much better mnemonics, and helps to understand the principles of it and its subtests.

4.1 Tester Limitations and Test Equipment Simulation

Usually, the generated test program is downloaded to the tester from the host, where it was developed, for debugging. Besides the use of fault simulators to validate the test vectors and testability analyzers to estimate the testability of a design, test development requires engineers to deal with the test equipment that is going to be used for prototype verification and production test. Some test equipment constraints and their effects on the testability of a particular circuit are shown in the following list:

1. The maximum number of DUT input pins that can be supported
2. The maximum number of DUT output pins that can be supported
3. The number of timing channels available
4. The physical range and diversity of stimulus and response signals
5. The range of complexity of stimulus and response waveform timing
6. Accuracy of signals versus simulation waveforms, due to limited timing resolution
7. Minimum pulse width limitation to stimulus waveforms
8. Minimum strobe width limitation of DUT responses observability
9. Pattern burst memory depth limits

Neglecting such tester specifications may be the reason if the test program fails. To support engineers in testing their complete test programs without having to run them on the tester on line (thus hindering production tests), Teradyne Inc. uses its package Testsim to make a LASAR simulation of the test environment to check a test program for the J941 tester [CHES85], and TSSI uses its Test(Sim) program to simulate the tester behavior and analyze its performance without needing expensive run time capacity the actual tester.

4.2   Test Generation from Simulation

There are several kinds of techniques to develop tests. A test may be generated by a test synthesis algorithm, such as to derive a test from a gate level description [AKE80 (survey), HIHU77], or to derive tests from a RT level description of the CUT [SM85, SMP86] (for more references see [SUL87]). Tests may also be written 'manually' where test development is a creative engineering process with human beings using test languages. A third technique derives tests from the simulation process. Test generation algorithms go beyond the scope of this paper.

Implementation techniques for generating tests from simulation are the following ones: (1) direct generation of a tester-specific test program, (2) conversion via a tester-independant intermediate form (in one case an event notation) and (3) conversion via a high level language used as an intermediate form. Techniques for developing tests by human beings or by automatic test pattern generators (ATPGs) use as an object language either a tester-specific language (a), or a (high level) tester - independent language (b).

The following examples use test derivation from simulation. (1) Calma's CATscope translator [EVAN86] derives tests for the IMS Logic Master verification tester from simulation in the TEGAStation (Calma Co). (2) Tests for Sentry testers may also be derived from CADAT-5 or TEGAS-5 simulators using the CADDIF format from Fairchild as a relatively tester-independant intermediate form [BEED86]. (3) Also the Test Development Series (TDS) software package from Test Systems Strategies Inc. (TSSI) [DENBE86, EVAN86, GOER85] derives tests from a simulation process. Simulator-dependent file translators convert simulation waveforms into a standard events format.

From this intermediate form the Test Resource Allocation Manager (TRAM) analyzes the waveforms with respect to their suitability for the tester. After this check the program Pattern Bridge (P Bridge) of TSSI generates a functional test program equivalent to the original waveforms. It uses auto-correlation analysis to detect the existence of multiple clocks within a pattern set and to determine the required number of timing sets. (4) A postprocessor by Honeywell Avionics Inc. drives tests from S-LASAR simulation. It transforms the simulation output into a high-level language is described in [CAMP83]. S-LASAR is a simulator from Scientific Machines Corporation (SMC). Another postprocessor converts Teradyne LASAR Version 5.3 simulation behaviour into ATLAS programs.

4.3 A Few Language Examples

The portation of a test program to a different test system mostly results in either a loss of information, or, in a loss of optimality and requires human beings, having been trained on the detailed hardware and software architecture of the particular test environments. The best solution to this problem would be a standardized tester independent problem-oriented language such as ATLAS [MAHO82, LIGU74, LOWW85, MELE84, LOWI85, LOWI86, ANMC81]. Also Ada has been proposed [ANMC81], however, the authors of this paper have the opinion, that there are still some major disadvantages in this approach (also compare [HA87]). A much better idea would be the extension of more comprehensible and more widely spread languages, such as Pascal, or, Modula-2.

FACTOR. Languages having been developed by FAIRCHILD to program testers of the SENTRY family are: the elder FORTRAN-like FACTOR language for all Sentry testers, the new PASCAL-like TDL language for the Sentry 50 [KURA86, LOWI85]. The M3 FACTOR (Fairchild Algorithmic Compiler - Tester ORiented) programming language, first developed for the Sentry V/VII and VIII, supports all general-purpose digital test systems Sentry V/VII/VIII, Series 20 and Sentry 21 equipped with the MASTR Modular Monitor (M3) operating system and also the Sentry 50 tester. A FACTOR program is compiled into object code which is interpreted by the testers.

FACTOR statements that cause a value to be loaded in a tester register are normally executed in the so-called "direct memory access (DMA) mode". The tester executes the test program directly until an escape instruction is encountered which transferes execution back to the system software. Execution in purely DMA mode is more efficient. This fact and many other programming features imply a deep knowledge of the device under test and of the tester.

TDL. Another Fairchild language is TDL [PREP83]. The new Program Enhancement Processor (PREP) for the Sentry 50 test systems runs on VAX/VMS computer systems available with the Test Area Manager (TAM) [TAM84] operating system. It features a new PASCAL-like Test Description Language TDL with interface specifications in terms of the semiconductor device rather than in terms of the specific tester configuration [MAHO82]. The test program contains default parameters, and it is developed and debugged off-line making the tester available for other tests. The output of PREP is formulated in the FACTOR source language of the tester. The input to the PREP is given either menu driven or in tabular form shown in the following table:

- PINDEFinition Table
- TIME PARAMETER Table
- TIME CYCLE Table
- FUNCTIONAL EXPANSION Table
- LEVEL ASSIGNMENT Table
- DCTEST Table

- PINGROUP Table
- TIME RELATION Table
- VECTOR GLOBAL Table
- FUNCTION LEVEL Table
- DC Parameter Table
- EVENTS Table

Pascal-T. Teradyne Inc. has developed two high-level test programming languages for the test equipment of the J900 test equipment series [TERA86]: PASCAL-T, and T900. PASCAL-T, a PASCAL extension, is used for the medium size tester J941. T900, a C language extension running on UNIX 4.2 BSD tester operating system has been created for VLSI testing. It is used on the high-end test systems J983 and J967.

T900. Another Teradyne language is T900 which also supports the interfacing to device handlers and features window techniques. Non-tester-specific test program modules can be written in C, FORTRAN 77, PASCAL, or the Motorola 68000 assembly language. The design engineer has to know up to two other activation languages for simulation (LASAR Version 6) and test (the tester object code) generated automatically generated from LASAR simulation data.

ATLAS and its dialects. ATLAS (Abbreviated Test Language for All Systems) [ATLA78] is a test programming language having been standardized by IEEE standard no. 416. Unlike most test languages, ATLAS features a tester-independent description. Because of several approved change proposals ATLAS has become a rather baroque language difficult to implement [LUBE82]. As a result, numerous different dialects like EQUATE/ATLAS [KETO72] have been specified and implemented. Later on the IEEE Std 716 C/ATLAS (Common ATLAS [CATL82]) subset of ATLAS 416 has been created. This subset was specifically designed for automatic compilation and execution and contains almost only the constructs that are mostly required in actual practice [MCGA82]. Another dialect is the ATLAS/UUT Description Language [ATLA83].

The CAE system FutureNet DASH-1 is used as the front end of schematic entry for the ATLAS system [MILL84]. This system supports several simulators (AAIDS [ROLY84], LASAR [LASA85], HITS [CORA84,HOMO83], and their automatic pattern generators provide the test patterns which are converted to a ATLAS program by a post processor in a final step. Furthermore a combination of the LASAR (Logic Automated Stimulus And Response) [LASA85] simulator with ATLAS [CAMP83] combines the widely used simulator activation Language of LASAR with ATLAS .

Since 1982 at least three special menue-driven user interfaces have been developed for ATLAS test program development: ALEX, the ATLAS LEXicographer software system, has been developed 1982 by the Grumman Aerospace Corp.[LUBE82] and supports a wide range of ATE stations of the company accepting ATLAS test programs. The Automatic ATLAS Program Generator (AAPG) from ITT Avionics [CROS83], having been developed in 1983, supports different ATLAS extensions used at ITT Avionics. The Interactive Development Environment for the ATLAS Language (IDEAL) from Harris-GSSD Orlando Operation [LAC85] features a syntax-directed editor.

4.3.1 Graphic Languages

The most innovative user interfaces especially for DC and AC testing in an instrumentation cluster testing environment are those published in the first half of 1983 (also see [HILL83]). They use powerful personal computers like the IBM PC-AT, the Apple Macintosh and the Hewlett-Packard Vectra Personal Computer with their high resolution bit-map graphics and with interfaces to the IEEE-488 bus to support the user in set-up of the testing configuration, test program generation and controlling the instruments.

TestStation. Summation Inc. delivers the TestStation test equipment [MONT86] with the TestWindows software running on a IBM PC AT that automates test programming tasks. All the standard setup and control functions are entered graphically by using windows. Individual statements that implement these instrument settings are generated automatically and inserted to the test program. One or more windows display the system configuration, and one or more windows are attached to the instruments to show their status and to allow step by step operation. The control of the system is simplified by icons and pull-down menus accessed with a mouse.

(HP). Hewlett-Packard [SISM86] offers interactive computer graphics similar to the last one to control instruments such as oscilloscopes, counters, function generators, or multimeters, manually through the personal computer. The interactive use is supported additionaly by function keys with changing meanings.

LabView. National Instruments Corp. offers LabView (the Laboratory Virtual Instrument Engineering Workbench) [WOLF86], whose icons stand for the front panels of lower-level virtual instruments. Due to the system's hierarchical structure, any virtual instrument can be put into the block diagram of another instruments. Controls, indicators, data paths, or functions can be added, deleted, or modified using pull-down menus. LabView makes possible to implement usual programming concepts by an instruction set of graphical elements. Thus, not only instrumentation clusters, but also test programs can be easily generated using block diagrams and icons.

5. TEST SPECIFICATION LANGUAGES

As has already been mentioned the term 'test description language' is used for a commercial product. So we use the less catchy term 'test specification language' (TSL). Test Pattern Description Languages (TSL) integrate simulation and test by providing a high level, common source language. Through the use of specific code generation modules, which directly support the architecture of the target simulator or tester, TSL's can be compiled to input for various simulators and testers. The compilers , and not the users, are responsible for the correct formatting of the ports and the patterns in the output. This chapter deals with test pattern description languages like ICTEST or VTL and with simulator- and tester-independent data formats used to exchange test data between various testers.

5.1 Common Features

The process of debugging the test program described in a TSL with the simulator contributes to design verification, since the test is already run during simulation, and, since only a single test specification is used for both, simulating and testing. An alternative approach to generating the test would be to apply the stimuli in such a way to the simulator, that it serves as a response generator for test derivation. This approach, however, lacks in validation. TSL's have language structures to describe the following information:

- Test name
- Packaging and socket type
- Port declarations
- Port type: input, output, bidirectional, tristate, pull-up, pull-down, clock, power, ground
- Formatting of pins (threshold, delay, width, format type)
- Correspondence of the circuit port name to the tester pin name or number
- Test statements (time of drive or sensing w.r. to the test cycle or to signal edges)
- Control statements

The ports of the highest module in the hierarchy of the design are assumed to be the pins of the IC, and a list of the corresponding identifiers is created. The physical characteristics of the IC (number of pins, used technology) and pin identification statements are included to the test program. The last ones assign identifiers to the pins, and help identify a pin's type (input, output, bidirectional or power). The generation of device-oriented test vectors instead of tester-oriented ones, replaces confusing tables of long binary 1 and 0 chains with more readable descriptions of circuit operations. Futhermore it requires very little effort to move a test program among various testers because the patterns are portable , so that the same file can be used to produce vectors for many target testers.

The individual code generation modules of the TSLs contain all the specific knowledge about the target simulator or tester, e.g. in the case of STL [IVIE86]. Moreover, the routines can identify any portions of the test language that cannot be executed by the simulator or tester, e.g. in case of VTL [EGMO84]. It is the author's opinion that in the future instruments characterization languages like ROLAIDS [SOLL82, SOLL83] can be used to describe the knowledge about the configuration and the facilities of an automatic test equipment.

It is common to all TSL's, that they don't correspond to specific hardware description netlists or languages, and, that they are not directly executable. After translation into simulator or tester commands they run in batch mode. However, TSLs allow to supply simulator or tester commands as literals to be passed to the produced output file, e.g. to meet memory length requirements during testing, or to interrupt the simulator to go into interactive mode.

Comparing the TSL's to the high-level tester activation languges (TAL) of different vendors, it is interesting to note that the last one are not used for functional testing, but only for the overall test program control, DC parametric testing, etc. The functional testing of the complete, tester specific test programs is still described in the vendors low level tester activation language [IVIE86]. Comparing the TSLs to the high-level simulator activation languges (SAL), TSLs don't have the possibilities to define breakpoints and associated action lists, like the THEMIS simulator command language.

ICTEST. ICTEST [MANEWA, DOWA82] is an extension of the C programming language for describing functional tests of digital circuits, especially tailored for two-phase, synchronous designs. ICTEST supports the integration of testing and simulation: the same test program may be compiled to run with the ESIM simulator [BATE80, BYLI84, HILL83, FREY84], the TSIM switching simulator [BATE80], functional simulator, and different testers (TEKTRONIX S-3260 systems, MINIMAL and MEDIUM testers [MATHEW]). The C extensions of ICTEST have three classes of language features: Port declarations (1) specify the clocking and formatting associated with each logical input or output, for example:

'input' A 'serial' 'lsb' N 'valid' PHI1

indicates that A excepts N-bit numbers serially and starts with the least significant bit first, all bits valid at the falling edge of PHI1 (by definition). Because of its excellent mnemonics ICTEST has a good readability which helps a lot in teaching designers not being testing specialists, but being aware of design for testability. Test statements (2), like the *drive* ('='), *sense* ('=?') and *store simulation result* ('=>') statements, operate on ports. Compound statements (3) are used to express concurrency and pipelining [MANEWA].

VTL. The software package called VTItest [VTIT85], which works with the test pattern description language VTL, enables to develop a circuit and its test program simultaneously, and notifies in the early design stages of the simulator- or tester-specific details that affect the circuit. Through VTL, designers create a file describing the physical characteristics, timing, stimuli patterns, and expected responses of a circuit, indepedent from simulator and tester characteristics. VTItest translates the high level description into commands to drive specific simulators like the logic and timing simulator VTIsim [VTI86, VTIS85], or different popular testers, such as the Sentry testers. In the last case, the generated FACTOR program contains all specifications for the timing generators, strobes, and registers, besides the test vectors.

STL. STL, the Simulation and Test Language STL of SDA Systems [IVIE86] is a high-level programming language which is an extension of the IL interpreter [IVIE86] and programming language, used by SDA systems for a number of CAD tools, by incorporating a number of high-level structures needed to program functional tests. STL uses two-pass compilers for the SILOS logic simulator, the SPICE circuit simulator, Sentry testers and other targets.

An STL program can produce a simulation for a circuit simulator. The *defformat* structure defines (by name) the order of the signals that correspond to the functional data specified by a special command, which defines the actual test vector (stimuli and responses) at various simulation or test times. Commands for repeating a single vector or calling a sequence of patterns, respectively, take advantage of the test pattern compression hardware of a target tester, if available. STL allows to write functional test programs including nested procedures. In general, STL allows to combine algorithmic pattern descriptions with similar methods used in the compact simulator stimulus descriptions discussed earlier. The language looks C-alike, although it is based on LISP concepts.

5.2 Test Description Exchange Formats

EDIF. Exchange formats are data formats for information exchange between different environments, or, between different tools. The most popular one of them is EDIF (Electronic Design Interchange Format) [EDIF86, CRAW85], an IEEE standards proposal supported by industry. Also within EDIF test description frame formats may be defined as part of the so-called behavioral view and associated to EDIF interface statements. The current EDIF version (1.1.0) lacks to efficiently support the transfer of other important test data like socket type, pin number, or test channel specifications adjustments. To solve this testing interface problem EDIF extensions have been proposed [EDIF85, SHRO86]. Maybe the new version 2 0 0 [EDIF87] will be helpful.

CADDIF. Some sort of EDIF-like formats for exchange of test and simulation data and a fault dictionary is CADDIF [BEED86] having been developed by Factron Schlumberger because of the same reason. A converter, called CADPROT (CAD to automatic test equipment PROcessor) [CADP84] uses CADDIF as a link between different CAE systems and Sentry test equipment.

SDIF. The SDIF format [PIE86] is the Stimulus Data Interchange Format of the GE Calma Co. Its syntax is similar to that of EDIF. SDIF also permits power supply definitions, assignments, format generator assignments, device driver and receiver assignments, and test control instructions. In addition to the VIEWs such as defined in EDIF, a new VIEW called TEST and three subVIEWs are available within SDIF: the FUNCTIONAL VIEW (which specifies a block of functional test commands), the DC VIEW and the AC VIEW. Part of the SDIF syntax is also shown in [MAV87].

6. SIMULATION AND TESTING LANGUAGES

In previous sections of this paper some commercially available languages have been mentioned, which are SALs and TSLs at the same time. I would like to call them SATLs (simulation and testing languages). The goal of implementing such languages is to support synergisms between simulation and test development. Examples of such languages are ICTEST, STL, and VTL (see previous sections of this paper). But not all of the following synergistic goals have been reached by such languages and their environments:

- use <u>high level language</u> for simulator activation and test development
- use the <u>same language</u> for simulator activation <u>and</u> testing
- use the same language for simulator activation and <u>interactive</u> testing
- replace simulation models by physical models (e.g. standard chips)
- use same equipment for <u>prototype</u> testing and simulation acceleration
- use same equipment for simulation and low quantity <u>production</u> testing

It is well known, that most libraries of simulation models are much more expensive than the simulator itself. That's why the replacement of simulation models by hardware is very important (physical models: circuits to be used within the system under design, microprocessors, co-processors, other supporting standard chips, and/or parts of the design having been already completed), since simulation models are by orders of magnitude more expensive than off-shelf chips and require highly qualified man power which is short. Of course also a common high level notation for both areas helps a lot. But it would be much more helpful, if such a language could be used for interactive simulation <u>and</u> prototype testing as well as physical modelling. I would like to call a SPTL (simulation and prototype testing language). None of the systems commercially available or having been published has that feature. All tests used in simulators and using an interactive command language have to be compiled such, that testing is a batch job. So it seems that no SPTL has been implemented up to now.

One exception is SCIL-III [BAS87, MAV87, SW87] currently being implemented in Kaiserslautern, the grammar of which is shown in appendix F. It is a successor of SCIL-II, the SAL of the KARL3 simulation system [HLW86]. Also SCIL-III may be used for the KARL-3 system, running under AEGIS, UNIX, VMS and others on VAX-11/7xx, VAX-11/8xxx, microVAX, Apollo workstations, SUN workstations, Atari ST, IBM PC-AT and (in preparation) Apple Macintosh. It may also be used for the advanced KARL-4 simulation system, currently being implemented in Kaiserslautern [ESP87]. SCIL-III and KARL-3 with its combined physical modelling and testing feature reach all the goals listed above.

7. CONCLUSION

Simulator activation as well as tester activation languages are known for much more than a decade. In the early eighties several test pattern description languages appeared for simulation and testing and to ease the portation of simulation and test data between design stations and testers. Since about 1983 all major vendors of simulators have developed physical modelling systems to accelerate simulation. Since physical modelling is closely related to testing this further closes the design/testing gap. A number of languages support the integration of simulation and production testing. However, none of them supports synergisms between simulation and <u>prototype</u> testing. SCIL-3 is the first real language which integrates simulation and not only test development, but also production testing and prototype testing. So it is the first really combined Simulation Activation and Testing Language.

8. REFERENCES

[AKE80] S. B. Akers: Test Generation Techniques; Computer, March 1980.

[ANMC81] R. E. Anderson, R.L. McGarvey, L.A. Zeafla: ADATLAS - The Test Language of the Future?; Proc. IEEE International Automatic Testing Conference, Orlando, Florida, 1981.

[ATLA78] IEEE Standard ATLAS Test Language, ANSI / IEEE std 416 - 1978.

[ATLA83] ATLAS/UUT Description Language; Proc. IEEE ATPG Workshop, San Francisco, 1983.

[BARB80] M. R.Barbacci: The Symbolic Manipulation of Computer Descriptions: An ISPS Simulator, Technical Report, Department of Computer Science, Carnegie-Mellon University, 1980.

[BARB81] M. R. Barbacci: Instruction Set Processor Specifications (ISPS): The Notation and Its Applications; IEEE Transactions on Computers, Vol. C-30, No. 1, Jan. 1981.

[BAS87] K. P. Bastian: Implementation of the SCIL-III Incremental Compiler/Interpreter; Diploma Thesis, Universität Kaiserslautern, 1987.

[BATE80] C.M.Baker and C.Therman: Tools for Verifying Integrated Circuit Designs; Lambda, 4th Qu. 1980.

[BEED86] Mitch Beedie: Data format links ATE to CAE workstations; Electronic Design, January 9, 1986.

[BOOT81] A.W. Booth: Computer Generated Timing Diagrams to Supplement Simulation; Proc. CHDL 1981 - 5th Int'l Conf. on CHDL, Kaiserslautern, F.R.G., 1981, North Holland, 1981.

[BYLI84] C.T. Bye, M.R. Lightner, D.L. Ravenscroft: A Funtional Modeling and simulation Environment based on ESIM and C; Proc. ICCAD, Santa Clara, California, Nov. 1984.

[CADP84] CADPROT; Users Manual Rel.1.0, Publ. No. 57510034, Fairchild, June 1984.

[CAMP83] T. E. Campbell: LASAR to ATLAS Post Processing; Proc. IEEE Int'l Automatic Testing Conference, Fort Worth, Tex, 1983.

[CATL82] IEEE Standard C/ATLAS, IEEE Std 716-1982.

[CHES85] M. Chester: Bridging from CAD to ATE over the VLSI ravine; IEEE Design & Test, Aug 1985.

[CORA84] V. Coletti, S. Raudvere: HITS Modelling Language - A Simple Solution to Complex Modelling Problems; Proc. IEEE Int'l Automatic Testing Conference, Washington, DC, 1984.

[CRAW85] J.D. Crawford: EDIF: A Mechanism for the Exchange of Design Information; Design&Test, 1/1985.

[CROS83] O. Cross, J. Gerg: Automatic ATLAS Program Generator (AAPG) for the Advanced Electronic Warfare Test Set; Proc. IEEE Int'l Automatic Testing Conference, Fort Worth, Texas, 1983.

[DAIS85] N.N.: Daisy Verification Tools; Daisy Systems Corp., 1985.

[DENBE86] W. E. Den Beste: Tools for Test Development; VLSI Systems Design, July 1986.

[DOSS84] M. H. Doshi,R.Sullivan, D.M. Schuler: THEMIS Logic Simulator - A Mix Mode, Multi-Level, Hierchical, Interactive Digital Circuit Simulator; Proc.21st Design Automation Conference, 1984.

[DOWA82] I.M.Watson, J.A.Newkirk, R.Mathews, D.Doyle: ICTEST: A Unified System for Functional Testing and Simulation of Digital ICs; Proc. 1982 IEEE International Test Conference, 1982.

[DOWN83] A. Downey: A 'Three Mode' Command Language for ATE; Proc.IEEE Int'l Test Conf. 1983.

[EDIF85] EDIF Steering Committee: EDIF Proposals, ICCAD EDIF Workshop, Santa Clara, CA, Nov.1985.

[EDIF86]   EDIF Steering Committee: EDIF Electronic Design Interchange Format; Version 1.10, 1986.

[EDIF87]   EDIF Steering Committee: EDIF Electronic Design Interchange Format; Version 2.0, 1987.

[EGMO84]   K. van Egmond: Software unites test program development with circuit design; ED, Nov.15, 1984.

[ESP87]   Bonomo et al.: Specification of the CVS_BK language; CVS report (ESPRIT), CSELT, Torino, Italy / Kaiserslautern University, 1987.

[EVAN86]   S.Evanczuk: IC Prototype Verification Test and Simulation; VLSI Systems Design, April 1986.

[FREY84]   E. J. Frey: ESIM: A Functional Level Simulation Tool; Proc. ICCAD, Sta.Clara, CA, 1984.

[GLAD85]   B. Gladstone, T. Westerhoff: Powerful PC-based tools close the loop between designing and testing ICs; Electronic Design, June 1985.

[GOER85]   R. Goering: CAE and ATE Vendors Tighten Link between Design and Test; Comp.Des. Oct.1, 1985.

[GUSC84]   S.Gunther, et al: CAE workstation sets up direct connection to board design system; ED, Nov.15,'84.

[HA87]   R. Hartenstein: The Classification of HDLs; within this volume.

[HADA85]   R.M.Haas and D.B.Day: The Integration of Design and Test, Part 2; VLSI Design, April 1985.

[HALE86]   R.W. Hartenstein, K. Lemmert: Test-Freundlicher und Strukturierter Architektur-Entwurf bei VLSI-Bausteinen: Benutzer-Fuehrung durch ein CAD-System; Proceedings 2.EIS-Workshop "Entwurf Integrierter Schaltungen"; 1986, GMD MBH, Bonn, March 1986.

[HAR77]   R.W.Hartenstein: Fundamentals of Structured Hardware Design - A Design Language Approach at Register Transfer Level; North-Holland Publ.Co, Amsterdam, 1977.

[HAR87]   R. Hartenstein: Simulation von VLSI-Schaltungen; CAD/CAM report; Febr. 1987.

[HAVP79]   R.W. Hartenstein, E. v. Puttkamer: KARL - a Hardware Description Language as Part of a CAD Tool for VLSI; Proc. CHDL 1979 - Int'l Conf. on CHDL, Palo Alto, California, 1979.

[HAWE86]   R.W. Hartenstein, U. Welters: Higher Level Simulation and CHDLs; Proc. Summer School on VLSI Tools and Applications, Beatenberg, Switzerland, 1986 (eds. M. Morf, W. Fichtner), Kluwer, 1986.

[HAWO85]   R. Hartenstein, A. Wodtko: Automatic Generation of functional test patterns from an RT language source; Proc. EUROMICRO Symp., Brussels, Belgium, 1985 (ed.: K. Waldschmidt) North Holland, Amsterdam, 1985.

[HIHU77]   F. J. Hill, B. Huey: SCIRTSS: A Search System for Sequential Circuit Test Sequences; IEEE-TC, May 1977.

[HILL83]   Dwight D. Hill: Edisim and Edicap: Graphical Simulator Interfaces; Proc. 20th DA Conf. 1983.

[HILO84]   N. N.: HILO-2 User Manual; Doc# 2521-0100/1, GenRad, Santa Clara, June1, 1984.

[HLW86]   R.W.Hartenstein, K.Lemmert and A.Wodtko: KARL-3 Ref. Manual; Univ. Kaiserslautern, 1986.

[HOLL83]   Y.Hollander: Using an RTL Simulator to Simplify VLSI Design, VLSI Design, Sept.1983.

[HOMO83]   Loring Hosley, Mukund Modi: HITS - the NAVY's New DATPG system; Proc. 1983 IEEE Int'l Automatic Testing Conference, 1983.

[IVIE86]   J.Ivie and K.Lai: STL- A High Level Language for Simulation and Test, Proc. 23rd DA Conf 1986.

[KETO72]   J. Kelly, P. M. Toscano: EQUATE - New Concepts in Automatic Testing; Proc. Automatic Support Systems for Advanced Main- tainability, Philadelphia, Pennsylvania, 1972.

[KURA86]   K.S. Kurasaki: TDL - A Tester Independent Test Language; Proc. IEEE 1985 Workshop on Simulation & Test Generation Environments, San Fransisco, 1985.

[LAC85]   D. Lacasia, B.J. Rosenberg: IDEAL = Interactive Development Environment for the ATLAS Language; Proc. IEEE Int'l Automatic Testing Conference, Long Island, N.Y., 1985.

[LASA85]   N. N.: LASAR Version 6, Design simulation and test generation for VLSI circuits, Publ.Nr 04641-0485, Teradyne, Inc., 1985.

[LEIM84]   W. Leimbach: Simulation von digitalen elektronischen Schaltkreisen; CAE Journal, 1984.

[LENE87]   K. Lemmert, W. Nebel: Conceptual Design based on CHDL Use; (within this volume).

[LIGU74]   F.Liguori (ed.): The Test Language Dilemma; Automatic Test Equipment, IEEE Press, 1974.

[LIMI86]   K.L Kindsay, J. Miller: Tester guarantees ASIC performance; Digital Design 5 / 1986.

[LOWI85]   A. Lowenstein,G.Winter: The TDL and Tester Independent Translation Problem Solved, Proc. IEEE International Automatic Testing Conference, Long Island, N.Y., 1985.

[LOWI86]   A. Lowenstein,G.Winter: VHDL's Impact on Test', IEEE Design & Test, April 1986.

[LOWW85]   A. Lowenstein, G. Winter: Importance of Standards; Proc. IEEE 22nd DAConference, 1985.

[LUBE82]   Kathryn A. Luber: Inherent Implementation Problems of an ATLAS Statement Interpreted System, Proc. IEEE International Automatic Testing Conference, Dayton, Ohio, 1982.

[MAHO82]   R.C.Mahoney: A Common Pascal Test Language: Reality or Pipedream; IEEE Int'l Test Conf.1982.

[MANEWA]   J.Newkirk,R.Mathews,I.Watson: Testing Chips using ICTEST Version 2.5, Information Syst. Lab, Stanford Univ., VLSI File #012782, 1982.

[MATHEW]   R. Mathews: The MINIMAL and MEDIUM testers; Stanford Univ., VLSI File No. 010882, 1982.

[MAV86]   A. Mavridis: A Physical Model Extension for KARL Simulators, Proc. ABAKUS workshop ABL and KARL User Group Workshop, Passau, June 1986.

[MAV87]   A. Mavridis: SCIL-III - A Simulator and Tester Activation Language for VLSI Circuits; Dissertation; Kaiserslautern, 1987.

[MCGA82]   R.L.McGarvey: A Critique of IEEE-716 ATLAS; IEEE Int'l Aut. Testing Conf., Dayton, OH, 1982.

[MELE84]   E.M.Melendez: Emerging ATE Engineering Design Tools; Int'l Aut. Testing Conf,Wash.DC 1984.

[MILL84]   G. A. Milles: Schematic to ATLAS; IEEE Int'l Aut. Testing Conf., Wash., DC, 1984.

[MONT86]   S. Montgomery: Measurement system's buses, windows, and icons simplify programming and operation; Electronic Design, Apr. 3, 1986.

[PIE86]   Chris Pieper: Stimulus Data Interchange Format (2 parts); VLSI Systems design, 7+8 / 1986.

[PREP83]   Program Enhancement Processor (PREP); User Manual Rel.2.0, Publ.No.57510023, Fairchild 1983.

[ROLY84]   Robinson, R.L., Lyon, J.E.: AAIDS as a Comprehensive Transportable Digital CAD Package; Proc. IEEE International Automatic Testing Conference, October 1984.

[SCHI84]   M.Schindler: Advances in software let system engineers take charge of IC design; ED Nov. 15, 1984.

[SCIL85]   A. Mavridis: SCIL-II Instant; CVT report, Univ. Kaiserslautern, 1985.

[SERT86]   N. Sertl: Einbindung der Hardware in die Logiksimulation; Design & Elektronik, No. 9, 1986.

[SHRO86]   U.Schroeder: A General Interface for Design and Test of Integrated Circuits; Proc.2nd E.I.S. Workshop, Bonn, F.R.G., 1986.

[SISM86]   R. Sismilch, W. Walker: Interactive Computer Graphics for Measuring Instrument Control; hp Journal, May 1986.

[SM85]   I. Stamleos, M. Melgara, CVT TIGER: a RT level test pattern generation and validation environment; CVT report; CSELT, Torino, Italy, 1986.

[SMP86]   I. Stamelos, M. Melgara, M. Paolini, S. Morpurgo, C. Segre: A multi-level test pattern generation and validation environment; Int'l Test Conf. 1986.

[SOLL82]   L. C. Sollman; First Sequel to roll-your-own automatic test system (The ROLAIDS connection); Proc. Autotestcon Conference, 1982.

[SOLL83]   L. C. Sollman; First Sequel to roll-your-own automatic test system (The Information connection); Proc. Autotestcon Conference, 1983.

[STEV82]   A. K. Stevens: Structured Programming and the I.C. Test Engineer; IEEE Int'l Test Conf. 1982.

[SUL87]   S.Y.H. Su, T. Lin: Test Pattern Generation from HDLand Similar Sources; in this volume.

[SW87]   Schwarz: Implementierung von Standardfunktionen und Prozeduren für SCIL-3; Diploma Thesis, Universität Kaiserslautern, 1987.

[TAM84]   N. N.: Test Area Manager (TAM); Product Description, Publ. No. 57140002, Rev. 3, Fairchild Camera and Instr. Corp., March 1984.

[TEK83]   N. N.: DAS 9100 Series Operators Manual; Tektronix, Inc., 1983.

[TERA86]   N.N. (Teradyne): Uniting the worlds of design and test; Electronics, June 9, 1986.

[THOM86]   D.Thomas: RTL Simulation makes a comeback for complex VLSI; Computer Design, Feb. 1, 1986.

[VALID84]   N. N.: CAE tool set offers hardware modelling on as VAX; Computer Design, 7 / 1984.

[VALID86]   N. N.: SCALD Analysis Tools: SCALD Logic Simulator; Valid Logic Systems Corp., 1986.

[VTI86]   N. N.: IC Design Software; Publ. No. 408-942-1810, VTI, Febr. 1986.

[VTIS85]   N. N.: VLSI Design System: VTIsim User Manual; VLSI Technology Inc., 1985.

[VTIT85]   N. N.: VLSI Design System: VTItest User Manual; VLSI Technology Inc., 1985.

[WEB81]   B.Weber: Pascal-Implementierung eines Simulators auf KARL-II Basis; Diploma Thesis, Univ. Kaiserslautern, F.R.G., 1981.

[WERO83]   J. Werner, et al.: Comparing the Computer-Aided Engineering Systems in Action; VLSI Des., 11'83.

[WIHA84]   L.C.Widdes, W.C.Harding: CAE station uses real chip to simulate VLSI-based systems; ED, 3/1984.

[WOLF86]   R. Wolfe: Blockdiagrams and Icons alleviate Customary Pain of Programming GPIB Systems; ED, April 17, 1986.

Appendix A (Hierarchy of Activation Languages):

AL	Activation Language
SAL	Simulator AL
TL	Testing Language
TSL	Test Description Language
TVL	Test Vector Language
TPL	Test Programming Language
TEL	Test Equipment Language
TAL	Tester AL
ESL	Equipment Set-up Language
SATL	Simulation Activation and Testing Language

Appendix B (Synopses of SCIL-2 Commands):

execution command:

step	execute one micro step

assignments: >(execution will be evoked by next step command)<

init value	all switches,registers and delays are set to this value
assign [*name value* // ',']+	assignment of values to words
seq [*name i* [*value*]+ // ',']+	--> assign // i times step
set *name* ([*bitnr* ['..' *bitnr*] // ',']+)	set specified bits to '1'
reset *name* ([*bitnr* ['..' *bitnr*] // ',']+)	set specified bits to '0'
invert *name* ([*bitnr* [',' *bitnr*] // ',']+)	invert specified bits
lock [*name* // ',']+	freeze values until encounter of **release** command
keep [*name value* // ',']+	overriding freeze until encounter of **release**
release [*name*]+	see **keep** and **lock**

block definitions: >(never use DEF and ENDDEF within loops!)<

loop *n* [*command*]+ endloop	loop n times the command sequence
repeat [*non-def-command*]+ until *condition*	(test at end of the loop)
while *condition* [*command*]+ endwhile	(test in front of the loop)
def *name* [*command*]+ enddef	define command macro
delete	delete all commands back until most recently executed one
skip	jump to first command behind end of outermost block
escape	during definition mode: escape jump behind the block

conditional commands   >is is different from '=' !<

if *condition* then [*command*]+ [else [*command*]+] endif	
condition ::= name { is I isnot } value	see **value** syntax
is	1 only if operands (0,0), (1,1), (?,?), (*,*)
isnot	result bit is inverted **is** result

multiple occurrance of identifiers: >instant selection: also see *name syntax* <

in *modname* [*command*]+ endin	only module's local names are referenced

print commands:

print *name*	current value printed together with name
prntact	print all those values which changed at last **step** execution
prntrtc	print RT-code description

prntunit   see prntrtc, however, also most recent values shown
prntbit
sense [*name value* // ',']+   whenever values are unequal print "sense error", name and
                                           both values

plot commands:

plotall [*name* ([*bitnr* ['..' *bitnr*] // ',']+ ':' *size* // ',']+ plot-like print of all specified bit values,
                                                     repeated after each **step**
plotnew   see plotall, however, plot suppressed whenever all values are unchanged
endplot   all plotting terminated so far

file management commands:

stdin,stdout   simulator input, output, resp. switched over to terminal
newin,newout   enter dialogue mode to assign new file for input or output
oldin,oldout   switch back to input or output used before **newin** or **newout**
startprt,endprt

interrupt and termination of simulation run:

endsim   terminate simulation run
return   interrupt and enter KARL system dialogue mode

miscellaneous commands:

settime   start counting CPU time having been used
gettime   print CPU time used after **settime** execution
help   dump command interpreter state
edit   enter editor

comment commands:

comment " ' " *text string* " ' " *blank*   inserted into SCIL listing
write   see **comment**, however: inserted in simulator output

memory managment commands:

loadmem   loads ROM or RAM from a file
dumpmem   dumps RAM or ROM to standard out (file or terminal)

Appendix C (Synopses of ESIM Commands):

Command	Description
h *node*	force *node* to logic high
l *node*	force *node* to logic low
x *node*	unsticks *node*
pn *node*	prints value of *node*
dv *name node node* ..	defines a vector with *name*
av *name value*	assigns *value* to vector *name*
pv *vector* / *format*	prints value of nodes on *vector*
	(formats: d -decimal, o -octal, x -hexadecimal)
iu *stepcounts*	tries to initialize the circuit *stepcount* times
st	steps the simulator
ig *node*	prints the state of *node* and all transistors that it affects
in *node*	prints the state of *node* and all transistors that determine the node
do *n commands* od	execute *commands* between do and od *n* times
:*label*	destination *label* of jump commands
go *label*	jump to *label*
macro-name .sim	read simulator macro from file with the same *name*
ws *filename*	write values of all nodes into *file*
rs *filename*	read values of all nodes from *file*
!*command*	executes operating system commands
q	quit the simulator

Appendix D (Synopses of main VTIsim Commands):

Menu	Command	Action
Browser menue	load	load the selected cell
	load opt	load the selected cell with options
	browser	browser specific commands
Wave pane menue	osc	prepare an OSC command file
	set	set waveforms options
	display	table of displayed nodes
	zoom	zoom commands
	ruler	measure the time between indicated points
	save wf	save the waveform as a cell
Transcript pane menu	cycle/phase/time	current stepping mode
	single	simulate one cycle/phase/time period
	multi	simulate a number of cycles/phases/time periods
	set	set simulation options
	show	show parameter settings and status information
	stop	stop the simulation
	watched	write names of watched nodes to transcript pane
	write	write transcript pane to file
	erase	erase transcript pane
	search	search for string in the transcript pane
	misc	miscellaneous commands
Popup Button menu	set charged	set selected node charged
	set forced	set selected node forced
	set input	set selected node input
	show after	show after selected node
	show before	show before selected node
	show syn	show synonyn for selected node
	unwatch	unwatch selected node
	watch	watch selected node
	view	view selected node
	display	display selected node in wave pane
	undisplay	undisplay selected node in wave pane
	copy	copy text into paste buffer
	enter	copy text into command pane
	read	read cell into transcript pane
VTIshell commands	bus	create bus
	capacitor	declare a capacitor between two nodes
	check	run several static checks
	show	show parameter values or status information
	test	execute com'd conditionally on node or vector value
	vtitest	enter VTItest

Appendix E (Keywords for DSL pattern descriptions):

ALL	DELAY	HZ	MEND	OUTPUT	TO
BIDIR	DO	INCLUDE	MHZ	PERIOD	TRI
BITSTREAM	DRIVE	INPUT	MONITOR	PINS	UN
BY	END	KHZ	MS	PS	UNMASK
CEND	FAMILY	LO	NEG	S	US
CIRCUIT	FOR	MACRO	NOT	SENSE	VALUES
CONST	FREQ	MASK	NS	TIMEDEF	VECTORS
CYCLE	HI	MCALL	OD	TIMES	WAIT

Appendix F (SCIL-3 Grammar):

DECLARATION-GRAMMAR

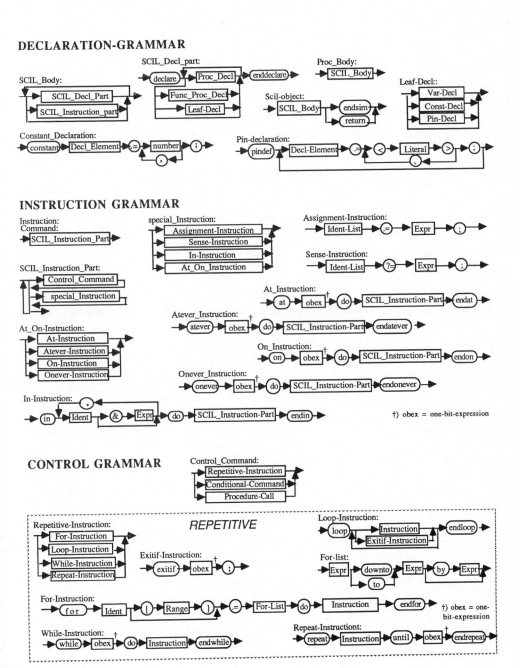

INSTRUCTION GRAMMAR

CONTROL GRAMMAR

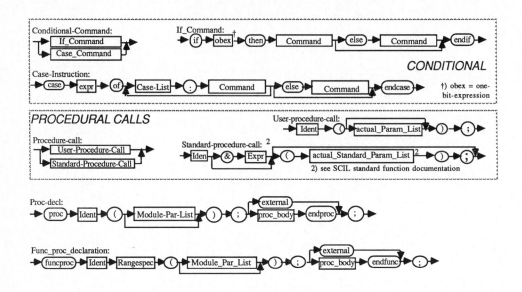

BASIC PRODUCTION RULES

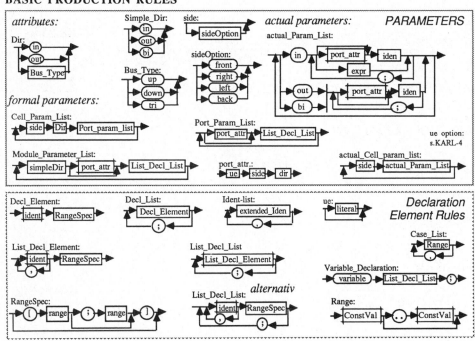

EXPRESSION-GRAMMAR

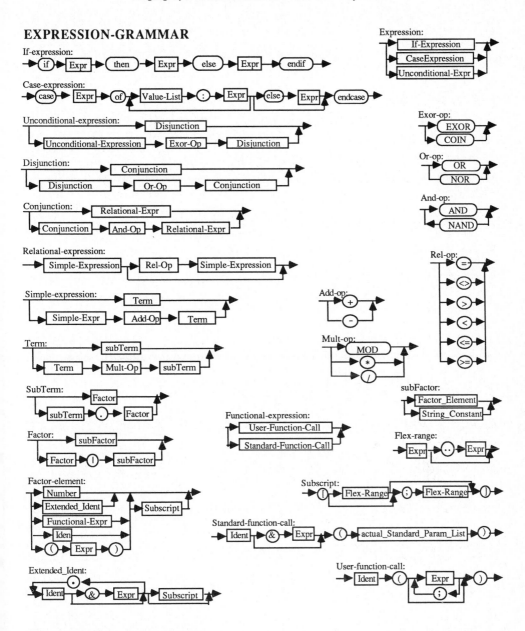

STANDARD   FUNCTIONS
STANDARD PROCEDURES

tester adjustment and driving commands	
user extensions	*also commands from appendix B*
other environment implementations	*for upward compatibility of SCIL-3 to its predecessor SCIL- 2*

4. HDL Applications

HARDWARE DESCRIPTION LANGUAGES
R.W. Hartenstein (Editor)
© Elsevier Science Publishers B.V. (North-Holland), 1987

4.1. CONCEPTUAL DESIGN BASED ON HDL USE

Karin LEMMERT and Wolfgang NEBEL

Department of Computer Science, University of Kaiserslautern
Postfach 3049, D-6750 Kaiserslautern, F.R.G.

This text describes a design methodology, which was enabled by a demystification of LSI design and by the wider acceptance of structural hardware development. The methodology encourages the architectural designer to design his circuit at all levels of abstraction using CAD-tools for specification, refinement, and verification of circuit designs. We show the role of hardware description languages at all steps of the design cycle.

INTRODUCTION

The process of digital hardware design consists of successive refinements of functional circuit specifications into silicon structures. Usually this process is executed by several steps. The designer generates descriptions of the hardware designed at the several design levels of integrated circuits. Generally there are five design levels with different detail stages:

- Register Transfer level (RT):

  At the RT level hardware is specified in a mixed behavioral/structural way. The components of purely behavioral descriptions are algorithmic variables, whereas structural ones use hardware elements like multiplexer, registers, nodes etc..

- Gate level:

  Gate level decriptions specify the function of the hardware in terms of combinational logic or sequential logic networks.

- Switch level:

  At the switch level hardware is specified by technology independent models of connectors, transistors, resistors, pull-up and pull-down loads.

- Circuit level:

  At the circuit level the technology independent models of the switch level elements are substituted by the corresponding of diodes, transistors, resistors etc..

- Layout level:

  Definition of the hardware in silicon structures.

The purpose of hardware description languages is to facilitate the design of integrated circuits. In the past mainly lower level HDL's have been integrated into design tools. The main application of high level description languages at the RT level has been to document user's manuals of circuits. In the middle of the seventies RT languages have been developed as input formats to design automation systems, like ISP at the Carnegie Mellon University [44] and APL used in the Alert system developed by IBM [19]. Rapid progress in VLSI technology and increasing complexity of hardware units caused a great increase of the designer's productivity. The consequence was a bottle neck in the design of integrated circuits. Solutions to the problem are the development of new design facilities and methods.

The new design strategies based on a method propagated by Mead and Conway [38] lead the designer to the conceptual design of digital hardware. The conceptual design process can be characterized by the following features:

- It starts the description of digital hardware at the register transfer (RT) level aided by RT languages, since with respect to the increasing complexity of integrated circuits a description at logic or circuit level would be too extensive and confusing.

- It supports the hierarchical design of integrated circuits, leading to the conception of key cells with array capability and their placement on the chip, the floor plan.

- It guides the designer not only to capture the design problem by developing algorithms suitable to be implemented in silicon structures, but also to think about the testability of design [60].

- At each abstraction level the design is simulated to check whether it meets the design problem [37].

- The steps of refinement are computer aided and the results should be automatically verified.

- Automatically generated test patterns valid for the test at RT level as well as at other design levels and for circuit test equipment.

The technologically possible circuit complexity tended to grow faster than the power of CAD tools for automatic circuit design (silicon compilers) during the last years. So correctness by construction is only today available for some special types of architecture. This implies the need for powerful and sophisticated design verification tools to validate the correctness of each design step. Current standardisation efforts enforce electronic design interchange formats, which are some kind of a hardware description language [17]. Since the designer will no longer be forced to use the CAD tools of 'his' vendor, these languages also support the acceptance of a wide range of CAD tools. The second part of the text will describe some of their algorithms and how they are related to hardware description languages at different levels of abstraction.

CONCEPTUAL DESIGN PROCESS OF INTEGRATED CIRCUITS

The conceptual design process of integrated circuits can be partitioned into two main parts, the top down design and the bottom up extraction. The goal of the first part consists of transforming hardware descriptions at a high design level (RT level) into lower design level descriptions with more and more implementational details until the circuit is described in silicon structures. The second part can be considered as some kind of hardware verification. Thereby the designer passes the design levels in a reverse order. He applies hardware extraction systems to get out of lower level description the corresponding specification at the next higher level until he again gets an RT level description. At every design stage he simulates the description generated by the extractor and compares the simulation results with the corresponding one during the design phase. In the following paragraphs the reader will find a description of the two subprocesses.

TOP DOWN DESIGN PROCESS

As mentioned above the conceptual design process is based on the new design philosophy propagated by Mead and Conway [38]. Following the above mentioned design method the conceptual phase of the design process mainly takes place during the specification at the RT level. Already at the high RT level the designer may specify the behavior as well as some implementational details of the circuit, like the topology of the floor plan of the composed system and of the input-, output- and bidirectional ports of the subcells. This requires an RT level language, which provides features to express behavioral and structural informations. Using such a HDL the conceptual design phase at RT level passes in the following steps:

STEP1: Divide the function of the whole hardware system to be designed into several subfunctions and search for algorithms suitable to be implemented in silicon.

STEP2: Generate a functional description of every subfunction and simulate them to validate their behavior towards the requirements of the design problem.

STEP3: Specify an early floor plan by the design of the topologies of the interfaces of each subunit and generate the corresponding structural and topological descriptions. The main goal thereby is that most of the cell interconnections can be achieved just by wiring by abutment without additional wiring cells.

STEP4: Design to every subunit (cell) the corresponding key cell with array capability so that the whole data path can be formed just by iterative abutment of the key cells, if possible.

STEP5:   Consider again the several key cells (one bit wide) and describe complex RT
         elements by lower levels elements with the same I/O structure to maintain the
         consistency of test pattern. For example:
         • Replace multiplexer constructions by the corresponding combinational
         expressions or at switch level by models of passtransistor networks.
         • Replace operators (arithmetic, logical and relational) of the HDL by the
         corresponding user defined sliced operators.
         • Describe registers by catenation of dynamic storage cells with a 1 bit path
         width and more differential timing and control signals.

STEP6:   Now, partition the arrays of the cells into subarrays separated by a single key cell
         which may be described at a lower design level (STEP3). Following this strategy
         describe the whole circuit and simulate it, so that the key cells may be validated in
         an RT level context, which is much more efficient than a simulation of a purely
         switch level description.

STEP 1 and STEP 2 are the phases of capturing the design problem and validating the specification.
During the executing of the following steps the conceptual top down design phase takes place. Thereby
the designer generates a combination of behavioral and topological description of the circuit. During
STEP 5 the designer describes some RT elements like multiplexers, registers etc. at different
abstraction level, since sometimes it may be necessary for the designer to plunge down to switch level
to see whether his design concept may also be realizable at lower levels. After the conceptual design
phase at RT level the designer has a validated combined structural and behavioral description.

The following designs at the circuit and layout level may pass in two differrent ways. One
possibility may be, that the designer has used for the hardware specification at RT level a HDL being
the input format to design automation systems, like ISP at the Carnegie-Mellon University [44]
and APL of the ALERT system developed by IBM [19]. In this case the RT level description is
translated automatically to an intermediate code being the input format to other parts of the design
system generating the corresponding designs at the lower levels. The further design process could
be influenced by the designer giving input parameters specifical to the circuit. If, however, the designer
has chosen a hardware description system at RT level, which is not integrated to design tools at
lower levels, then he has manually to make the transformation between the different levels to
generate input formats to hardware simulation tools at lower design levels like DOMOS [47] and
SPICE2 [40]. The disadvantage of this design method is the occurrence of errors during the
transformation of the descriptions. However, with respect to the state of the art of existing design
automation systems the resulting layout influenced by the designer's experience is more efficient.
The design process at the logic, circuit and layout level is treated exhaustively in literature, so this
article releases a detailed description.

CHARACTERISTICS OF AN RT LANGUAGE SUITABLE FOR CONCEPTUAL DESIGN

Considering the conceptual phase of the design process of integrated circuits described above the RT language used for the specification of hardware should have the following characteristics:

- The language should be easy to learn.
  It should describe hardware elements and their interconnections in a natural way, since it is not only used to communicate between humans and machines but als between the members of a design team.

- The language should be nonprocedural.
  Control informations of procedural languages, like ISP, suitable for some simulation purpose, need additional mechanisms to get real hardware structures, either design automation systems or second descriptions with implementational details. Structural nonprocedural base descriptions are closer to real hardware and therefore more understandable to the programming shy hardware designer.

- The language should support the philosophy of hierarchically constructed hardware systems.
  In that case the integration of predesigned circuit descriptions stored in a libary or the design of standard cells will be also possible.

- The language should be capable to describe just in this early specification phase some topological aspects of the hardware, like floor plan topology of compound hardware cells and the topology of input, output and bidirectional ports.

- The language should be capable to describe hardware at several abstraction levels minimizing the number of languages necessary througout the design system.

- The high level description should be technology independent.

- The RT language should be the input format to several design tool, such as simulators, test pattern generator and test evaluation tools as well as verification systems.

Considering the traditional HDL's at RT level like AHPL [27], CDL [13], DDL [15] and ISP [48] one can say that they are all behavioral. They are well suited to generate procedural functional descriptions, but they don't provide concepts to describe structural and topological details. Surveys of such design languages can be found in [7], [14], [36], and [46].

In parallel to the advent of VLSI technology and new design methods HDL's at RT level are developed capable to express both behavioral and structural as well as topological informations of hardware. Such HDL's are the model of the USC Information Sciences Institute [34], FDL [29], HARPA [53] and KARL-III [24]. Although those languages don't meet all the above mentioned characteristics, they all support the philosophy of hierarchically designed circuits.

In the following paragraph the conceptual design of an incrementer/decrementer circuit is executed. The incrementer/decrementer circuit is one part of an address generator designed within the EIS project at Kaiserslautern University. The description language used is KARL-III, because of the authors familarity with the language.

CONCEPTUAL DESIGN OF AN INCREMENTER/DECREMENTER CIRCUIT

The address generator is one part of a design rule check machine developed within the EIS project at Kaiserslautern University. It generates the addresses of the layout information to be loaded into the cache for the design rule check. Thereby the cache simulates a window of size 4x4 raster elements to be moved over the layout during the design rule check is executed. Therefore it is necessary to increment or decrement the actual address of the layout information by 1, 2, or 3. In the following the design of a single incrementer/decrementer circuit is executed according to the above mentioned design steps. At each step a KARL-III description and the corresponding ABL-diagram is presented. ABL stands for A Block diagram Language [20], [58]. It is the graphical version of the textual language KARL-III. With respect to the length of this article the authors release the representation of the corresponding simulation files [37]. Base to the design is the following informal description:

Informal incrementer/decrementer description:

INPUT: data signal DATA_IN [7..0],

                bidirectional control signal ACT_BI, INC_BI (single bits).

OUTPUT: data signal DATA_OUT [7..0].

ACT_BI	INC_BI	actions
0	*	DATA_OUT <-- DATA_IN
1	0	DATA_OUT <-- decrement(DATA_IN)
1	1	DATA_OUT <-- increment(DATA_IN)

Design during STEP 1:

Since the incrementer/decrementer circuit is one subfunction of the adressgnerator, the design starts at STEP 2.

Design during STEP 2:

Fig. 1a/b show the ABL-diagram and the corresponding functional KARL-III description of the incrementer/decrementer circuit. 'inc' and 'dec' are KARL-III standard functions.

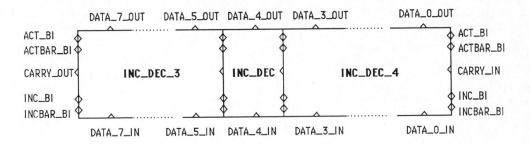

Fig. 6a: Floor plan of incrementer/decrementer circuit

```
cell INC_DEC_8
  ( back in DATA_7_IN, DATA_6_IN, DATA_5_IN, DATA_4_IN,
          DATA_3_IN, DATA_2_IN, DATA_1_IN, DATA_0_IN
   front out DATA_7_OUT, DATA_6_OUT, DATA_5_OUT, DATA_4_OUT,
          DATA_3_OUT, DATA_2_OUT, DATA_1_OUT, DATA_0_OUT
    right bi ACT_BI_R, ACTBAR_BI_R
        in CARRY_IN
        bi INC_BI_R, INCBAR_BI_R
    left bi ACT_BI_L, ACTBAR_BI_L
       out CARRY_OUT
        bi INC_BI_L, INCBAR_BI_L );



cell INC_DEC_3 ( back  in DATA_IN [2..0]
             right  bi ACT_BI, ACTBAR_BI
                in CARRY_IN
                bi INC_BI, INCBAR_BI
             left  bi ACT_BI, ACTBAR_BI
               out CARRY_OUT
                bi INC_BI, INCBAR_BI
             front out DATA_OUT [2..0] );

external;


cell INC_DEC_4 ( back  in DATA_IN [3..0]
             right bi ACT_BI, ACTBAR_BI
                in CARRY_IN
                bi INC_BI, INCBAR_BI
             left  bi ACT_BI, ACTBAR_BI
               out CARRY_OUT
                bi INC_BI, INCBAR_BI
             front out DATA_OUT [3..0] );

external;
```

```
cell INC_DEC ( back in DATA_IN;
              right bi ACT_BI, ACTBAR_BI
                    in CARRY_IN
                    bi INC_BI, INCBAR_BI
              left  bi ACT_BI, ACTBAR_BI
                    out CARRY_OUT
                    bi INC_BI, INCBAR_BI
              front out DATA_OUT    );
external;

node DATA_7_OUT, DATA_6_OUT, DATA_5_OUT, DATA_4_OUT,
     DATA_3_OUT, DATA_2_OUT, DATA_1_OUT, DATA_0_OUT,
     DATA_OUT_75[2..0]; DATA_OUT[3..0];
     CARRY_OUT;
tribus ACT_BI_R, ACTBAR_BI_R,
       INC_BI_R, INCBAR_BI_R,
       ACT_BI_L, ACTBAR_BI_L,
       INC_BI_L, INCBAR_BI_L;

begin
make INC_DEC_3 : INC_DEC : INC_DEC_4
  ( back in DATA_7_IN.DATA_6_IN.DATA_5_IN.DATA_4_IN,
        DATA_3_IN,
        DATA_2_IN.DATA_1_IN.DATA_0_IN
    front out DATA_OUT_75, DATA_4_OUT, DATA_OUT_30
    right bi ACT_BI_R, ACTBAR_BI_R
        in CARRY_IN
        bi INC_BI_R, INCBAR_BI_R
    left bi ACT_BI_L, ACTBAR_BI_L
        out CARRY_OUT
        bi INC_BI_L, INCBAR_BI_L );

terminal DATA_7_OUT .= DATA_OUT_75 [7];
         DATA_6_OUT .= DATA_OUT_75 [6];
         DATA_5_OUT .= DATA_OUT_75 [5];
         DATA_3_OUT .= DATA_OUT_30 [3];
         DATA_2_OUT .= DATA_OUT_30 [2];
         DATA_1_OUT .= DATA_OUT_30 [1];
         DATA_0_OUT .= DATA_OUT_75 [0];
end.
```

Fig. 6b: KARL-III description of incrementer/decrementer circuit

LOGIC LEVEL REFINEMENT OF CONCEPTUAL DESIGNS

Up to this point the design has been structurally refined within the RT level.

During the logic design two strategies will be used joinedly. The first approach is the top down strategy. In this strategy the structural elements of the RT level description will be substituted by logic equations defining the input/output relation of the elements. For example these equations can be computed by classical minimisation techniques from the truthtables of the RT level elements. This refinement is almost independent of the technology the circuit will be physically realized in.

However during logic level refinement, the target technology will already have some influence on the logic description of the design. This is caused by the second strategy of refinement, the bottom up design. In this approach the designer already has some knowledge of the technology and knows realisations which best fit the technology. For example, if NMOS or CMOS technology is employed, single stage compound gates will be extensively used. These are logic AND and OR relations with one inversion at the output.

From this we see that the logic level description of a design is influenced by the technology used, but the language, the logic design is described in, may still be technology independent, since the primitives of the language are Boolean operators and functions.

Usually the description of an entire design is not transferred to the logic level in one step, but by a set of small steps. In each of these steps only a small part of the design, e.g. a single cell, is modified. So in order to simplify the design process the HDL used should provide RT level primitives and logic level primitives. In this case the designer has to learn just one language, and he can simulate a mixed RT logic level description with one single simulator. The simulation results can be compared to the RT level simulation results in order to have a first check of each refinement step.

PHYSICAL REFINEMENT OF CONCEPTUAL DESIGNS

The next step in the design cycle is the development of a circuitdescription of the design. The transition from logic to circuit level is characterized by two main features, the first of which is the technology dependency of the circuit design. The second is the difference in modeling logic and electrical behavior of the elements of logic and the circuit description.

These two points are the reason why most HDLs for RT level do not cover circuit level descriptions. The technology dependency would require a large set of primitives in the HDL and the different modeling would require a different simulator. So in most cases the designer will use a different description at circuit level than at logic or RT level. The most commonly used language is the SPICE2 input format [40], [54]. Circuit level descriptions are usually simple netlist formats, which are either device (SPICE, DOMOS [47]) or net oriented (EDIF). Some of them support hierarchical designs (SPICE, EDIF).

In order to force a standardisation in the area of circuit descriptions the EDIF steering committee has published the 'Electronic Design Interchange Format' EDIF [17]. It is to be hoped that it will become a standard for low level design descriptions. EDIF is an interchange format covering a wide range of representation levels. In was developed by a commitee of representatives of the biggest American data processing companies.

The development of a circuit description for a given logic description is almost simple, since the logic gates can usually directly be substituted by equivalent transistor circuits. More difficult is the correct dimensioning of the circuit elements, specially if timing or powerdissipation constraints are given. After all elements have been correctly sized a first transient simulation can be performed in order to verify the performance of the circuit.

A detailed simulation can be performed after the layout generation of the design, which is not subject of this text. For this detailed simulation an exact parameter extraction from mask artwork has to be done to compute the paratisics of the circuit.

LOGIC DESIGN OF THE INCREMENTER/DECREMENTER CIRCUIT

Having a close look at the INC_DEC cell as designed during step 4 of the RT level refinement mentioned above, we find that this description still has some RT level elements, such as 'if', 'encode', 'coin'. These RT level elements have to be substituted by logic level elements during this step.

At this point some bottom up decisions influence the design. So for example the multiplexer described by the 'if' statement in the RT level description can be replaced by an enables statement or by a Boolean expression at logic level. There are two multiplexers used within the incrementer/decrementer cell (Fig. 7).

```
CARRY_OUT .= (if encode (INC_BI.INCBAR_BI)
        then not (DATA_IN)
        else DATA_IN
    endif)
            nor
            CARRYBAR;
DATA_OUT .= if encode (ACT_BI.ACTBAR_BI)
        then (DATA_IN coin CARRYBAR)
        else DATA_IN
    endif;
```

Fig. 7: Part of RT Level Incrementer/Decrementer Description

The first 'if' statement in the CARRY_OUT assignment is one argument of the 'nor' operator. So if NMOS or CMOS technology is employed the entire CARRY_OUT assignment can be realized by just one single stage compound gate plus an inverter for the DATA_IN signal. The condition of the if statement is true, if INCBAR_BI is 0 and INC_BI is 1. In all other cases the condition is false. By definition INCBAR_BI is always not (INC_BI). So the logic expression for CARRY_OUT can be formulated according to fig. 8.

The second 'if' statement is different, since the result is a direct output of the cell. Another important feature of the DATA_OUT assignment is the fact, that again the control variables are ACT_BI and ACTBAR_BI, with ACTBAR_BI equals not (ACT_BI). So the designer will immediately think of a passtransistor realisation using NMOS or CMOS technology. If the output DATA_OUT were a bus, the most natural description of the assignment would be in terms of the 'enables' statement. In our example the DATA_OUT signal is a node and the assignment can be done using the case statement, which is equivalent to the 'if' statement, but allows more than two inputs, or by the 'if' statement itself. The condition of the 'if' statement is true, if ACT_BI is 1. So the logic expression of the DATA_OUT assignment can be simplified according to fig. 8.

```
        .
        .
        .

    CARRY_OUT .= ((INC_BI and (not (DATA_IN))) or
           (DATA_IN and INCBAR_BI))
               nor
               CARRYBAR;
    DATA_OUT .= if ACT_BI
           then (DATA_IN = CARRYBAR)
           else DATA_IN
         endif;

        .
        .
        .
```

Fig. 8: Part of Logic Level Incrementer/Decrementer Description

CIRCUIT DESIGN OF THE INCREMENTER/DECREMENTER CELL

The next step in the methodological refinement of the incrementer/decrementer is the circuit design. In our example we will use NMOS technology [38]. In this technique we will need five subcircuits to ensure the required function. The multiplexer can be realized by two passtransistors (fig. 10), T1 and T2, which are controled by the signals ACT_BI resp. ACTBAR_BI. The equivalence gate is realized by a tricky NMOS circuit, built of the transistors T3, T4 and T18. The two inverters are designed using the NMOS standard inverter, built of transistors T7 and T11 resp. T16 and T17. The rest of the circuit is realized by a compound gate built of transistors T8, T9, T13, T14, T15 and T12.

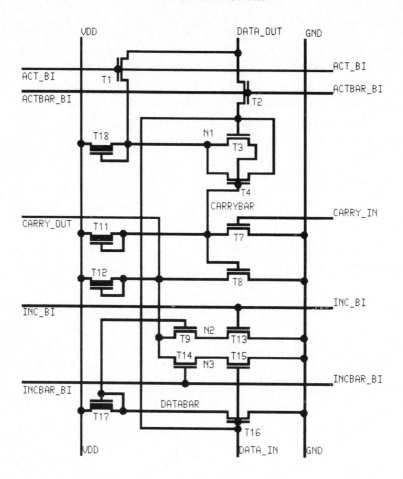

Fig. 9: Circuit Diagram of the Incrementer/Decrementer Cell

Although electronics designers are used to think and document their work in schematics most of the circuit simulators only accept a textual description of the circuit. So the designer has to translate the circuit diagram into a netlist description. Fig. 10 shows the EDIF netlist description of the incrementer/decrementer cell.

```
(edif example
 (status (edifversion 1 0 0)
  (ediflevel 1)
  (written (timestamp 1985 6 13 14 52 0)
       (accounting author "W. Nebel, Kaiserslautern University")
  )
 )
 (external TRANSISTORLIB)
 (comment "TRANSISTORLIB is an external library holding the models
       of the used transistor types")
 (design INCDEC_CELL (qualify EXAMPLELIB INCDEC))
 (comment "This states that the root of the design INCDEC_CELL is
       the cell INCDEC within the library EXAMPLELIB")
 (library EXAMPLELIB
  (cell INCDEC
   (view netlist TRANSISTORNET
    (interface
     (define inout  port (multiple ACT_BI ACTBAR_BI INC_BI
                    INCBAR_BI VDD GND))
     (define input  port (multiple DATA_IN CARRY_IN))
     (define output port (multiple DATA_OUT CARRY_OUT))
    )
    (contents
     (define local signal (multiple N1 N2 N3))
     (instance
      (multiple
       (qualify TRANSISTORLIB NMOS_DEPL) TRANSISTORNET T11)
       (qualify TRANSISTORLIB NMOS_DEPL) TRANSISTORNET T12)
       (qualify TRANSISTORLIB NMOS_DEPL) TRANSISTORNET T17)
       (qualify TRANSISTORLIB NMOS_DEPL) TRANSISTORNET T18)
       (qualify TRANSISTORLIB NMOS_ENHA) TRANSISTORNET T1)
       (qualify TRANSISTORLIB NMOS_ENHA) TRANSISTORNET T2)
       (qualify TRANSISTORLIB NMOS_ENHA) TRANSISTORNET T3)
       (qualify TRANSISTORLIB NMOS_ENHA) TRANSISTORNET T4)
       (qualify TRANSISTORLIB NMOS_ENHA) TRANSISTORNET T7)
       (qualify TRANSISTORLIB NMOS_ENHA) TRANSISTORNET T8)
       (qualify TRANSISTORLIB NMOS_ENHA) TRANSISTORNET T9)
       (qualify TRANSISTORLIB NMOS_ENHA) TRANSISTORNET T13)
       (qualify TRANSISTORLIB NMOS_ENHA) TRANSISTORNET T14)
       (qualify TRANSISTORLIB NMOS_ENHA) TRANSISTORNET T15)
         (qualify TRANSISTORLIB NMOS_ENHA) TRANSISTORNET T16)
```

```
  )
  )
  (joined VDD     (qualify T18 DRAIN) (qualify T11 DRAIN)
          (qualify T12 DRAIN) (qualify T17 DRAIN))
  (joined GND     (qualify T7 SOURCE) (qualify T8 SOURCE)
          (qualify T13 SOURCE) (qualify T15 SOURCE)
          (qualify T16 SOURCE))
  (joined DATA_IN  (qualify T16 GATE) (qualify T15 GATE)
          (qualify T3 GATE) (qualify T4 SOURCE)
          (qualify T2 SOURCE))
  (joined CARRY_IN (qualify T7 GATE))
  (joined ACT_BI   (qualify T1 GATE))
  (joined ACTBAR_BI (qualify T2 GATE))
  (joined INC_BI   (qualify T13 GATE))
  (joined INCBAR_BI (qualify T14 GATE))
  (joined DATA_OUT (qualify T1 DRAIN) (qualify T2 DRAIN))
  (joined CARRY_OUT (qualify T12 SOURCE) (qualify T8 DRAIN)
          (qualify T9 DRAIN) (qualify T14 DRAIN)
          (qualify T12 GATE))
  (joined CARRYBAR (qualify T3 SOURCE) (qualify T4 GATE)
          (qualify T7 DRAIN) (qualify T8 GATE)
          (qualify T11 SOURCE) (qualify T11 GATE))
  (joined DATABAR  (qualify T9 GATE) (qualify T16 DRAIN)
          (qualify T17 SOURCE) (qualify T17 GATE))
  (joined N1      (qualify T1 SOURCE) (qualify T3 DRAIN)
          (qualify T4 DRAIN) (qualify T18 SOURCE)
          (qualify 18 GATE))
  (joined N2      (qualify T9 SOURCE) (qualify T13 DRAIN))
  (joined N3      (qualify T14 SOURCE) (qualify T15 DRAIN))
  )
  )
  )
  )
  )
```

Fig. 10: EDIF Netlist Description of Incrementer/Decrementer Cell

Starting from this netlist or circuit diagram the designer will develop the layout of the integrated circuit. This design step is not subject of this text, but for the sake of completeness the layout of this circuit is given in fig. 11. The presented layout is just a demonstration example and far from being optimal with respect to area optimisation.

Since the layout is usually handmade, it has to be extensively checked for topographical and topological errors. The topological check is the first step in a set of design verifications. This step is called circuit-extraction from mask artwork and will be subject of the following subchapter.

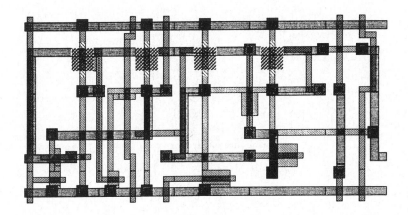

Fig. 11: Layout of Incrementer/Decrementer Cell

FUNCTIONAL VERIFICATION OF THE DESIGN

In order to verify the function of a design a number of verification steps have to be performed. These steps usually prove the correctness of one methodological refinement step, e.g. proving the transition from logic to circuit design. This verification step will be performed after each refinement step to close the design cycle, it may, however may be done covering several steps of refinement, e.g. to close the design cycle between RT level and layout level design (fig. 12). Figure 12 does not contain the chip fabrication and the test of the chip. Since fabrication is expensive and time consuming all design errors should be detected and corrected before the prototype fabrication in order to avoid a redesign. To detect all design errors in an early stage of the design requires high performance and sophisticated verification tools [41].

CIRCUIT EXTRACTION

The task of circuit extractors is to retrieve component and connectivity data from mask layout. A more detailed extraction recognizing parasitic components is called parameter extraction. The latter is necessary for timing performance validation. Circuit and parameter extractors, both of which will be called circuit extractors in the following text, have to detect devices and their interconnections from mask artwork.

Devices in all planar technologies are defined by a vertical sequence of different masks. Fig. 13 a shows the layout of a bipolar transistor. The transistor is defined by the vertical sequence of the masklayers n-epitaxy, p-diffusion and n-diffusion. The base terminal of the transistor is defined by

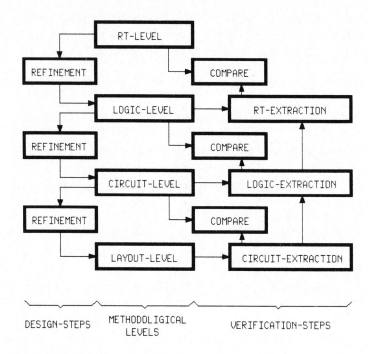

Fig. 12: Design and Verfication Loops

the vertical sequence of the masklayers n-epitaxy and p-diffusion and not n-diffusion. In the NMOS example (fig. 13 b) the channel of the NMOS transistor is defined by the vertical sequence of the masks n-diffusion and poly. So devices and their terminals can be found by Boolean 'AND' resp. 'AND NOT' operations on the mask layout.

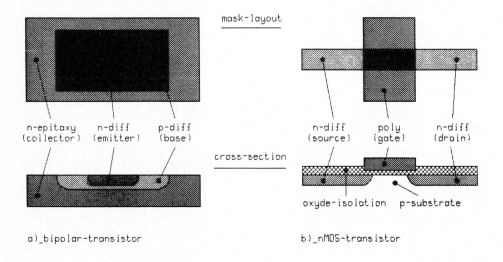

Fig. 13: Planar Semiconductor Devices

On the other hand two nodes of a circuit are connected if there exists at least one path of at least one of the connectivity layers, e.g. metal, between both points. This path may lead through different materials connected by vias.

Depending on the layout representation and the used algorithms for this task today's circuit extractors can be divided into three different classes:

1. polygon oriented extractors
2. pixel oriented extractors
3. extractors aided by a special input datastructure

The first class may be regarded as the classical extractors. In their case the layout geometry is defined by boxes or polygons. The layout data are represented by the edges of the layout elements.

The basic computational problem of the polygon oriented extractor is the intersection and touching point detection of polygon edges. These points can be the corners of devices or abuting polygons forming a common electrical node. In case of a bipolar circuit transistors can be found if a n-diffusion area is completely within a p-diffusion area, which itself is within a n-epitaxy area (fig. 13). Also this 'IN' function can be answered by the polygon intersection algorithm. A polygon A is completely within polygon B if no edge of A intersects an edge of B and if at least one point of A is within B. The latter question can be answered by computing the number of intersections of a straight line between the point of A and a point which is surely outside B. If the number of intersections is uneven, the point of A is within B.

We can state that the polygon oriented circuit extractor can be reduced to a set of polygon intersection problems. Algorithms have been reported that solve this problem in a complexity of $O(n \log n)$ [9] with n representing the number of polygons.

The pixel oriented extractors use another layout data representation. In their case the layout is rasterized into a grid. Each grid element is called a pixel. The grid width equals the tolerance distance lambda [38]. A dataword is associated with each pixel and the bits of the dataword correspond to the mask layers. So Boolean operations on the layout can be performed by computing Boolean operations on the bits of the pixels. This dataformat is used for a wide range of algorithms, e.g. pixel oriented design rule check [26], [41].

If neighboring pixels are examined, connectivity of these pixels can be found. In case of connectivity the pixels will be labeled with the same node number (fig. 14a). Devices are detected when the characteristic device material combination is found next to the device terminal combination (fig 14 b).

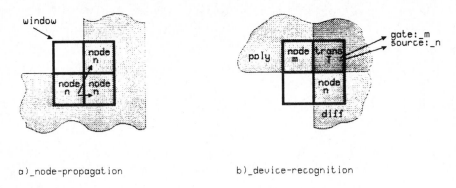

a)_node-propagation   b)_device-recognition

Fig. 14: Pixel Oriented Circuit Extraction

The two by two pixel window is moved (one) column by (one) column through each line of the layout, and each pixel is examined. By counting node pixels and perimeter information parasitic capacitances can be computed. By counting lines and columns of transistors the length and width of devices can be estimated. These simple algorithms are even well suited for hardware accelerators [26]. A more detailed description of pixel oriented extraction algorithms is given in [41]. Since each pixel is processed just once the time complexity of the pixel oriented extractor is O (n) with n being proportional to the layout area. The required memory however is larger than for a polygon oriented extraction.

As an example of the third class of extractors John Ousterhout's MAGIC [43] can be mentioned. MAGIC is a complete VLSI design system with layout editor, online design rule check, circuit extractor and compaction and routing features. The system is based on a unique data structure called corner stitching. The layout elements are rectangles placed on one of the layout planes. The remaining are of the layout plane is then filled with empty rectangles. Each rectangle is linked together by four pointers with its neighbour rectangles at the upper right and lower left corner (fig. 15 a). If you follow these pointers the direct environment of a rectangle can be found. This avoids the need for search algorithms and so connectivity can efficiently be derived (fig. 15 b).

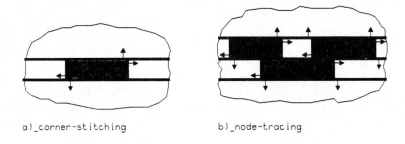

a)_corner-stitching   b)_node-tracing

Fig. 15: Corner Stitching Datastructure

Extractors of each of these classes can work in a hierarchical way to exploit structural information in order to minimize computation time [3], [10], [12], [21], [28], [35], [42], [49], [51], [55]. In this case each layout cell descripition is examined just once. The results are circuit descriptions of cells, which can be linked together according to the design hierarchy and topology.

The output of all circuit extractors is a netlist description. Common formats are SPICE2, DOMOS or EDIF, which were mentioned above. These descriptions can be used as input for a circuit simulation in order to analyse the timing behavior of the circuit. They can also be input for a logic simulator like MOSSIM [11] to validate the logic function of a circuit. A real functional verification of a circuit cannot be performed by simulation, because the device count of real circuits does not allow the simulation of all possible state transitions. So a verification can be performed either by proving isomorphism between the retrieved circuit description and a specification circuit description [1] or by further logic and functional block extraction and subsequent verification of the functional description.

LOGIC AND FUNCTIONAL BLOCK EXTRACTION

Logic extraction is the opposite process to circuit level refinement. It can be undertaken during two steps of the design cycle. After the circuit level refinement the designed circuit level description has to be verified. This can be done by a functional equivalence prover [22], which compares the specification description with the extracted logic description of the circuit. The VERENA system for example compares two KARL descriptions, one at RT level and one at logic level, with respect to functional identical behavior. Instead of using a functional equivalence prover, the extracted description could also be validated by simulation with test patterns generated manually or automatically [60] from the functional description of the design. Because of the smaller number of logic devices and the simpler simulation models used in comparison to the circuit description of the design, even a complete simulation of appropriate designs could be performed. Logic extraction is also undertaken to prove the correctness of an extracted circuit description, if no isomorphism check with a specification circuit description is performed.

A distinction between logic and functional block extraction can be made with regard to the algorithms used. Logic extractors analyse a circuit description technology dependent. They trace paths from output nodes to Vdd and Gnd nodes in order to compute a logic expression for the behavior of the output nodes. Functional block extractors work almost technology independent. They try to substitute parts of the circuit description by an equivalent functional description.

Usually both approaches use a circuit representation in form of a weighted graph, called a net (fig. 16). The weight of the relations is the terminal type of the associate devices.

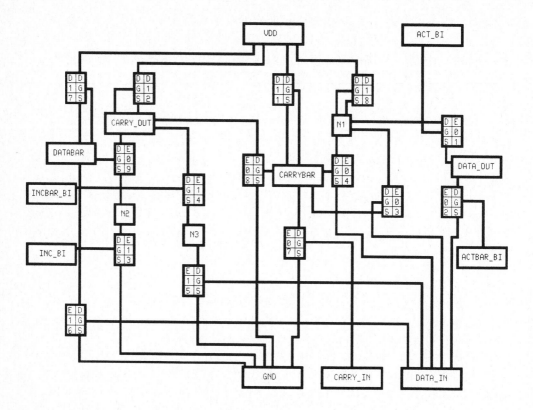

Fig. 16: Circuit Net of Incrementer/Decrementer Cell

LOGIC EXTRACTION

In this subchapter a simple algorithm is presented, which can be used to extract the logic function from NMOS subcircuits. The algorithm can easily be adapted to CMOS technology.

Typical NMOS logic gates are composed out of a pull up transistor and a pull down network. The output of the circuit is 1 unless at least one conducting path exists from the output node to the Gnd node. So the algorithm has to find out the logic condition for closing each path from output to Gnd (fig. 17).

If we apply the algorithm to the carry_out node of the incrementer/decrementer cell, the first parameter of the function, the node would be carry_out. The second parameter will be the pullup transistor T12. This second parameter is necessary to ensure that the algorithm does not trace the old path back. By this recursive function an AND/OR expression is generated for the parallel and series transistors of the pulldown net. The logic function of the NMOS circuit is the inversion of the AND/OR expression. In our example the AND/OR expression is the following:

CARRYBAR OR DATABAR AND INC_BI OR INCBAR_BI AND DATA_IN

and the logic expression for the CARRY_OUT node is:

CARRY_OUT =
  NOT (CARRYBAR OR DATABAR AND INC_BI OR INCBAR_BI AND DATA_IN)

This logic expression can obviously be formulated with the aid of a HDL, e.g. KARL-III:

TERMINAL

CARRY_OUT .=
  NOT ((CARRYBAR **OR** DATABAR **AND** INC_BI) **OR** INCBAR_BI **AND** DATA_IN);

Despite of the rather simple algorithm for logic extraction this approach has some vital disadvantages. First it is strongly technology dependent. Second feedback loops and transmission gate recognition will be necessary extensions to the algorithm and make it more complex. These additional features are partially described in [2], [4]-[6], [23], [30]-[33], [39], [45], [50], [56], [59].

```
FUNCTION gndpath (node_X, oldtransistor);
BEGIN
 IF node <> Gnd
   THEN
     FOREACH (Transistor T with its drainnode
         connected to node_X <> oldtransistor) DO
       downpath := downpath OR
                   (gatenode (T) AND gndpath (sourcenode (T), T));
       gndpath := downpath;
END;
```

Fig. 17: Logic Extraction Algorithm for NMOS Logic Gates

FUNCTIONAL BLOCK EXTRACTION

Functional block extractors [8] try to find subgraphs in the analysed circuit graph. These subgraphs correspond to functional blocks, e.g. registers, multipliers etc.. The subgraphs and the corresponding functions are user defined. So each user will either use a standard library of subgraphs and associate functions or he will create his own library matching his own design style.

The computational problem of finding subgraphs is the problem of subgraph isomorphism [52]. Appropriate weighing of device terminals allows the use of fast matching algorithms. In large designs, however, or if extensive reference libraries are used, there will be more than one possible solution for partitioning the circuit graph into functional blocks. This problem can be solved, if either heuristic approaches are used, or by user interaction. The result of the functional block extraction is a graph of functional blocks and logic nodes. Optionally this graph may again be compared with a reference library of high level functions in order to substitute the functional blocks by high level functional blocks. If this technique is used, the device count of the design could be minimized. This would also minimize the design verification cost.

The retrieved graph can be transformed into a functional netlist, which will be input of a translater, which transforms the netlist description into a HDL. This HDL description can later be functionally compared with a specification description using for example the VERENA system [22].

CONCLUSION

We have shown in which way the conceptual design of the 'long thin man' is guided by a set of CAD tools and methodologies based on hardware descriptions at various levels of abstraction. Using these tools the designer refines his architectural concept from his inspiration through RT, logic and circuit level down to to the physical layout. On the other hand we have demonstrated the great importance of a closed verification cycle. Again during this cycle hardware description languages are the carriers of the input information of the verification tools.

REFERENCES

[1]   Ablasser, I.; Jaeger, U., Circuit Recognition and Verification Based on Layout Information, IEEE 18th DAC, Nashville (1981).

[2]   Ackland, B.; Weste, N., Functional Verification in an Interactive Symbolic IC Design Environment, Caltech Conference on VLSI, Pasadena, CA (1981).

[3]   Annevelink, J; Dewilde, P.; Fokkema, J.T., A Hierarchical Layout to Circuit Extractor using a Finite State Approach, IEEE ICCD '83, VLSI in Computers, Port Chester, NY (1983).

[4]   Apte, R.; Scheffer, L., LSI Design Verification Using Topology Extraction, Twelfth Asilomar Conference on Circuits, Systems and Computers, Pacific Grove, CA (1978).

[5]   Apte, R.; Chang, N.; Abraham, J. A., Logic Function Extraction for NMOS Circuits, IEEE ICCC '82, New York, NY (1982).

[6]   Apte, R. M.; Chang, N., Consistency Checking for MOS/VLSI Circuits, IEEE 20th DAC, Miami Beach (1983).

[7]   Barbacci, M.R., A comparison of register transfer languages for describing computers and digital systems, IEEE Transactions on Computers vol. c-24 (1975) 137-150.

[8]   Barke, E.; Hoepken, T.; Luellau, F., A Technoilogy Independent Block Extraction Algorithm, IEEE 21st DAC, Albuquerque, NM (1984).

[9]   Bentley, J. L.; Ottmann, T. A., Algorithms for Reporting and Counting Geometric Intersections, IEEE Transactions on Computers, Vol. 28, No. 9 (1979).

[10]   Bello, S. E.; Hoffman, J. L.; McMillan, R. I.; Ludwig, J. A., VLSI Hierarchical Design Verification, IEEE ICCC '82, New York, NY (1982).

[11]   Bryant, R. E., An Algorithm for MOS Logic Simulation, Lambda, Vol. 1, 4th Quarter, Palo Alto, CA (1980).

[12] Chao, S.; Huang, Y.; Yam, L., A Hierarchical Approach for Layout   Versus Circuit Consistency Check, IEEE 17th DAC, Minneapolis (1980).

[13] Chu,Y., Introducing CDL, IEEE Computer vol. 7 no.12 (1974) 42-44.

[14] Cleemput van, W.M., Computer hardware description languages and their applictions, IEEE 16th DAC, San Diego, CA (1979) 554-560.

[15] Dietmeyer,D.L., Introducing DDL, IEEE Computer vol. 7 no. 12 (1974) 34-38.

[16] Dudani, S. and Stabler, E., Types of hardware descriptions, in: Uehara, T. and Barbacci, M. (eds), 6th Int. Sym. o. Computer Hardware Description Languages and Their Applications (North-Holland, Amsterdam, 1983).

[17] EDIF Steering committee, EDIF Electronic Design Interchange Format Version 100 (Pro Print & Services, Sunnyvale, CA, 1985).

[18] Eveking, H., The application of register transfer languages in the design of real hardware, in: Proc. of the 4th Int. Sym. on Computer Hardware Description Languages (1979).

[19] Friedmann, T.D. and Yang, S.C., Methods used in   an automati logic design generator [ALERT], IEEE Trans. on Comp. vol. C-18 no. 7 (1969) 593-614.

[20] Girardi, G., ABL data structure, CVT-report, CSELT Torino, Italy (March 1984).

[21] Gupta, A.; Hon, R. W., Two Papers on Circuit Extraction, Report CMU-CS-82-147, Carnegie Mellon University, Pittsburgh (1982).

[22] Grass, W.; Schielow, N., VERENA - A Program for Automatic Verification of the Refinement of a Register Transfer Descrition into a Logic Description, IFIP International Symp.on Computer Hardware Description Languages and their Applications, CHDL '85, Tokyo (1985).

[23] Greenberg, S.; Buazza, M., Logic Recognition in the SAVVY Timing Verification System, Digest of Technical Papers, IEEE ICCAD '84, Santa Clara, CA (1984).

[24] Hartenstein, R.W., KARL-III Reference Manual, Department of Computer Science, Kaiserslautern University (1985) (in work).

[25] Hartenstein, R.W. and Borrmann, B., HYPERKARL-III Language, Reference   Manual, Department of Computer Science, Kaiserslautern University (March 1985).

[26] Hartenstein, R. W.; Hauck, R.; Hirschbiel, A.; Nebel, W.; Weber, M., PISA, a CAD Package and Special Hardware for Pixel-Oriented Layout Analysis, IEEE ICCAD '84, Santa Clara, CA (1984).

[27] Hill, F.J., Introducing AHPL, IEEE Computer vol. 7 no. 12 (1974) 28-30.

[28] Hon, R. W.: The Hierarchical Analysis of VLSI Designs, Dept. of Computer Science, report CMU-CS-83-170, Carnegie-Mellon University, Pittsburgh (1983).

[29] Kato, S. and Sasaki, T., FDL: A structural behavior description language, in: Uehara, T. and Barbacci, M. (eds), 6th Sym. on Computer Hardware Description Languages and their Applications (North-Holland , Amsterdam, 1983).

[30] Kawamura, M.; Hirabayashi, K., Logical Verification of LSI Mask Artwork by Mixed Level Simulation, IEEE International Symposium on Circuits and Systems (1982).

[31] Kawanishi, H.; Kishimoto, A.; Kani, K., An Automati Layout-Logic Verification Algorithm for VLSI, ECCTD (1980).

[32] Kishimoto, A.; Kawanishi, H.i; Yoshizawa, H.; Ohno, H.; Fujinami, Y.; Kani, K., An Interconnection Check Algorithm for Mask Pattern, IEEE International Symp. on Circuits and Systems (1979).

[33] Kishimoto, A.; Mori, K.; Takahashi, S.; Kawanishi, H., A Logic Function Extraction Algorithm for MOS VLSI, IEEE ICCAD '83, Santa Clara, CA (1983).

[34] Klein, S. and Sastry, S., Parameterized modules and interconnections in unified hardware descritpions, in: Breuer, M. and Hartenstein, R. (eds), Int. Conf. on Computer Hardware Description Languages and their Applications (North-Holland, Amsterdam, 1981).

[35] Lauther, U., Simple but Fast Algorithms for Connectivity Extraction and Comparision in Cell Based VLSI Designs, ECCTD (1980).

[36] Maissel, L.I. and Ofek, H., Hardware design and description languages in IBM, IBM Journal of Research and Development vol. 28 no. 5 (1984) 557-563.

[37] Mavridis, A., Languages for Simulator Activation and Test Description, Hartenstein R. (ed.), Advances in CAD for VLSI: Vol. 7 Hardware Description Languages (North-Holland, Amsterdam, 1987).

[38] Mead, C. A.; Conway, L. A., Introduction to VLSI Systems (Addison-Wesley Publ. Comp., 1980).

[39] Munch, P. H.; Munch, K. H., A General Solution for Design Verification, IEEE ICCC '82, New York, NY (1982).

[40] Nagel, L. W., SPICE2: A Computer Program to Simulate Semiconductor Circuits, Dissertation, ERC-M 520, University of California, Berkeley, CA (1975).

[41] Nebel, W., CAD-Entwurfskontrolle in der Mikroelektronik -mit einer Einfuehrung in den Entwurf kundenspezifischer Schaltkreise (german) (B. G. Teubner Verlag, Stuttgart, 1985).

[42] Newell, M. E.; Fitzpatrick, D. T., Exploitation of Hierarchy in Analyses of Integrated Circuit Artwork, IEEE Transactions on Computer-Aided Design of Integrated Circuits and Systems, Vol. CAD-1, No. 4, (1982).

[43] Ousterhout, J. K.; Hamachi, G. T.; Mayo, R. N.; Scott, W. S.; Taylor, G. S., Magic: A VLSI Layout System, IEEE 21st DAC, Albuquerque, NM (1984).

[44] Parker,A., Thomas,D., Siewiorek, D., Barbacci, M.R., Hafer, L., Leive, G. and Kim,J., The CMU Design Automation System, IEEE 16th DAC, San Diego, CA (1979) 73-80.

[45] Saab, D.; Hajj, I., A Logic Expression Generator for MOS Circuits, ICCC '82, New York, NY (1982).

[46] Shiva, S.G., Computer Hardware Description Languages - A Tutorial, Proceedings of the IEEE vol. 67 no. 12 (1979) 1605-1615.

[47] Sibbert, H., DOMOS: A Nonlinear Transient Simulation and Optimisation Program for Integrated MOS Circuits (Version 7/5), User Manual, Dortmund University, Lehrstuhl Bauelemente der Elektrotechnik, Dortmund, FRG (1982).

[48] Siewiorek, D.P., Introducing ISP, IEEE Computer vol. 7 no. 12 (1974) 39-41.

[49] Son, K.; Kishimoto, Z., A Formal Verification Method for Hierarchical VLSI Designs, IEEE ICCD, VLSI in Computers, Port Chester, NY (1983).

[50] Takashima, M.; Mitsuhashi, T.; Chiba, T.; Yoshida, K., Programs for Verifying Circuit Connectivity of MOS/LSI Mask Artwork, IEEE 19th DAC, Las Vegas (1982).

[51] Tarolli, G. M.; Herman, W. J., Hierarchical Circuit Extraction with Detailed Parasitic Capacitance, IEEE 20th DAC, Miami Beach (1983).

[52] Ullmann, J. R., An Algorithm for Subgraph Isomorphism, Journal of the ACM, Vol. 23, No. 1 (January 1976) pp. 31-42.

[53] Veiga, P.M.B. and Lanca, M.J.A., Harpa: A hirarchical multi-level hardware description language, IEEE 21st DAC, Albuquerque, NM (1984) 59-65.

[54] Vladimirescu, A.; Zhang, K.; Newton, A. R.; Pederson, D. O.; Sangiovanni-Vincentelli, A., SPICE Version 2G.5 User's Guide, Dept. EE/CS, University of California, Berkeley, CA (1981).

[55] Wagner, T. J., Hierarchical Layout Verification, IEEE Design & Test of Computers, Vol. 2, No. 1 (1985).

[56] Watanabe, T.; Endo, M.; Miyahara, N., A New Automatic Logic Interconnetion Verification System for VLSI Design, IEEE Transactions on Computer-Aided Design of Integrated Circuits and Systems, Vol. CAD-2, No. 2 (April 1983).

[57] Waxman, R., The Many Languages of Electronic Computer-Aided Design, IEEE ICCD: VLSI in Computers (1984) 74-77.

[58] Welters, U., Graphic Hardware Description Languages, in: Hartenstein R. (ed.), Advances in CAD for VLSI: Vol. 7 Hardware Description Languages (North-Holland, Amsterdam, 1987).

[59] Williams, L., Automatic VLSI Layout Verification, IEEE 18th DAC, Nashville (1981).

[60] Wodtko, A., RT-Languages in Goal-Oriented CAD Algorithms, in: Hartenstein, R. (ed.) Advances in CAD for VLSI, Vol. 7, Hardware Description Languages (North-Holland, Amsterdam, 1987).

HARDWARE DESCRIPTION LANGUAGES
R.W. Hartenstein (Editor)
© Elsevier Science Publishers B.V. (North-Holland), 1987

4.2. SILICON COMPILATION FROM HDL AND SIMILAR SOURCES

M. GLESNER, H. JOEPEN, J. SCHUCK and N. WEHN

Fachgebiet Halbleiterschaltungstechnik, Technische Hochschule Darmstadt
D-6100 Darmstadt, F.R.G.

Abstract

Using the implementation of ALGIC silicon compiler system as a demonstration example this work tries to give an overview on the topics related to the development of "Silicon Compilers": Description Languages/Compiler Construction, Architecture Construction, Floorplanning, Automated Layout Generation, Cell-Libraries and Verification. Components necessary to build such a general system are described. Special emphasis is given to the description of digital signal processing systems.

1. Introduction

The design of an integrated circuit may be specified on different levels of hierachy spanning from high level representation down to low level mask geometry. In the traditional design strategy the designer specifies the function of the chip on a higher level. By a series of successive refinements the design is ultimately transformed to a low level format. If the successive transformation steps are performed by hand, the design procedure is tedious, error prone and time consuming. Silicon compilers will in the future be capable to automate the design process thereby reducing the costs and time of a design dramatically. In its ultimate form Silicon compilers will accept a behavioral specification and generate the geometrical representation of a high performance chip layout. Such tools allow the highest degree of automation in the design process.

N.B. The work on the ALGIC Silicon Compiler System has been sponsored by the European Commission, Task Force Information Technology, under the Microelectronics Regulation in the CAD-VLSI for Telecommunications project (CVT project, subtask 1.8)

A classification of a Silicon compiler system can be done according to the following criteria which every general automatic layout genereation tool has to consider:

- Chip description level and the corresponding input language
- Target architecture
- Transformation strategy from the description level to topological level
- Component layout generators and predesigned cells
- Placement and routing strategies
- Data representation
- Design rules and technology independence.

The Bristle Block System /1/, orginating from the Ph.D.-thesis work of D. Johannsen at Caltech, is one of the earliest approaches on Silicon-compilation. Johannsen introduced a block-concept, where cells are treated in a first phase as hierarchical structured black boxes and where only sizes and external connection positions (bristles) are known. A fixed target architecture has been chosen and the user has to describe his system in terms of target architecture components.

A more powerful approach is the MACPITTS-system, a layout-generation tool developed by Siskind, Southard and Crouch at MIT Lincoln Laboratory /2/. MACPITTS and its successor METASYN use a LISP-based functional design language which is interpreted by a LISP program.   The system consists also of several component generation tools and is able to produce layout on a relative flexible target-architecture floorplan.

ARSENIC, an experimental Si-compiler study at the University of Illinois /3/ uses a hierarchical recursive method to derive the complete system layout by stepwise refinement. For example missing informations on a system block level below and - if it is also not specified in the next level below - the reference mechanism will descend in the tree-structure until the derived information is found.   Decisions between alternative refinements are taken by a look-ahead strategy.

The FIRST Silicon-compiler /4/ has been developed by Denyer, Renshaw and Bergmann at University of Edinburgh and is a tool constrained to the field of bit-serial architectures in digital signal processing.   The input language consists of a fixed set of operators and produces layouts on a fixed floorplan.

Although today still in a development phase some Si-compilers have moved out of the research laboratories into commercial application. However acceptance from industry still remains low.

2. Overview of the ALGIC Silicon Compiler System

The structure of silicon compiler systems can be splitted into three fundamental functional units:

- the input language processor
- an "intelligent" control module and
- the layout generation module(s).

The power of the system basically depends on the complexity of the implementation of these blocks. Usually the Si-compiler systems are either restricted to special fields of applications and thereby realizing efficient layouts, or they are trying to cover a more general scope of application resulting in poor performance chips.

In the ALGIC system /5,6/ we try to overcome with these restrictions on the one hand by using a general PASCAL-like algorithmic language as an input of the system and on the other hand by using sophisticated tools like an optimizing PLA-generator or parameterizable operator cells (multipliers /7/, adder, etc.) to produce dense high performance layouts.

The ALGIC input language offers to the user the full range of algorithmic statements and of block structuring known from PASCAL combined with a declaration part specific for digital systems. So the behaviour of a system can be specified by writing a PASCAL-like program, taking no care of hardware implementation constraints. Only a declaration part forces the user to specify his system in terms of ports, registers or terminals and their input/output attributes. The hierarchical modularization of complex system is supported in the language using block structuring and procedure/function concepts. The set of language constructs of the ALGIC silicon compiler is summarized in table 1.

DEF ... ENDDEF	GOTO
LABEL	IF ... THEN ... ELSE
CONST [ARRAY]	CASE ... OF ... END
PORT [IN,OUT,INOUT,TRISTATE] [EXT]	WHILE ... DO
REG [ARRAY]	REPEAT ... UNTIL
PROCEDURE [LIB]	FOR ... (TO||DOWNTO) ... DO
FUNCTION [LIB]	Operators:
BEGIN ... END	relational: = <> < > <= >=
PARBEGIN ... ENDPAR	logical: OR NOR AND NAND NOT EXOR NEXOR
CONBEGIN ... ENDCON	arithmetic: + - *

Table 1: Summary of ALGIC language constructs

Especially for the description of dedicated VLSI digital signal processing circuits the design system offers to the user special features to support an efficient generation of the signal and control path of the circuit. This includes the possibility to reference external procedures and functions (modules from the flexible blocks library), e. g. to select between several forms of multiplication, to specify the corresponding circuit implementation (e. g. pipelining) and the wordwidth of signals to be processed . Furthermore a separate set of statements can be used to include user specified controller into the system to support the description of dedicated control modules for special DSP architecture concepts.

The overall structure of the system is shown in fig. 1. It mainly consists of the language processing block (scanner, parser), the synthesis tool block (PLA generator, Weinberger-array generator, cell library), the floorplanning block (placement, routing, floorplan compaction), the verification block and the central architecture construction block which controls the actions of the subsiding tools.

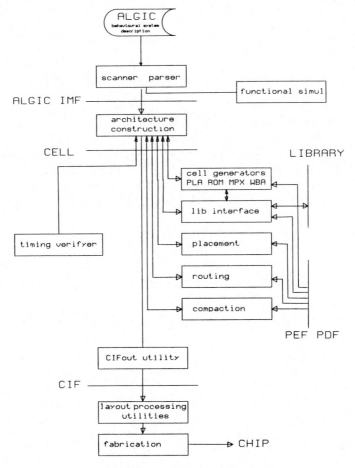

Fig. 1: Structure of the ALGIC System

3. Specification Languages

3.1 Levels of Chip Description

To implement the front-end of a layout generation tool we may choose between several possible levels of chip description. The optimal description level should allow the description of detailed chip-characteristics on the lower end and should also be powerful enough to describe complex systems in an efficient way.

Basically, we can find the following possible description levels for VLSI-systems:

- functional
- algorithmic
- behavioural
- architectural (structural)
- logical
- topological.

Clearly, the logical or topological level are not suited for complex system descriptions. A functional description specifies the output of the system to be designed in terms of its inputs. An example for functional description is the class of PLA and finite-state-machine languages, describing systems in terms of its inputs, outputs and state-transitions. Only relative simple transformations are necessary to derive the structure and the components of a system from the functional level.

An algorithmic description specifies a system in a "computable" form (e. g. programs in block-structured languages like PASCAL,ADA, etc.). From this abstract form we have to derive a system that performs in a way (or can replace) the given program or routine. However, there are several degrees of freedom in the way of system realization (e. g. parallel, serial or in a mixed form) and difficulties in assigning system inputs and outputs to program-variables of abstract types.

A structural (architectural) description describes the system as a set of components (modules or blocks) of known functional description and their interconnection. Of course the components themselves may be sub-systems, described again by components of lower levels. Each component-level can be characterized by the functional level of its components.

For our purpose we want to define the behavioural form as an algorithmic description which is interpreted in a application-specific context as it is quite common today. For example the FOR-statement, generally interpreted as a fixed number of repetitions of a statement or a sequence of statements by incrementing a loop-variable, may describe in a layout generator a special hardware structure like a counter, whose overflow-detect is controlling a part of the system. Thereby the degrees of freedom contained normally in algorithmic descriptions are reduced.

According to this we have defined the ALGIC language which consists out of a general behavioural part and the DSP specific part which allows a functional description.

When we are defining the ALGIC-language as close to PASCAL as possible we especially hope to achieve an easy acceptance of the new language because the learning effort could be kept on a minimum.

3.2 General Behavioural Part of the ALGIC Language

In a conventional computer programming language we can find two essential language functions: a description of the data types and elements in a declaration part and a description of actions which are to be executed after compilation on the underlaying machine. In contrast we can locate in the language for Si-compilation the following functions: the description of a system-frame, the actions on the data held in the frame and the control of these actions.

In ALGIC we want to support the description of fully synchronous digital systems and therefore we include registers and arrays of registers (RAM's) in the chip-frame declaration. Other explicitly declared frame-elements can be internal or external ports, constants and arrays of constants (ROM's).

The actions on data in the frame are described by forming expressions and assigning them to declared registers or connection points. The repetetive and conditional statements used to specify the algorithm are implicitly containing the information how to control the actions on data. This control information is used to construct the controllers in the system realization.

In the following a list of the language constructs and their features is given:

PROGRAM
    program identifier and global parameters input; opens a
    chip frame;

DEF...ENDDEF
    declaration-part enclosure.

PORT [IN/OUT/INOUT/TRISTATE] [EXT]
    definition of global and local bus-connections will be
    helpful to route busses in an efficient manner.

REGISTER [ARRAY]
    definition of internal storage-elements; the array-
    attribute will be used to define relative small RAM's e.g.
    used for storing microprogram instructions.

CONST [ARRAY]
>   definition of constant values; depending on the program
>   context two kinds of constants are possible:
>>   a) physical constants to control execution; these constants are
>>   used as binary input-values on the generated chip (realized
>>   as a combination of VDD- and GND-lines);
>>   b) algorithmic constants to control execution; these constants
>>   will implicitly be contained in the extracted controllers
>>   (e.g. number of states in a state diagram).
>   The ARRAY-attribute in this case will be used
>   to define small ROM's,e.g. necessary to store
>   coefficients to perform digital filter algorithms.

BEGIN...END
>   execution block; it will also in the floorplan be realized in a local
>   block with its own controller and data-path.

PARBEGIN...ENDPAR
>   this construct is used analog to the MACPITTS par-construct if it is
>   necessary to express explicitely parallel execution;   in this way
>   together with the statement sequencing it is possible to include explicit
>   control information e.g. concerning external lines (read-/write-cycles
>   of an external memory) in the description.

<statement> ; <statement>
>   sequence of statements; it implies the execution-sequence (→ control -
>   extraction);   if   the   analysis of the internal data-flow detects
>   operations which could be executed in parallel, this sequence might be
>   changed.

IF...THEN...ELSE...
>   comparison of two binary-vectors in the control-expression generates a
>   control-signal (inverted and non-inverted) which activates the local
>   controllers of the alternative execution-blocks.

CASE...OF...ELSE...
>   analog to IF-statement with an n-input-comparator;   parallel or serial
>   comparator-realizations are possible.

WHILE...DO...
>   internally the WHILE clause can be transformed into a IF-GOTO-loop
>   structure.

REPEAT...UNTIL...
>   internally the REPEAT-clause can be transformed into a IF-GOTO-loop
>   structure.

GOTO

    branch to the state marked by <label> in the actual block (→ no branches between different controllers).

FOR...TO...DO

    The general form of the FOR-loop (variable boundaries) installs two additional registers to be loaded with the actual values of the upper and lower bound. In a simplified form: utilizing a counter to start an execution-block n times (→ decrement counter and restart controller until counter equals zero).

PROCEDURE

    in fact it installs a new frame for a subsystem; calling mechanism (formal contents, ports and terminals); in contrast to a chip-frame on the program declaration- level no EXT - attributes must be used and so path analysis methods will be applicable over the local frame borders.

FUNCTION

    analog to PROCEDURE; the function result port attributes are implicitly PORT OUT; predeclared functions are: lshift, rshift, neg, abs,...; external parameterizable function blocks can be referenced by using the keyword LIB.

3.3 DSP-Specific Part of the ALGIC-Language

    A special part of the ALGIC-language is intended for the description of dedicated DSP circuits.

    Starting from the algorithm which specifies a certain DSP function or filter important constraints are involved for the choice and definition of the architecture for the customized DSP circuit. In fact a lot of constraints and important influences have to be observed like

- flexibility (special or general purpose)
- optimability (max. throughput min. delay, heuristics)
- data structure (bit-seriell/bit-parallel)
- data processing (pipelining, parallelism, sequential)
- architectural units
- speed (depends on task),etc..

Because of these contraints on automatic translation from any signal processing algorithm to the corresponding architecture is restricted. Therefore the designer has to specify the architecture with regard to these constraints and this requires the full creativity and experience.

The translation of the architecture into the corresponding DSP circuit layout is the task of the ALGIC system. For that task the architecture and the DSP circuit has to be described on a hardware description level as an input to the silicon compiler system. Therefore a special part of the ALGIC language is intended for a structural and functional description of dedicated DSP circuits.

For DSP-specific application the following additional language constructs are offered:

TERMINAL
   The TERMINAL declaration part allows the declaration of local single lines and vectors. In this context a couple of lines or vectors can be referenced by one single name and an index selection. In this way the cell interconnections of the data path can be described in a simple manner.

SET....TO/DOWNTO....DO
   This statement is used to build up complex cells by abbutting simple cells like registers or 1-bit adders. The actual values of the loop variable reference to certain local lines and interconnections of the cells.

IF....THEN....ELSE
   This statement is used as a perfect mechanism for specifying the synchronisation of the system or for describing the function of multiplexers or demultiplexers. The statement only allows the description of transmission gates. For example, the statement IF a THEN b:=c specifies a transmission gate with source c, drain b and gate a. Optionally the ELSE part can be included or not. If the ELSE statement part is used it is assigned to the most inner IF statement.

ON....DO....WITH CLOCK
   The ON statement can be used if more than two alternatives must be specified. The ON statement allows the specification of a list of alternatives each alternative marked by one or more ON selectors. The ON statement is highly suitable for describing the states of a ROM to be programmed or the function of a PLA and finite-state machine (FSM). The ON conditions on the left side of the statement indicate the input values of the controller. If the function of a FSM is described, actual states are assigned, too. The actions on the right side of the statement indicate the output values (including the next states, if the function of a FSM is described). For those conditions which are not specified the ELSE clause determines the actions. Optionally the WITH CLOCK clause of the ON statement can be used. It specifies the clock phase of the inputs and outputs of the controller.

4. Architecture Construction

The result of the evaluation of the input language text generally will be the description of the system-architecture in an intermediate form.
But there are several restrictions which make it impossible to generate VLSI-systems of any arbitrary architectural structure:

- limited description possibilities from restricted input
  language grammars,
- insufficient, fixed transformation algorithms from special
  purpose input languages to a system topology,
- limited number of basic operator cells and component generation tools,
- difficulties in implementation of general placement- and
  routing-algorithms.

Therefore, most Si-compiler approaches meet these limitations by choosing a target architecture either fixed or built up on a special scheme.

Up to now only very few attempts are made to give a formal definition for the transformation from a behavioural description level to the topological (architectural) level.

For most of the considered concepts this formal definition would be quite simple because every input language construct denotes a special component of the target architecture (e. g. FIRST). So the number and types of components are fixed and the only degree of freedom will be the placement of the components on the floorplan scheme to minimize wiring effort and to increase performance. Difficulties arise in finding a formal transformation definition, when several degrees fo freedom are introduced:

- free placement of architecture components (floorplan construction);
- derivation of a control-part, implicitly contained in the system
  description (e. g. state-allocation);
- alternative realization possibilities for a system component
  (e. g. implementing Boolean expressions in a Weinberger-array or
  in a PLA);
- no direct semantic context existing between input language constructs
  and target architecture components.

With respect to these demands we have developed and implemented a general architecture construction module for the ALGIC system which is able to derive very flexible architectures (fig. 2).

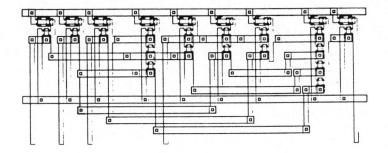

Fig 3:   Generated example Weinberger array

5.1.3 ROM Generator

The ROM generator is capable to generate the memory devices according to the
programmed states which are user defined. Furthermore the partition of the ROM
structure to be generated can be controlled, e. g. word-organized, bit-organized
or mixed-mode ROM arrangement.

To increase the performance and to minimize the power consumption and silicon
area the ROM structure has been realized as a dynamic arrangement with two-phase
non-overlapping clock-rates Phi1 and Phi2. Figure 4 shows the transistor circuit
of a part of the ROM. The decoder bases on a simple NOR-structure. With clock
Phi 1 the word-lines are grounded. In addition the bit-lines are precharged to
3.5 V. During clock Phi 2 the selected word-line is loaded to more than 5 V which
is enabled by the bootstrap circuitry . The bit-transistors are activated by the
relevant word-line and the corresponding bit-lines are discharged.

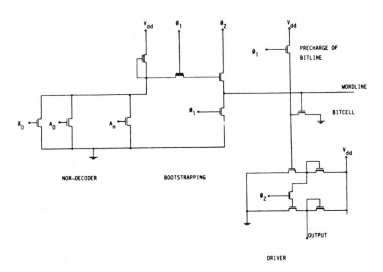

Fig. 4:   Reduced ROM-circuit

5.1.4 Multiplexer Generator

Another important component generator within the silicon compiler system ALGIC is the multiplexer generator. The switches of the multiplexer are realized by a passtransistor logic. The layout generation is controlled by a set of parameters which specify

- the number of input busses
- the number of bits per bus
- the input-to-output switch control
- the input drives for the control-lines
- Optionally the output drives to refresh the output signal

5.2 Parameterised Cells

In the past a lot of parameterised cells have been designed, for example simple gates, different adders, registers (static, dynamic), double pointer hardware stack, sign inverters, buffers, counter, RAM, etc.. Especially for high performance custom design DSP circuits different multipliers have been designed for a wide range of applications. Multipliers are required for bit-serial and bit-parallel processing, for any precision and speed depending on the application. Other variables are the desired wordwidth of signals, pipelining processing, rounding or truncation of the product. Based on a hierarchical cell design methodology parameterised multipliers for the automatic layout generation have been designed for various applications /12/:

a) For bit-serial processing a parameterised fully pipelined bit-serial multiplier has been developed. Its most important features are fully pipelined processing, bit-parallel input of the coefficient, bit-serial processing of signal, wordwidth n of signal independent of wordwidth k of the coefficient ($n=k$, $n<k$, $n>k$), rounding of the product during processing, result and input signal represented in two's complement, in-to-out delay equal (k+1) clock cycles, computation of the product in n clock cycles.

b) For bit-parallel processing a fixed coefficient multiplier, a pipelined guild array multiplier and a bit-parallel multiplier have been designed as parameterised cells:

- The first one multiplies a two's complement input signal with a fixed coefficient by adding and hardware-shifting the input signal, CSD coding is used to reduce the number of additions.
- The most important features of the parameterised pipelined guild array multiplier are a very efficient floorplan for the realization of bit-parallel pipelining multiplication, two's complement representation signal and coefficient, various amount of pipelining (latch layers) in the array, optionally addition of other functions (e. g. pipelined adder) and a small number of elementary cells.

- The last one, the two's complement bit-parallel multiplier, has been
  developed with respect to the following demands: Selection of an efficient
  multiplication algorithm for two's complement representation input vectors,
  selection of a suitable type of parallel array multiplier, definition of an
  efficient architecture and a minimization of the number of elementary cells.

As an example figure 5 shows the block structure of a 6x4 bit-parallel
multiplier. It consists of only 5 different kinds of elementary cells (exor-gate,
super- buffer, half- and full-adder, and-gate) which are assembled to the
sub-cells PMi and PMKj (1≤i, j≤12) to achieve an efficient parameterised
multiplier structure for a simple layout generation. The sub-cell representation
is shown in figure 6. The general construction rules for the automatic layout
generation of a n x k bit-parallel multiplier can be described line by line:

```
PMK1        (n-4)PM2                                   PM3
------------------------------------------------------------

            for i = 1 to k - 3 do

PM4         (i-1)PM5        PMK5        (n-i-4)PM5   PM6
------------------------------------------------------------

PM7         (k-3)PM8        PMK10       (n-k-1)PM8   PMKX
```

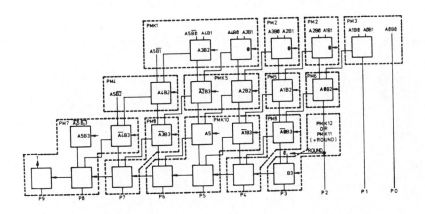

Fig. 5: Building Block of a two's complement 6 x 4 Bit Parallel Multiplier

By abbutting the sub-cells according to the construction rules the layout of
two's complement bit-parallel multipliers can automatically be generated for any
desired wordwidth of input signals without using placement and routing techniques.
There is no restriction with regard to wordwidth of multiplicand or multiplier
(n=k, n≠k; n,k=wordwidth of the two input signals). Additional the product can
be rounded to the wordwidth n or k respectively. After generation the final
layout will be a good result with regard to silicon area, power consumption and
computing rate. As an example figure 7 shows a generated layout of a 7x5 bit
two's complement multiplier.

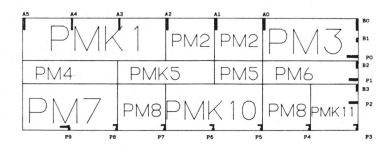

Fig. 6: Sub-Cell Representation of a Two's Complement 6 x 4 Bit Parallel
        Multiplier

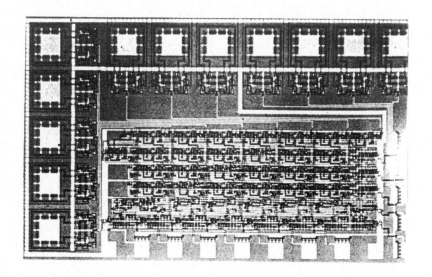

Fig. 7: Chip photograph of parallel multiplier

6. Cell Library and Process Definition File

One essential idea of the ALGIC-system is based on the fact that the geometric layout of the elementary or sub-cells and all parameters depending on the technology and actual process are strictly seperated from the ALGIC-compiler tools like the cell layout generation, verification and floor- planning tools. This seperation has the evident advantage that new designs or redesigns of cells and any change of the electrical process parameters, geometric design rules etc., does not lead to a modification and recompiling of the implemented software. The separation has been realized by the

- implementation of a process definition file (PDF)
- implementation of a powerful cell library.

In the process definition file all necessary parameters of the actual process are stored like e. g. the number of mask levels and the geometric design rules. The floorplanning tools and the timing verifier have access to the PDF.

In the cell-library all the geometric and electrical data of cells of all hierarchy levels are stored to provide the cell generation and verification tools with the necessary information. In the cell library we can distinguish between 4 types of data:

- geometric layout of the elementary and sub-cells
- construction rules for the generation of the parameterised and macro-cells
- electrical properties of the cells
- cell organisation

The geometric layout is represented in a simple CIF-like format. The electrical properties of the cells include the delay time of the cell, the power consumption, the output drive capability, the input capacitance, etc.

The cell organisation includes cell type, size, pin position, clocking etc.. All the cell data are available to the cell operator handler, the component layout generators, the floorplanning tools and the timing verifier. All these tools communicate with the cell library via the library interface.

7. The Timing-Verifier

Locating the critical pathes concerning the timing of a large digital systems is a great problem to the designer due to the complexity of the circuit and the often unforeseen influence of the delaytime behaviour of the interconnections between its elementary cells. The SCAT timingverifier helps the designer doing so by estimating the necessary minimum clock period of the circuit for the worst and nominal case and by identifying those pathes determining the clock period.

The basic algorithm of SCAT is blockoriented and uses two passes for the handling of the internal graph. In the first pass, it is traversed in the direction of the signal flow, calculating the arrival- time of the signals at each internal note. With this information, the verifier is now able to calculate the slacktime of each node traversing the graph in the opposite direction. The slacktime is hereby defined as the amount of time by which a single node can be delayed without affecting the minimum clockperiod. This approach is commonly known as the "Critical Path Method", which has been used for the first time in an Operations Research program named PERT /13/ but it has been extended in several points. The first difference between PERT and SCAT is that SCAT is able to model the effects on the delaytime elements, which occur if the underlying elementary cells of the circuit are serially or parallely connected. The second difference is that SCAT is able to handle clocked system.

8. Floorplanning in a Silicon Compiler

Due to the different demands to a floor-planner in a Silicon-Compiler environment a very flexible approach is used. First of all different placement methods in a hierarchy structure are provided by the system (bottom-up, top-down methods) to allow a flexible placement process. In contrast to other floor-planners our system doesn't require a fixed set of module dimensions, modules can be specified with a set of possible rectilinear dimensions giving more flexibility to the module generators.

During the placement process relationships among area, connections, module dimensions and orientations are considered. The fundamental key to this approach is a binary tree, in which every node represents a cut in the layout where the leafes of the tree represent modules. This datascheme leads to a slicing structure of the overall layout. A sequence of different traversals are performed on the tree during the placement and routing process in order to determine the possible module dimensions, positions and routing spaces. By using weights, certain nets can be specified to be critical indicating that this length should be taken preferable into account during the floor-planning process i.e. for electrical reasons.

Furthermore it is possible to define relations between modules (left, right, top, bottom) which are also considered during the floor-planning. The advantages of the binary tree approach are significant: Layouts based on this structure can easily be routed since cyclic conflicts are eleminated. Routing channels correspond to inner nodes of the tree, thus channel widths can be accurately calculated via global routing. The channel order is determined by the depth of the corresponding node in the tree.

Structure manipulations can be performed by fast treetraversals. Before starting the placement process, a cluster generator is invoked which produces module clusters according to the net weights and relation constraints. The system provides two different methods to place modules inside a cluster and the clusters itself which are also treated as modules.

First we employ a min-cut approach which produces for each partitioning step two nodes in the binary tree.At the beginning the tree consists only of one node - the root containing all modules. Therefore this approach differs from previous efforts. The bipartioning step is repeatedly applied to the tree with changed direction until each remaining subset contains only one module and therefore forming the leafs of the tree. For the partitioning of a set in two subsets, a heuristic scheme according to Fiduccia and Mattheyses is used in order to minimize the number of nets crossing the virtual cutline which is represented by a node. Normally this partitioning leads to a local optimum with respect to the two subsets being partitioned. In our approach based on the treescheme a global optimum can be achieved since there is exact information about the location of modules which are not included in the two actual partitioned subsets via modules included in these subsets. Therefore the bipartitioning can be performed in an environment represented by the tree which reflects its actual location on the floorplan. As a last step an improvement process is performed which tries to optimize the overall area by rotation and local module changes.

As second method a bottom up placement scheme is provided by the system which places 'best' module by 'best' module successively on the 'best' place. The criteria if a module or a place is 'best', are controlled by a parameter set allowing a flexible placement process. These two placement-schemes can be mixed applied to the cluster levels. Summarized the placement step results in a binary tree which provides for every module a possible shape. If the dimensions of a module is exactly specified, a treetraversal is performed to specify the correct position of a module in the floorplan.

The routing process of the floor-planner is divided into four steps. First a channel definition is performed by a trivial treetraversal which assigns a channel to every node. As next step the VDD and GND lines are routed. This step uses also the advantages of the tree scheme. The line widths are varied according to the current-flow on these lines. Furthermore VDD/GND connections inside of modules are considered to allow regular structures i.e. in register chains. The global routing step is also simplified by the tree because each channel corresponds to a node.

Due to the NP-complexity of this problem we employ a method which is based on a heuristic Steiner tree search to determine minimum net lengths. Based on a stack technique, nets with equal netweigths are processed simultaneously in the Steiner tree search. Penalty functions for every channel avoid bottlenecks and provide equilibral distribution. As last step a greedy channelrouting scheme is performed on every channel, where the channelorder is determined by the depth of the corresponding node in the tree.

9. Example on ALGIC

To illustrate the use of the behavioural part of the ALGIC-language we introduce a short example program "magnitude approximation". This algorithm has also been described in terms of input languages of the MACPITTS and the FIRST silicon compiler system. Altough this example demonstrates only very few language elements it is simple enough to show the complete construction process from the ALGIC description to the topological optimized controller layout.

The hardware generated from this program should calculate an approximation for the expression $\sqrt{a^2+b^2}$ using the following algorithm:

1. Set g = max (abs(a), abs(b))
   and l = min (abs(a), abs(b))
2. Then a + b will be approximately
   max (g, 7*g/8, 1/2)

The following program is the complete ALGIC-description of the GENMAG example:

```
PROGRAM genmag;
DEF
    CONST
        msb = 3
    ENDCONST;
    PORT
        IN EXT      a [0..msb],
                    b [0..msb];
        OUT EXT     result [0..msb]
    ENDPORT;
    REG
        aab   [0..msb],
        bab   [0..msb],
        g     [0..msb],
        l     [0..msb],
        sqs   [0..msb]
    ENDREG
ENDDEF
BEGIN
    IF a[msb] THEN
        aab := -a
    ELSE
        aab := a;
    IF b[msb] THEN
        bab := -b
    ELSE
        bab := b;
    IF aab > bab THEN
        PARBEGINN
            g := aab;
            l := bab
        ENDPAR
    ELSE
        PARBEGIN
            g := bab;
            l := aab
        ENDPAR;
    sqs := g - lshift (3,g) + rshift (1,1);
    IF sqs > g THEN
        result := sqs
    ELSE
        result := g
END.
```

First in the program declaration part a constant "msb" was declared to indicate the wordlength. Note that constants can be used in two different contexts:

- computational: the constant "msb" is used only to
    substitute the approxriate wordlength anywhere
    in the source program text.
- generative: if e.g. a statement
    <u>IF</u> a = msb <u>THEN</u> <u>BEGIN</u>...<u>END</u>;
    occurs in the program the constant "msb" would
    refer to the generated system and for this
    example the binary value of "msb" will be connec-
    ted together with port "a" as inputs to a
    comparator physically.

The constant declarations are followed by the port and the register declaration part. The ports are declared with the appropriate signal direction attribute and are marked as lines running from or to the chip by using the keyword <u>LIB</u>. Internal registers are declared to buffer the intermediate results between the single operations resulting in a sequential architecture.

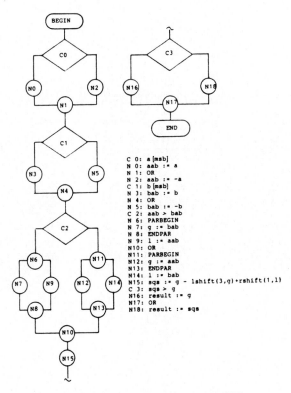

```
C  0: a [msb]
N  0: aab := a
N  1: OR
N  2: aab := -a
C  1: b [msb]
N  3: bab := b
N  4: OR
N  5: bab := -b
C  2: aab > bab
N  6: PARBEGIN
N  7: g := bab
N  8: ENDPAR
N  9: l := aab
N10: OR
N11: PARBEGIN
N12: g := aab
N13: ENDPAR
N14: l := bab
N15: sqs := g - lshift(3,g)+rshift(1,1)
C  3: sqs > g
N16: result := g
N17: OR
N18: result := sqs
```

Fig. 7a: Precedence-graph of GENMAG

The program body consists out of a sequence of IF-statements and will need no
further comments. As shown in the example possible control-expressions are
testing of single lines and magnitude comparison of line vectors. Comparing
results of arithmetic expressions is also possible. Note also the
parallel-statement (PARBEGIN...ENDPAR) which is used to load the g- and l-register
in the same clock-cycle.

Figure 7a shows the precedence graph generated from the input program. From
this structure the state tables of the transition diagram in figure 7b have been
evaluated by the controller construction algorithm. Putting in the state-tables
to the ACARF PLA/FSM generator results in the layout shown in figure 7c.

A remarkable feature of our algorithm is the partitioning of the control path
into one 20-state main controller and 11 small subcontrollers having only 3 to 5
states.

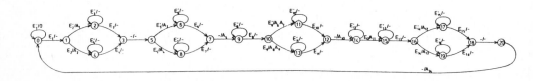

Fig. 7b: Transition diagram of the GENMAG main-controller

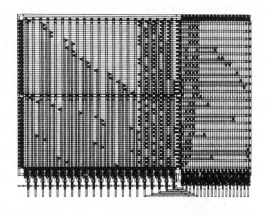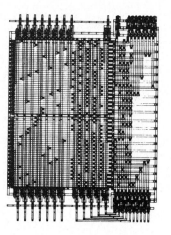

Fig. 7c: Resulting PLA/FSM layouts

10. Summary and Outlook

We have extensively discussed the basic concepts and ideas of a flexible Silicon compiler.

Starting from a behavioural input description level we have shown the different modules which are necessary to transform the source text down to the layout level. The ALGIC system developed at TH Darmstadt has been introduced to illustrate the transformation process and the whole range of layout generation tools.

The ALGIC system is already running in an experimental version and is able to generate chip-layouts for several examples both from DSP-specific and general descriptions.

References

/1/   D. L. Johannsen:
      "Bristle Blocks: A Silicon Compiler",
      Proc. 16 th Design Automation Conference June 1979, pp. 310-313

/2/   J. Siskind, J. R. Southard, K. W. Crouch:
      "Generating Custom High Performance VLSI-Design from Succinct
       Algorithmic Descriptions",
      Proc. Conference on Advanced Research in VLSI, MIT, January 1982, pp. 28-40

/3/   D. Gajski:
      "The structure of a Silicon-Compiler",
      IEEE Proc. on Design Automation Conference, 1982, pp. 272-276

/4/   P. Denyer, D. Renshaw, N. Bergmann:
      " A Silicon Compiler for VLSI Signal Processors",
      ESSCIRC Proceedings, Brussels, 1982, pp. 215-218

/5/   H. Joepen, M. Glesner:
      "Architecture Construction for a General Silicon Compiler System"
      Proc. of ICCD 85-Conference, Port Chester, N.Y., October 1985

/6/   M. Glesner, J. Schuck, H. Joepen:
      " A Flexible Silicon Compiler for Digital Signal Processing Circuits",
      ICCD, New York, October 1984

/7/   M. Glesner, J. Schuck:
      " Layout Generation for Multipliers in VLSI-Digital Signal Processing",
      ESSCIRC, Sept. 1984, Edinburgh

/8/   M. Wei, H. Sholl:
      " Extraction and Parallelism in Control Structures"
      IEEE Trans. on Computers, Vol. C-31, No. 9, Sept. 1982

/9/   W. Grass:
      "Steuerwerke: Entwurf von Schaltungen mit Festwertspeichern"
      Springer, Heidelberg, 1978

/10/ J. Southard:
      " MACPITTS: An Approach to Silicon Compilation"
      Computer-Magazine, December 1984

/11/ N. Bergmann:
      " A Case Study of the First Silicon Compiler"
      Proc. 3rd Caltech Conference on VLSI, Pasadena, March 1983

/12/ S. Meier, H. Schlappner:
      " Entwurf von Multiplizierern fuer die Digitale Signalverarbeitung",
      Studienarbeit TH Darmstadt, Institut fuer Halbleitertechnik, December 1983

/13/ T. Kirkpatrick, N. Clark:
      "PERT as an Aid to Logic Design"
      IBM Journal, March 1966, pp. 135-141

4.3. TEST PATTERN GENERATION FROM HDL AND SIMILAR SOURCES

Stephen Y.H. SU and Tonysheng LIN

Research Group on Design Automation and Fault-tolerant Computing
Thomas J. Watson School of Engineering, Applied Science and Technology
State University of New York, Binghamton, New York 13901, U.S.A.

Basic concepts in testing and its relation to hardware description
language (HDL) are introduced. The relationship between physical and
functional faults is presented. Four HDL-driven test generation methods are described and illustrated with simple examples. A brief
discussion on some related recent work in VLSI testing is given.
Current trends and future development on test pattern generation
are addressed with open problems.

4.3.1 INTRODUCTION

The increasing complexity of integrated circuits has made fault testing one of
the most important problems in digital system area. The major difficulty is that
conventional gate level testing techniques are not suitable for testing VLSI (very
large scale integration) circuits. Functional testing using test pattern generated
from HDL is one of the promising solutions for today's and future VLSI testing problems.

Digital systems can be tested at various levels. In a typical top-down Computer-Aided Design (CAD) procedure, a digital system is designed starting from the
specification of system behavior. Then, repetitive refinements of the higher-level
specification are performed level by level down to the level of actual implementation. Testing is a procedure for finding out whether the system has any abnormal
behavior. It is an important step which can be (and should be) considered during
each level of system design. A good designer always considers the testing issue in
the very beginning of his design not only because testing is a difficult job and
very expensive after manufacturing, but also that an earlier consideration on testing can show the quality of the present design (e.g., whether the system-under-design
has low testability or even untestable) with valuable clues for a better redesign.

In a CAD process, test patterns corresponding to each level of refined functions can be generated with respect to a pre-established fault model long before
the system is manufactured. The simulation results using such test patterns may
assure that the system under development is really well-designed with good testability.

In reality, testing is a generic term that covers multiple activities during
the life of a digital system. As far as design correctness is concerned, in design

stage, we may perform testing for <u>design verification.</u> During the production stage, we perform <u>fabrication testing</u> to discover any component defects or assembly mistakes. After the system is delivered and in operation, we may use <u>maintenance testing</u> to locate and repair physical faults in the fields.

Before we perform actual testing, we must prepare the tests. The process of preparing such tests is called <u>test generation.</u> Test generation involves two stages. The first stage generates the bare test patterns. These test patterns are then converted to a format suitable for actual testing of the hardware. For example, a format suitable for a tester to perform the actual testing.

Test pattern generation from HDL is a natural approach especially in VLSI testing to reduce the high complexity involved [1,2]. Since test patterns are derived from the primitive functions (e.g., register transfer operations) represented by HDL, they are called <u>functional test patterns</u>. In this section, we define <u>functional testing</u> as the process of testing digital systems using functional test patterns to make sure the systems behave correctly. Functional testing is quite different from gate-level testing not only because it uses higher level of system representation, but also because it tests functional faults (e.g., addition operation) rather than signal faults (e.g., a line stuck-at-0). Figure 4.3-1 shows conceptually the contrast between gate-level and functional-level testings. When conventional gate-level testing is performed, every line in the logic circuit must be considered. In Fig. 4.3-1, the stuck-at-0 fault on line 1, is an example of gate-level faults. Whereas, when functional testing is performed, only the functions (behavior) of modules (to be further defined on different levels of system specification i.e., RTL level, subsystem level, etc.) are tested. Faults f_1 (e.g., incorrect addition) and f_2 (e.g., incorrect control sequencing) are examples of functional faults for data operation module and control module respectively.

A complete functional testing effort for a system includes, in general, the following steps:

(1) Description - Describe the system-under-test (SUT) in terms of a HDL.

(2) Fault Modeling - Set up a desired functional fault model (or models).

(3) Test generation - Generate functional test patterns and the order of test sequences.

(4) Evaluation - Evaluate the test effectiveness (e.g., test generation complexity and fault coverage) usually by a fault simulator.

Readers may refer to [3] for background in testing area.

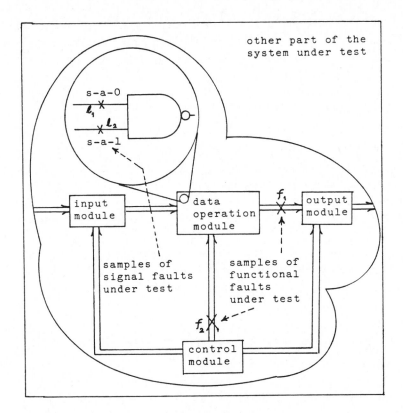

Fig. 4.3-1 Gate-level testing vs. functional testing

4.3.2 Functional Faults and Fault Models

According to the definition in 4.3.1, a functional test pattern is derived by considering a fault in a functional fault model. A functional fault model is the representation of the faulty behavior and it contains a set of functional faults to be tested. It is developed mainly for analytical purposes.

The following definitions are established.

Definition A fault refers to the malfunctioning of hardware within a digital system. A fault is permanent if it always exists. Otherwise, it is temporary.

Definition An error is the manifestation of a fault (or faults) which causes the wrong results observable at the output of a system.

Definition A functional fault refers to the malfunctioning of a basic functional primitive (or primitives) represented by an HDL used to describe a digital system.

Typical examples of functional faults in register transfer level (RTL) are register decoding faults, RTL instruction sequencing faults, etc. Here, we restrict
our discussion on the test pattern generation of permanent functional faults.

Definition Two functional faults F_1 and F_2 in a digital system are said to be
functionally equivalent if and only if their faulty effects to the HDL description
of the system are identical.

Definition If the changes of functions described by an HDL due to functional
fault F_1, is a subset of those changes due to F_2, then only F_2 needs to be considered in test generation. This reduction process is called fault collapsing.

Definition A functional fault is redundant if its existence in a system does
not affect the correct functional operations. Otherwise, it is non-redundant.

Definition A functional fault is testable if it is non-redundant and its faulty
effect is observable at the output of the SUT. Otherwise, it is untestable.

Definition A functional fault is detectable by a certain test set if it is testable and by applying at least one input pattern from the test set, its faulty
effect can be obserbed at the SUT's output. Otherwise, it is undetectable by that
test set.

It is important to note that based on the above definitions, a functional fault
is testable with respect to the system itself, and detectable with respect to a
certain test set.

Definition Fault coverage is the ratio of the number of faults detected with
respect to the total number of faults considered in a fault model.

Fault models may be developed at different levels of system representation. The
signal stuck-at fault model is most popular in gate-level representation. When the
system is described in an HDL, functional fault model are generally used. Although,
a deterministic mapping bewteen functional faults and lower-level faults (such as
gate-level stuck-at faults) is still an interesting research topic today, a general observation can be made on the size of the fault domain. It is generally true
that for the same physical faults, the domain of fault model established in functional level is smaller than that in gate-level. Figure 4.3-2 shows this concept.
The universal fault F_u represents the set of all possible physical faults in a digital system. After a gate-level fault modeling process, a gate-level fault model
containing those faults of interests can be established. The functional fault model on the other hand, is developed through functional fault modeling process. The
functional faults F_i's are originated from circuit/gate-level faults by fault e-
quivalence and/or fault dominance 3. Note that more than one circuit/gate-level
fault may contribute to one functional fault. The arrows from F_u to F_i's in Fig.
4.3-2 denote this basic idea.

The fault coverage is calculated based on the fault model used. Figure 4.3-3
shows the relation among all possible faults, faults under consideration, and faults

detected by testing. The fault coverage is F_d/F_c which is always less than or equal to 1.

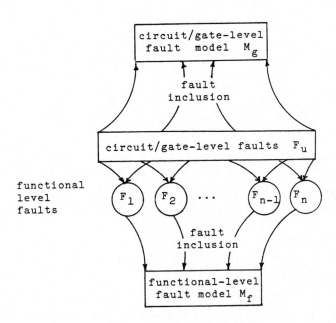

Fig. 4.3-2 The close relation between functional-level and circuit-level
fault models

Fault coverage serves as a reference to show how good a testing effort is. But, fault coverage alone can not determine the quality of a testing technique (also the test generation method) since the measurement is relative to the fault model used. The content of the fault model is an important criterion to judge the usefulness of a test method. For simplicity, most fault coverages are measured through simplified fault simulation which is usually based on single-fault assumption. As a consequence, the more functional faults included in a fault model, the more physical faults (e.g., bridging faults, multiple stuck-at faults, and technology-dependent faults) will be covered. Therefore, the fault coverage will reflect a measure closer to the actual faults detected.

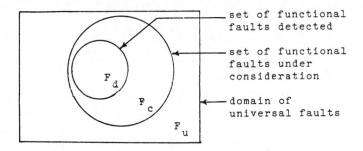

Fig. 4.3-3 Fault domain and fault coverage

4.3.3 Some Test Pattern Generation Methods

In this subsection, we briefly introduce four typical functional-level test pattern generation methods for digital systems proposed today. Interested readers are referred to the references listed at the end of this section for details [4-11].

4.3.3.1 RTL and S-algorithm

Su and Hsieh [4] and Lin and Su [5-7] developed a systematic functional test pattern generation technique for digital VLSI systems. A register-transfer-level (RT-level) HDL is defined and a comprehensive test pattern generation algorithm called S-algorithm ("S" for symbolic) is designed and implemented. The development of the algorithm is based on two foundations: the RT-level fault model and symbolic execution technique. The RT-level fault modeling and fault collapsing analysis are performed on the well-defined RT-language to lay an analytical foundation for the investigation of faulty behavior among RT-level faults. The RT-level symbolic execution technique is utilized to derive test patterns.

A complete RTL description (RT-description) of a digital system consists of two distinct parts: the declaration part and the execution part. The former contains declarations for input, output, and internal registers. The latter is composed of a series of RT-statements representing the functional behavior of the SUT.

A typical statement is given below:

$$k:(t,c) \quad R_d \texttt{<--} f(R_{s1}, R_{s2}), \texttt{-->} \quad n$$

In the above statement R_{s2} may be empty which means that function f performs on R_{s1} only. Symbol k is statement label, t (designed for future extension) and c are timing and condition respectively, R_d denotes destination register, f stands for an ALU operator, <-- represents the transfer path, and --> n indicates the next RT-statement to be executed.

Figure 4.3-4 shows a simple-calculator described in this RTL. Note that the function is described in a sequential execution style. A comment statement starts and ends with "%". "@" is register concatenation operator. "< >" represents "not equal" sign. Note that if the jump section is omitted, the next consecutive RT-statement is executed. ".." is bit extraction indicator. For example, DB (0..2) means bits DB(0), DB(1) and DB(2) of the data bus. This simple calculator is activated when the content of the 1-bit register START is set. It contains seven basic functions: ADD, SUBSTRACT, MULTIPLY, LOAD A, B, Q, SC.

```
% SIMPLE CALCULATOR %
% DECLARATION PART %
INPUT: START(1), DBS(1), DB(8), CP(3)
OUTPUT: AS(1), A(8), Q(8), F(1)
INTERNAL: BS(1), B(8), E(3), SC(3), QS(1), S(1)

% PROGRAM PART %
0:   S <- START
1:   (S=0) , ->0
2:   E <- CP , ->3
     % INSTRUCTION DECODING CHAIN %
3:   (E=001B) , ->11    % ADD %
4:   (E=010B) , ->12    % SUBTRACT %
5:   (E=011B) , ->13    % MULTIPLY %
6:   (E=100B) , ->21    % LOAD A %
7:   (E=101B) , ->23    % LOAD B %
8:   (E=110B) , ->25    % LOAD Q %
9:   (E=111B) , ->27    % LOAD SC %
10:  S <- 0 , ->0       % OTHERWISE %
11:  F@A <- A+B , ->10
12:  F@A <- A-B , ->10
13:  AS <- BS XOR QS
14:  A <- 0
15:  F <- 0
16:  (Q(7)=1) F@A <- A+B
17:  F@A@C <- SHR F@A@Q
18:  SC <- SC-1
19:  (SC <> 0) , ->16
20:  S <- 0 , ->0
21:  A <- DB
22:  AS <- DBS , ->10
23:  B <- DB
24:  BS <- DBS , ->10
25:  Q <- DB
26:  QS <- DBS , ->10
27:  SC <- DB(0..2) , ->10
END
```

Fig. 4.3.3.-4   A simple calculator described in RTL.

<u>Definition</u> Each basic functional component in an RT-statement is called a
<u>RT-component.</u> That is to say, k, t, c, R_d, f, R_{si}, n, and are all RT-components.

<u>Definition</u> When an RT-component F becomes faulty, it is denoted by F/F'.

Based on the typical RT-statement syntax, nine RT-level faults are classified:

(1) k/k':   label fault
(2) --> n/--> n': jump fault
(3) t/t':   timing fault
(4) c/c':   condition fault
(5) (R)/(R)':   data storage fault
(6) <--/<--':   data transfer fault
(7) R/R':   register decoding fault
(8) f/f':   operator decoding fault
(9) (f)/(f)':   operator execution fault

Note that (R)' refers to the fault of data bits <u>within</u> a register which means
that the content of R is changed due to fault. R' refers to the decoding faults
<u>among</u> registers which means that a wrong register is chosen. Similar difference
also exists between f' and (f)' faults. The (f)' fault indicates that the result of
executing f (an ALU operator) is faulty. f/f' means that instead of executing f, f'
is executed.

Using reasonably analytical assumptions similar to gate-level stuck-at fault
model, the faulty behavior of these RT-level faults can be analyzed and RT-level
fault collapsing is conducted. The (f)/(f)' fault is related to Arithmetic Logic
Unit (ALU) which is a combinational circuit. Since this type of fault is heavily-
dependent on implementation, it seems more practical to process it by circuit/gate-
level test generation technique. Some theorems are given below. The proof can be
found in [7].

<u>Theorem</u> In an RT-description, the k/k' fault is collapsed to the -->n/-->n'
fault.
<u>Theorem</u> The (R)/(R)' fault is functionally equivalent to the <--/<--' fault.
<u>Theorem</u> t/t' fault is collapsed to either -->n/-->n' fault or c/c' fault.

After fault collapsing using the above theorems, a reduced RT-level fault model is obtained. Among these reduced faults, the <--/<--' fault is classified as non-
enumerable since there are numerous numbers of such faults and we don't intend to
enumerate each of them. Its test generation is performed by a path-oriented heuristics. Other fault types are enumerable and their faults are enumerated one at a
time.

The test pattern generation is performed on the reduced fault model using RT-level symbolic execution method. The symbolic execution technique is tractable in hardware description language mainly due to the simplicity of HDL and the simple philosophy used behind hardware testing. The basic idea of symbolic execution is to manipulate <u>symbolic values</u> (or symbols) instead of <u>actual values</u> during program execution. A <u>Symbolic value</u> is an expression of constants and variables (in RT-level, a read/write register is a variable) whose contents are fixed but unknown.

For example, the symbolic value of R_2 after the RT-statement $R_2 \leftarrow R_1+1$ is executed is $\$R_1+1$, where $\$R_1$ is the current symbolic value of register R_1. The basic idea in the test pattern generation algorithm is first to compute the symbolic execution results of both fault-free and fault-injected machines, and then compare them to obtain a feasible input test pattern which distinguishes the output of the faulty machine from that of the good one. Note that simplification of symbolic expression is crucial for complex systems.

The overall test generation method is divided into three stages: the pre-process, the S-algorithm, and the post-process stages, where the S-algorithm is the main stage. Figure 4.3-5 shows the simplified test generation flow.

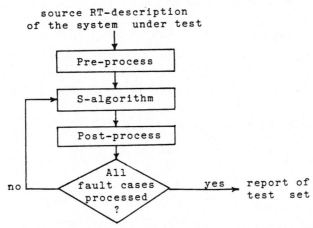

Fig. 4.3-5 The simplified execution flow of the overall
test generation algorithm

The summarized test generation steps are listed below:

Step 1   Perform syntax checking of the input RT-description representing the SUT.

Step 2   Partition the SUT into a set of function submodules (FS).

Step 3   Compute the test generation order of each FS.

Step 4   Find the path(s) which covers all distinct RT-components for the currently chosen FS.

Step 5   Perform machine symbolic execution on the covering path(s) to obtain fault-free symbolic execution results and path constraints, where path contraints are conditions for the machine to execute through certain path.

Step 6   Find test patterns for <--/<--' fault.

Step 7   Choose and inject a not-yet-tested enumerable fault into the fault-free ma-
           chine and find a fault-injected symbolic execution path.

Step 8   Perform symbolic execution along the fault-injected path.

Step 9   Derive a test pattern for the currently enumerated fault by solving the
           symbolic inequalities between the fault-free and the fault-injected symbo-
           lic execution results. If no test pattern is obtained, return to step 7 for
           another fault-injected path if it exists. Otherwise, claim the fault-under-
           test to be a hard-to-test fault.

Step 10   Remove other not-yet-tested faults which are also tested by the test pa-
           ttern obtained in step 9.

Step 11   Return to step 7 until no more unprocessed enumerable faults left in the
           current FS.

Step 12   Return to step 4 until no more unprocessed FS left in the SUT.

Step 13   Clean-up the hard-to-test faults by analyzing their symbolic execution
           results.

Step 14   Generate test generation report.

In the above steps, step 1 to step 3 are the preprocess, step 4 to step 9 form the
S-algorithm, and step 10 to step 14 are the postprocess. Details can be founf in [7].
A simple example is given below to illustrate the main concept just described.

Example 4.3.1

Let us consider a fault case in the simple calculator example. Suppose the decoding of addition operator in the RT-statement 11: F@A<--A+B is faulty (i.e., f/f')
and subtraction operation is executed instead of the addition. The fault-free symbolic execution path in the addition submodule is: 0-->2-->3-->11-->10-->0, where,
the number represents the label of an RT-statement. For faulty machine statement No. 4
instead on No. 3 is executed. The fault-free and fault-injected symbolic execution
results are shown below step-by-step:

RT-statement No.	fault-free	fault injected
0	S=START	S_2=START
(1)	P.C.:START=1	P.C.:START=1
2	E=OP	E_2=OP
3	P.C.:OP=001	
4		P.C.:OP=010
11	F=carry(A,B)	F_2=borrow(A,B)
	A+B	A_2=A-B
10	S=0	S_2=0

Here, P.C. stands for path constraint, and (1) means that RT-statement 1 is not actually executed but a path constraint is established. We then solve at least one of the following two symbolic inequalities for a feasible test pattern for this fault.

$$F \neq F_2 \rightarrow \quad carry~(\$A,\$B) \neq borrow~(\$A,\$B)$$
$$A \neq A_2 \rightarrow \quad A+B \neq A-B$$

There are many test patterns satisfying the above conditions; one possibility is to choose

$$A = 10010000$$
$$B = 10001000$$

as a feasible test pattern which satisfies both inequalities.

The RTL and S-algorithm method is most appropriate for test generation at the early system design stage as an efficient tool for CAD. It has the following good features:

(1) For a testable RT-level fault, it systematically finds an input test pattern for detecting that fault.

(2) It uses an extensible analytical RT-level fault model.

(3) It applies the RT-level symbolic execution as a computation tool.

(4) The test generation results obtained may be used for testability analysis for a system under development.

There are rooms for improvement in this method. The present version of the register transfer language can not describe parallel operations. More sophisticated skill needs to be developed for cleaning-up hard-to-test faults. Discussion on other possible improvements can be found in [7].

4.3.3.2 AHPL and SCIRTSS

Hill and Huey [8] developed the SCIRTSS (Sequential Circuit Test Search System) for digital systems described by a hardware description language called AHPL. It uses tree search technique to derive test sequences for sequential circuits. There are four primary components of the system: a heuristic problem reduction graph search routine, a logic fault injection simulator, a search-simulation interface, and a D-algorithm implementation. The heuristic graph search routine manipulates the sequential circuit by searching over the state graph of the machine to find a path from a present state to any desired state. A graph representation is obtained through the use of a data transfer simulator based on AHPL. The fault injection logic simulator verifies the results of the graph search. For a particular fault, the D-algorithm routine generates a set of D-vectors consisting of external input and flip-flop state specifications, which will sensitize a path from the fault site to a circuit output or into a flip-flop.

To understand the operation of SCIRTSS, one must know the basic ideas in the D-algorithm. The D-algorithm is a multiple path sensitizing algorithm which was developed by Roth [9] of IBM.

It consists of three major operations: Fault injection (Primitive D-cube of failure), D-drive (forward drive sensitization) and backward trace (line justification). A brief discussion on them is given below:

(1) fault injection - For each faulty logic element, select a primitive D-cube of failure which specifies certain logic values of inputs of the logic element such that the fault in that element causes a faulty signal (represented by D or $\bar{D}$) at its output.

(2) D-drive - This faulty signal is then propagated towards the output of the circuit by intersecting the primitive D-cube of failure with the propagation D-cubes of all logic elements on a sensitized path. The propagation D-cube specifies the input constraints to a logic element that allows the faulty signal to propagate from at least one input of the logic element to its output.

(3) Backward trace - After a D-drive has succesfully completed, certain lines in that circuit have been assigned specific logic values. Now, values for primary inputs are derived to obtain the assigned values for lines. This is accomplished by propagating the line values backward to the primary inputs using the input/output relation of the associated logic element. During the D-drive and line justification steps, the effects of every sensitizing line assignment on the entire circuit are determined. Additional lines will be assigned using the logic values implied by each path sensitizing line assignment.

As an example, consider the simple logic circuit in Fig. 4.3-6. Suppose we want to test line 7 stuck-at-0, the following three steps are performed to derive a test pattern:

Step 1   Set up the primitive D-cube of failure.

line	1	4	7
value	1	1	D

where D = 1 for good circuit
      = 0 for bad circuit with line 7 Stuck-at-0

Step 2   D-drive: Assign line 5=0 and line 6=1 to propagate D to the primary output
          (line 9).

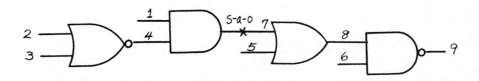

Fig. 4.3-6 A simple circuit.

Step 3   Backward trace: Trace from line 4 backward to line 2 and line 3 and assign
          both lines to 0's to obtain line 4=1.

Figure 4.3-7 shows an example of an APL description of a circuit. It's function is to determine whether the current four bits received on lines X_0 to X_3 match exactly any of the immediately previous eight four-bit characters. The first three lines are for registers and signals declaration. The actions in 1) are to set "ready" =1 and go to 2) if d(data received)=1 and r(reset)=0. Statement 2) initializes the 3-bit counter CT, move the content of input register X to INR, and reset Y and Z. If Y=0, it continues to 3), otherwise it returns to 1). In 3), the first character of the eight preceeding characters stored in the register file, denoted by FILE, is compared with INR and CT is incremented. Y is set to 1 if a match occurs. The other seven characters in FILE are individually compared in 4). In 5), the content of Y is moved to Z and the execution goes back to 1).

Registers: **FILE** (4,8) *INR*(4), *CT*(3), *y, z*.
Outputs: *z, ready*.
Inputs: *X*(4), *d, r*.
1) *ready* = 1
   $\rightarrow (1 \times (\bar{d} + r) + 2 \times (d \wedge \bar{r}))$
2) $CT \leftarrow \epsilon(3); INR \leftarrow X; y \leftarrow 0; z \leftarrow 0$
   $\rightarrow (3 \times \bar{r} + 1 \times r)$
3) $FILE^{3} \leftarrow INR; \alpha^{7} // FILE \leftarrow \alpha^{7} // \Uparrow FILE;$
   $y \leftarrow (\vee/FILE^{1} \oplus INR) \vee y; CT \leftarrow \text{INC}(CT)$
   $\rightarrow (4 \times \bar{r} + 1 \times r)$
4) $FILE \leftarrow \Uparrow FILE; y \leftarrow (\vee/FILE^{1} \oplus INR) \vee y;$
   $CT \leftarrow \text{INC}(CT)$
   $\rightarrow (5 \times (\bar{r} \wedge (\wedge/CT)) + 4 \times ((\wedge)/CT \wedge \bar{r}) + 1 \times r)$
5) $z \leftarrow y$
   $\rightarrow (1).$

Fig. 4.3-7 An APL description of a circuit

The fault model used is the conventional stuck-at fault of signals. Each faulty machine is taken into consideration for test generation and fault simulation
is performed to determine the faults detected by a given test pattern. The simplified flow diagram is shown in Figure 4.3-8, where the number on the upper left corner of each block is the step number.

In step 1, a homing experiment is perform to bring the network to a set of
known states. The uncertainty of the initial state of the good circuit can thus be
reduced. This step need not be actually applied if an embedded reset/preset circuit
is assumed. A fault is <u>stored</u> if the state of a faulty sequential network N_f is different from the state of the good network. If no fault is stored in step 2, then an
untested faulty network is selected along with a set of goal states Q_g and a single
input such that the faulty network and good network will go into a different next
state from Q_g. An example of goal states selection is shown at the end of this sub-
section.

The D-algorithm is applied directly in this step after the feedback loops are
cut at every memory cell. In steps 4 and 5, the search for an input sequence to
bring the network to Q_g is performed, and all inputs along this path are determined.
In step 6, the tree search for a propagation sequence from the distinguished fault-
injected state to output is performed. In steps 7 and 8, simulation is performed to
identify all faulty networks tested by each subsequences so that they can be deleted from the set of untested faulty networks. The whole process continues until no
faulty networks left unprocessed.

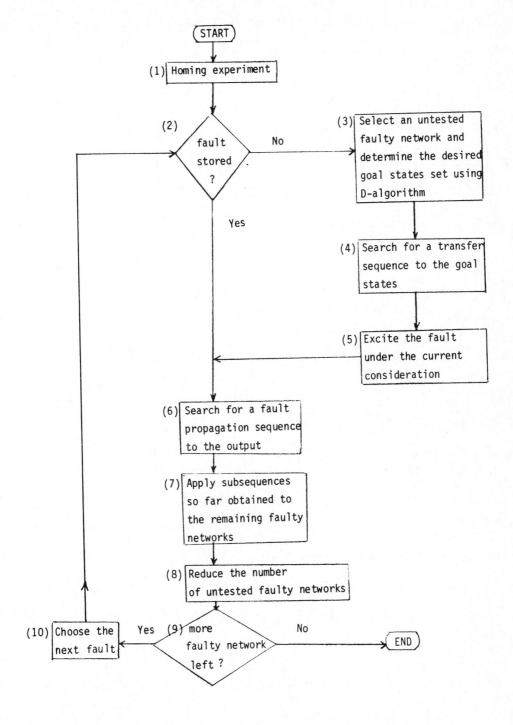

Fig. 4.3-8 Simplified flow diagram of SCIRTSS

The tree search is the major process during test generation. Its basic concept is illustrated by a simple example as follows.

Example 4.3.3.2

Consider the good circuit and a faulty circuit with the following state transition tables:

next state, output	input X	
	0	1
present state a	b,0	b,0
b	c,0	b,0
c	a,0	d,0
d	e,0	a,0
e	b,0	f,0
f	g,0	f,0
g	c,0	h,0
h	b,1	b,0

next state, output	input X	
	0	1
present state a	b,0	a,0
b	c,0	a,0
c	a,0	d,0
d	e,0	a,0
e	b,0	f,0
f	g,0	f,0
g	c,0	f,0
h	b,1	b,0

(a) Good sequential circuit   (b) Faulty sequential circuit

Suppose that initially both networks are in some states [b,c,e,g]. By inspection of the above state transition tables, it can be seen that [a,b] is a suitable set of goal states since after X=1 is applied, the next state of [a,b] is "b" for the good network while "a" for the faulty network. Therefore, a successor tree leading to this goal node (sensitization search) is given below:

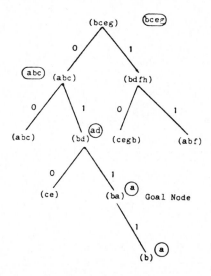

In this figure, the states enclosed by circles and elipses are states for the faulty circuit. After input sequence 0111 is applied, the good circuit will go to state b and the faulty circuit will go to state a. A propagation search for this faulty circuit is carried out as follows. Here, the states of the faulty circuit are placed below the good ones. It can easily seen that after input sequence 00101010 is applied, the good circuit will go to state a with output 0 while the faulty circuit will go to state b with output 1.

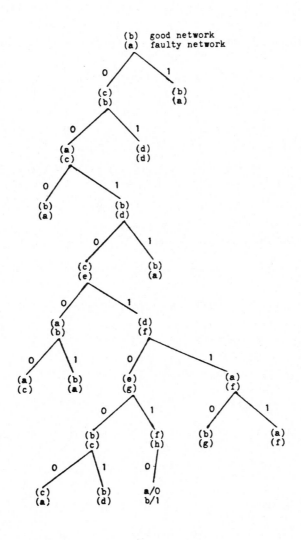

The SCIRTSS has been experimented on several sequential circuits with various complexity. Encouraging results of the prototype system are reported [8]. Two points are worthy of mentioning: (1) Without an RT-level fault modeling for test generation, the use of AHPL to describe a system looks less meaningful; (2) The use of state-transition information in tree-searching may cause storage problems in practical VLSI applications.

4.3.3.3 CHDL and Extended D-algorithm

Levendel and Menon [10] proposed an extension of the D-algorithm for test generation of digital systems described by a Computer Hardware Description Language (CHDL). Methods for D-propagation through the basic constructs of this language and the test pattern generation for circuits containing functions described in CHDL are discussed. The fault mode include: function variables stuck-at faults, control faults, and functional faults with user specified faulty behavior.

A CHDL is used to define and describe the functional elements in a digital system. Its description consists of a declaration section and a behavior section. The former defines hardware variables such as input, output, and internal variables. The latter consists of a collection of statements grouped in a sequence with logic meaning. An example of CHDL description is given below:

```
                  .
                  .
                  .
    After 2:      B<--A ;
    After 3:      C<--B ;
    If C=1    Then
    Begin
          After 2:    Z<--0 ;
          After 1:    Y<--1 ;

    End;
                  .
                  .
                  .
```

Here, "After 2: B<--A" means that after a delay of 2 units, the content of register A is transferred to B. As proposed in [9], the CHDL uses control constructs and functional operators to describe the behavior of a general logic network. Then switching algebraic expressions are used to derive D-cubes for each functional block. To derive tests for circuits containing CHDL blocks, one has to perform D-propagation from the outputs of a CHDL block to the inputs of next CHDL block. A CHDL block may contain different types of functional expressions such as shift, addition, etc., and cause-effect structures such as if-then-else, and memory elements. In such cases, a sequence of inputs to propagate a D or $\overline{D}$ to some of its outputs would be required.

The fault model considered here includes conventional data stuck-at faults and two functional faults - the control faults and the general faults. To test the control fault, each path in the control graph corresponding to the control construct must be tested. The general faults are defined as faults whose effect are known, but can not be modeled by stuck-at faults or control faults. This kind of faults can only be tested when their behavior is expressed explicitly.

The test generation algorithm contains the following basic steps:
(1) Describe the SUT as an interconnection of CHDL blocks.
(2) Insert a fault effect into a CHDL construct or a CHDL block containing CHDL blocks.
(3) Propagate the fault effect to an observable point.
(4) Justify all the decisions made in the above two steps by applying sequences at the primary inputs of the SUT.

The D-propagation is a major problem here. Both switching and non-switching algebraic expressions can be included as a part of CHDL constructs. Therefore, ways must be provided to propagate D variables through them. Also, the control constructs in the CHDL must also be able to perform D-propagation. Now, let us consider the following illustrative examples.

Example 4.3.3-3

To derive the set of D-cubes for a JK-flipflop, $Q=J\overline{q} + \overline{K}q$, where q is the present output and Q is its next output, we make use of the simple equations:

$$C^D = (a + b)^D = a^D\, b^0 + a^0 b^D + a^D b^D \qquad (1)$$

$$E^D = (ab)^D = a^D b^1 + a^1 b^D + a^D b^D \qquad (2)$$

By using the simple expansion technique of Boolean expression, we compute:

$$Q^D = (J\bar{q} + \bar{k}q)^D$$

$$= (\bar{J}q)^0(\bar{k}q)^D + (J\bar{q})^D(\bar{k}q)^0 + (J\bar{q})^D(\bar{k}q)^D \qquad \text{(by (1))}$$

$$= J^0K^0q^D + J^DK^0q^D + K^{\bar{D}}q^1 + J^0K^{\bar{D}}q^D \qquad \text{(apply (2) to each term)}$$

$$+ J^DK^{\bar{D}}q^D + J^1K^1q^{\bar{D}} + J^DK^1q^{\bar{D}} + J^Dq^0.$$

$$+ J^1K^{\bar{D}}q^{\bar{D}} + J^DK^{\bar{D}}q^{\bar{D}}$$

For non-swithching algebraic expressions, let us consider the addition operation in the following example.

Example 4.3.3-4

Analogous to the regular addition, we may use the SUM (EXclusive-OR) and CARRY (AND) tables shown below to derive its D-cubes:

SUM	0	1	D	$\bar{D}$	x
0	0	1	D	$\bar{D}$	x
1	1	0	$\bar{D}$	D	x
D	D	$\bar{D}$	0	1	x
$\bar{D}$	$\bar{D}$	D	1	0	x
x	x	x	x	x	x

CARRY	0	1	D	$\bar{D}$	x
0	0	0	0	0	0
1	0	1	D	$\bar{D}$	x
D	0	D	D	0	x
$\bar{D}$	0	$\bar{D}$	0	$\bar{D}$	x
x	0	x	x	x	x

where, x's in the table denote unspecified values. From the D outputs in the table, we obtain the D-cubes of SUM part as $A^0B^D + A^DB^0 + A^1B^{\bar{D}} + A^{\bar{D}}B^1$ and the D-cubes of CARRY part as $A^DB^1 + A^1B^D + A^DB^D$.

For the D-cubes of a CHDL control construct, let us consider the following binary cause-effect expression:

$$\text{If A Then } Z \longleftarrow B \quad \text{Else} \quad Z \longleftarrow C$$

where, A, B, and C are switching expressions. This statement is equivalent to the following switching expression: $Z \longleftarrow AB + \overline{A}C$. Its D-drive expression is derived as:

$$Z^D = (AB + \overline{A}C)^D = (AB)^D (AC)^0 + (AB)^0 (\overline{A}C)^D + (AB)^D (\overline{A}C)^D$$

The above expression may be expanded further as in example 4.3.3-3.

Other complex CHDL control constructs may be similarly handled. To be practical in complex CHDL functions, CHDL control construct may be transformed into binary graph representation and the D-propagation problem is treated as a graph traversal problem. The test generation steps involved in this method are essentially the same as those in gate-level D-algorithm: fault injection, D-propagation, and backward trace. Note that for control faults, every branch on that control path must be tested.

This method is interesting since it uses the existing popular gate-level technique at HDL level. However, the functional-level fault model has not been well-defined. Moreover, it is not easy to construct D-cubes for a functional block, mainly because even a simple functional blocks may need several different representations (e.g., the Boolean expression in Example 4.3.3-3 and the Tabular form in Example 4.3.3-4). Also, the efficiency of this technique needs to be demonstrated by an actual implementation of the test generation algorithm.

4.3.3.4 FTL and Path-sensitization Method

Song and Fong [11] presented a test generation method based on functional tables translated from a hardware description language called Function and Timing Language (FTL). FTL is a table-oriented language which describes mainly the functionalities extracted from SUT. A functional table T is defined as a data structure consists of two parts: one specifies the conditions for the corresponding actions to be executed and the other describes those actions in the form of algebraic expressions.

An FTL description contains the following four sections:
(1) Timing specifies the timing constraints of some input signals, such as clock pulses and enable lines.
(2) TABLE is the functional table corresponding to a functional block.

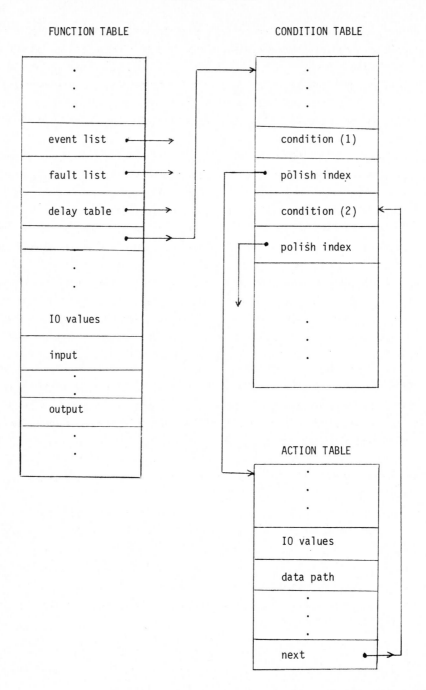

Fig. 4.3.3-10 Basic table structure of a function table

To illustrate the key idea of this technique, the test generation of a binary counter is given below:

Example 4.3.3-5

A presentable binary counter (SN74LS197) can be described in FTL similar to fig. 4.3.3-9. It has the following functionalities:

. It becomes a 4-bit binary counter if its least significant output is connected to the second clock output.
. It has a low-active direct input to clear register content.
. It has a count/load control input to preset its outputs to any states from data inputs.

. It may be used as a 4-bit latch by strobing a separate control line.

Suppose no user-defined functional fault is to be tested. The test generation for such a device may be carried out in the following steps after its functional table is established:

(1) Search tests to initialize all outputs to zeroes. To do this, one possible way is to set the direct-input clear to zero. The other way is to load all four data inputs with zeroes and set the count/load control input to load inputs.
(2) Search tests for exercising the load operations to test data faults on input and output lines. Under the load mode, 0000 and 1111 on data inputs may test all data faults in input and output lines.
(3) Test the control faults on resetting when counter limit is reached. To do this, we first load the counter with 1111 and then increment the counter in count mode.
(4) Search tests for operation faults to test the incrementing operation of counter. To do this, the counter is presetted to a value. Then the count mode is turned on, and the count triggering input is continuously applied until the counter goes back to the presetted value.

Generally speaking, this method is simple and application-oriented. Some improvements may make it more practical. First, more algorithmic development on searching technique needs to be done. The timing fault is implementation-dependent. Its faulty effect in functions is not obvious and needs more study. The automatic fault injection of user-defined functional fault is also a non-trivial problem in actual VLSI testing.

4.3.4 Other Related Works

Some other research and development in the functional testing area has been performed in recent years. Some of them are briefly described below. Su and Lin's overview paper [2] provides a little more detail.

Roboch and Saucier [12] proposed the "Abstract Execution Graph Technique" for testing microprocessors. Test programs of the microprocessor are generated based on the set of abstract execution graphs derived from its instruction set. This method is easy to understand but its fault coverage seems dependent on the exhaustive testing of particular functions and memory elements.

Thatte and Abraham [13] proposed another graph-theoretic approach to test micro-processors represented by a register-transfer graph. The register-transfer-level fault modeling and test generation are performed in a systematic way. Their work is valuable from analytical point of view.

Annaratone and Sami [14] proposed a testing technique for microprocessors with microprogrammed control. Testing of microprocessors is performed by testing each instruction at the microoperation level and each independent asynchronous signals observable at external pins. This method is attractive because microinstructions are basic to the functions of the control unit. Since no specific fault model is assumed, the fault coverage remains to be investigated. An algorithmic test procedure also need to be developed.

Breuer and Friedman [15] proposed a system which may generate tests for complex sequential circuits composed of functional primitive elements such as counters and shift registers. Similar to [10], it also directly applies D-algorithm for generating test sequences. More work is needed in practical applications.

Min and Su [16] presented a systematic way to generate tests for a given functional fault. Several procedures for testing microprocessors are developed. Saluja, Shen and Su [17] proposed a new approach to test instruction execution faults of microprocessors by measuring the cycle time used during instruction execution . Shen and Su [18] discussed the major issues involved in functional testing of microprocessors. The basic procedures for testing microprocessors were reported. Levendel and Menon [19] formalized the critical trace method originally developed for generating tests for gate level circuits, and extend it to the functional level. Recently, Shen and Su [20] utilized the critical path trace concept for VLSI functional testing. A systematic approach was proposed to test digital systems described by a hardware description language higher than register transfer language level. The method eliminated expensive fault simulation since in the process of generating tests, all faults detectable by a test are recorded.

4.3.5 Current Trends and Future Development

   With higher level of system representation, functional test pattern generation methods suggest feasible ways to solve the problem of increasing complexity in VLSI systems. Generating test patterns directly from a HDL is becoming more attractive than ever especially in CAD area for better efficiency and quality. The techniques of design-for-testability or built-in-self-test are also developed toward reducing the complexity problem in testing VLSI systems. Therefore, one possible trend in the future test technology is to combine functional test generation techniques with design-for-testability techniques for a better test efficiency. Another future trend is that, due to the various requirements set by the test quality, hierarchical testing or mixed-level test generation are also feasible ways for future testing.

   Here are some general criteria which may be used to judge a good functional test pattern generation method:
(1) From the practical point of view, every such technique should be designed with the ease of automation in mind. This suggests the use of an explicit, well defined HDL to describe the SUT, and the use of compact algorithmic expressions to describe the desired test generation procedure.
(2) Test generation algorithms should be proposed along with detailed analysis of computation complexity.
(3) A quantitative fault coverage and performance measures should be performed to demonstrate the effectiveness of the proposed method.
(4) The functional fault model should be explicitly established and should be extensible and general enough to cope with the future evolutions of VLSI technology.

   Fault modeling at functional level is a difficult problem in general. More studies are needed to formulate the general relation between gate-level faults and RT-level (or higher level) faults. The fault coverage measure can only reflect a restricted interpretation of the quality of a test generation method. Also, besides research papers, implementation of test generation methods and applications of these methods to practical VLSI circuits should receive more attention to make these methods of real value.

   Finally, future design automation should use the same data base generated by a HDL translator for automated logic design, hierarchical simulation and test generation [21].

ACKNOWLEDGEMENT:

The authors thank Mr. Michael Kessler of IBM-Böblingen Laboratory, West Germany for his administrative support to provide the final typing.

References:

[1]   S.Y.H., Su, "A Survey of Computer Hardware Description Languages in the U.S.A.", COMPUTER, December 1974, pp. 45-51.

[2]   S.Y.H. Su and T. Lin, "Functional Testing Techniques for Digital LSI/VLSI systems", (Invited Paper), Proceedings of 21-st Design Automation Conference, June 1984, pp. 517-528.

[3]   M.A. Breuer and A.D. Friedman, "Diagnosis and Reliable Design of Digital Systems Computer Science Press, 1976.

[4]   S.Y.H. Su and Y.I. Hsieh, "Testing Functional Faults in Digital Systems Described by Register Transfer Language", Journal of Digital Systems, Summer/Fall 1982. Also Digest of Papers, 1982 International Test Conference, October, 1982, pp. 447-457.

[5]   T. Lin and S.Y.H. Su", Functional Test Generation of Digital LSI/VLSI Systems Using Machine Symbolic Execution Technique", Digest of Papers, 1984 International Test Conference, Oct., 1984, pp. 660-668.

[6]   T. Lin and S.Y.H. Su, "The S-algorithm: A Promising Solution for Systematic Functional Test Generation", IEEE Transactions on CAD, July 1985.

[7]   T. Lin, "Functional Test Generation of Digital LSI/VLSI Systems Using Machine Symbolic Execution Technique", Ph.D dissertation, Computer Science Dept., State University of New York at Binghamton, May 1985.

[8]   F.J. Hill and B. Huey, "SCIRTSS: A Search System for Sequential Circuit Test Sequences", IEEE Transaction on Computers, May 1977, pp. 490-502.

[9]   J.P. Roth, "Diagnosis of Automata Failures: A Calculus and a Method," IBM Journal of Research and Development, Vol. 10, July 1966, pp. 278-291.

[10]   Y.H. Levendel and P.R. Menon, "Test Generation Algorithms for Computer Hardware Description Languages", IEEE Transactions on Computers, Vol. C-31, No. 7, July 1982, pp. 577-588.

[11]   K. Son and J.Y.O. Fong, "Automatic Behavior Test Generation", Digest of Papers, 1982 International Test Conference, October 1982, pp. 161-165.

[12]   C. Robach and G. Saucier, "Microprocessor Functional Testing", Digest of Papers, 1980 Test Conference, pp. 433-443.

[13]   S. Thatte and J.A. Abraham, "Test Generation of Microprocessor", IEEE Transactions on Computers, June 1980, pp. 429-441.

[14]   M. Annarartone and M. Sami, "An Approach to Functional Testing of Microprocessors", Proceedings of 12th International Symposium on Fault-tolerant Computing, June 1982, pp. 158-164.

[15]   M.A Breuer and A.D. Friedman, "Functional Level Primitives in Test Generation", IEEE Transactions on Computers, March 1980, pp. 223-235.

[16] Y. Min and S.Y.H. Su, "Testing Functional Faults in VLSI", Proceedings of 19th Design Automation Conference, June 1982, pp. 384-392.

[17] K.K. Saluja, L. Shen, and S.Y.H. Su, "A Simplified Algorithm for Testing Microprocessors", 1983 International Test Conference, Oct. 1983, pp. 669-675.

[18] L. Shen, and S.Y.H. Su, "A Functional Testing Method for Microprocessors", Proceedings of 14th International Symposium on Fault-tolerant Computing, June 1984, pp. 212-218.

[19] Y. Levendel and P.R. Menon, "The Algorithm: Critical Traces for Functions and CHDL Contructs", Digest of Papers, 1983 International Test Conference, October 1983, pp. 90-97.

[20] L. Shen and S.Y.H. Su, "VLSI Functional Testing via Critical Path Traces at a Hardware Description Language Level", Proceedings of 2nd GI/NTG/GMR Conference on fault-tolerant Computing Systems, Bonn, West Germany, September 1984, pp. 364-379.

[21] S.Y.H. Su, "IDAS - An Integraged Design Automation System," Proceedings of the 1984 National Computer Conference, June, 1984, pp. 143-150.

KEY WORDS

Testing, functional testing, behavior testing, fault testing, VLSI testing, fault detection, fault location, fault diagnosis, test generation, test sequence, hardware description languages, computer hardware description languages, design languages.

HARDWARE DESCRIPTION LANGUAGES
R.W. Hartenstein (Editor)
© Elsevier Science Publishers B.V. (North-Holland), 1987

4.4. HARDWARE VERIFICATION BASED ON HDL SOURCES

Masahiro FUJITA

Department of Electrical Engineering, University of Tokyo
7-3-1 Hongo Bunkyo-ku, Tokyo, Japan 113

Now at Fujitsu Laboratories Ltd. Nakahara-ku, Kawasaki, Japan 213

Abstract

This section explains formal verification methods for hardware logic designs widely used. Here, verification means logical correctness only, and does not necessarily mean any timing characteristics.

First basic verification techniques are classified and explained in section 4.4.1. Second several verification methods by logical reasonings are presented in section 4.4.2. Then, since hardwares work in parallel, some parallel execution models that have sufficiently mathematical backgrounds are necessary. In section 4.4.3, such models and verification methods based on them are presented.

In present computer aided design systems, the standard approach that is taken to verify the proposed logic designs is that of simulations [1]. Almost all the HDLs have their simulators, and some of them have tools for enhancing the man-machine interface such as, graphical input/output, interactive simulations, etc. However it is recognized that simulations have several drawbacks. As the logic network that is to be simulated becomes more complex, the number of possible inputs to the logic structure grows exponentially, so that a complete simulation becomes infeasible. Hence, only a representable subset of inputs is selected for simulations. This may lead a designer to fail to notice some errors.

In order to avoid failing to notice some bugs, formal verification methods, which can also be considered to be the methods that automatically generate all the simulation cases needed for verification, are proposed. Those methods analyze HDL sources, generate the conditions that must be satisfied, and then check them.

Verification is to check that a design satisfy a specification. Let S be logical expressions that correspond to specification and let D be logical expressions that correspond to designs. Then the following formula must be checked.

D → S,

that is, design logically implies specification. In verifying HDL sources, there are two possibilities: one is that D is described in a HDL and S is given by some special sources only
for verification, and the other is that both S and D are HDL descriptions. We here call
the former type of verification 'assertion verification', and call the latter type of
verification 'equivalence verification'. Of course the former is considered to be a special
case of the latter, and when the given specifications are sufficient, the former equals to the
latter.

Here first basic verification techniques are classified and explained in section 4.4.1.
Second several verification methods by logical reasonings are presented in section 4.4.2.
Then, since hardwares work in parallel, some parallel execution models that have
sufficiently mathematical backgrounds are necessary. In section 4.4.3, such models and
verification methods based on them are presented.

4.4.1 Overview

This section explains the basic and simple techniques for logic design verification, which are easily implemented on computers. Verification techniques are classified into several levels, depending on the deepness of analysis of HDL sources. Here examples of several levels are presented by using DDL [2] as a typical HDL.

Verifications of the easiest level are simple syntactic checks. For example, the following statement is not correct:

$$ADS(0:9) \leftarrow S(0:1)|D(0:8) + B(0:9),$$

where '|' means the concatenation operation. The above is incorrect, because the bit-width of the left hand side is shorter than that of the right hand side. Since syntactic verifications, such as above, are very easily done by computers, almost all HDL processors can do it. Of course, this level of verification also exists for software programs. Verification of hardware has the difficulty for verifying a sequential software programs, such as, Pascal, Fortran, etc. Moreover, it has more points to be considered. They are originated from the fact that hardware instinctively has parallelisms.

The next level is concerning mutual exclusions or physical constraints. For example, it is inhibited that a register is data-transferred from two sources at the same time. Therefore, the following statement is incorrect:

$$R(0:7) \leftarrow S1(0:7), R(0:7) \leftarrow S2(0:7).$$

(in DDL, the above two register transfer statements are executed in parallel) In this case, it is easily checked by syntactic analysis. However, in case of having conditional statements, the check is not easy. Consider the next statements. (|: or, &: and, ~ : not)

if $F1\&(F2|F3)$ then $R(0:7) \leftarrow S1(0:7)$,
if $\sim F1|(\sim F2\&\sim F3)$ then $R(0:7) \leftarrow S2(0:7)$,

In this case, exclusiveness of the two conditions, that is,

$$(F1\&(F2|F3))\&(\sim F1|(\sim F2\&\sim F3)) \equiv 0$$

must be checked. So in checking the exclusiveness of conditions for register transfer, terminal connection and state transition operations, it must be shown that every logical product of two conditions is identically false, which is usually done by symbolically manipulating boolean expressions. Therefore, it requires much time to do so especially in handling expressions that have many variables and logical connectives.

These verification methods are reported in [3]. Verifications in the most difficult level are formal reasonings on HDL descriptions. This is difficult because we must handle parallelisms. There are many ways of verification in this level, which can be viewed from the two points: 'what reasonings are made' and 'which model it is based on'. Section 4.4.2 and 4.4.3 are written from these two points of view respectively.

4.4.2 Verification by Logical Reasoning

Symbolic Simulation

The simpler and cleverer way than exhaustive simulation is symbolic simulation [4], that is, executing simulations symbolically. The principal feature of symbolic simulation is that it uses symbols rather than actual values for program variables or machine components. This causes important differences from running test cases on a real machine or on a simulator, but also makes possible a higher degree of confidence in machine design. In fact, if all variables are expressed with symbols, symbolic simulation has the same effects on verification as that of exhaustive simulation. On the other hand, if no symbolic values are involved, both the original and symbolic simulation would be the same results.

A design description will assign new values to components by combining values of other components. For example, part of a description may NAND two registers A and B to produce a new value for register C. A simulator has actual test case values for register A and B and produces an actual new value for C, that is, the negation of the conjunction of the two values of register A and B. A symbolic simulator has symbolic values, say $a and $b, for the registers and must produce a symbolic expression ~ ($a&$b) as the new value. Thus, as symbolic execution proceeds, the value of the components becomes more complicated. To provide meaningful information to the user, these symbolic values must properly be manipulated and simplified. This is one of the key points of symbolic simulation.

A simulator will need to simulate the action of branch instructions or other parts of the computer description requiring a choice. Again, this is no problem for test cases — the branch either is, or is not, made according to the value of the test on the test values. If the values are symbolic, such a branch often can not be resolved. Thus, a symbolic simulator must either follow all control paths from such a point, or allow the selection of a particular path by the user of the system. In the example above, a subsequent test $C = 0$ poses no problem in a test case on a simulator or actual hardware. In symbolic simulation, the test is whether the symbolic expression ~ ($ a&$ b) is zero. If this cannot be determined from known or assumed information about &a and &b, then both cases must be assumed and both paths must be followed. Execution sequences will, however, be infinite, whenever the program contains a loop for which the number of iterations is dependent on some procedure inputs. It is this fact that prevents symbolic execution from directly providing a proof of correctness.

```
<TIME> CLK<(10)>.
<AUTOMATON> COUNTER2:CLK:
  <REGISTER> C(2).
  <TERMINAL> START.
  <STATES>
    STATE: |* C=0 & START *| C <- 1.,
           |* C > 0 & C < 4 *| C <- C + 1..
  <END>.
<END> COUNTER2.
```

Figure 4.4.2.1 A DDL Description for a 2-bit Counter

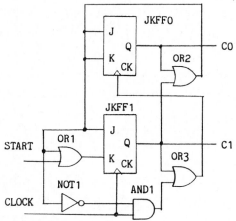

Figure 4.4.2.2 A Circuit Diagram for the 2-bit Counter

```
IDENT    :EXAMPLE;
VERSION :1.0;
DATE :84/12/16;
AUTHOR :M. FUJITA;
PROJECT :VERIFIER;
COMMENT :SAMPLE PROGRAM ;
NAME     :COUNTER2;
PURPOSE :LOGSIM ;
LEVEL    :MODULE;
EXT      :START,CLOCK,CO,C1;
INPUTS  :.START,.CLOCK;
OUTPUTS :.CO,.C1;
TYPES    :AND2,OR2,NOT;
AND2     :AND1;
OR2      :OR1,OR2,OR3;
NOT      :NOT1;
JKFF     :JKFF0,JKFF1;
START        =FROM(.START) TO(OR1.I2);
CLOCK        =FROM(.CLOCK) TO(JKFF1.C,AND1.I2);
CO           =FROM(JKFF0.Q) TO(.CO,OR2.I1);
C1           =FROM(JKFF1.Q) TO(.C1,OR2.I2,OR3.I1);
TT           =FROM(OR2.O) TO(JKFF0.J,JKFF0.K,JKFF1.J,OR1.I1,NOT1.I);
TS           =FROM(OR1.O) TO(JKFF1.K);
CT           =FROM(AND1.O) TO(OR3.I2);
NT           =FROM(NOT1.O) TO(AND1.I1);
CS           =FROM(OR3.O) TO(JKFF0.C);
END;
CEND ;
```

Figure 4.4.2.3 A HSL Description for a 2-bit Counter

In real use the symbolic simulation method is applied to prove that if the two machine descriptions are started in corresponding states, they will continue in corresponding states and therefore behave equivalently [4]. So, symbolic simulation belongs to 'equivalence verification'. A level where verification could certainly be helpful is in proving that an implementation in terms of network of primitive gates (e.g. flip-flops and gates) satisfies a higher level specification. This specification could be written in one of the register transfer languages, such as DDL. Figure 4.4.2.1 gives the specification of a 2-bit counter in DDL. A proposed implementation is shown in the circuit diagram (figure 4.4.2.2), which is in turn described in HSL as shown in figure 4.4.2.3. HSL is a hardware description language and shows structures of hardwares [5]. The two JK flip-flops are connected as a ripple-counter that is enabled when either of the bits is one. There is some extra logic to set the counter to one when the start signal is one. The symbolic simulation can prove that the two descriptions (figure 4.4.2.1 and figure 4.4.2.3) are behaved equivalently. The intended correspondence between the states of the two machine descriptions will be given by a user. To prove the equivalence of the two descriptions we will start the symbolic simulation of the two implementation in corresponding states, allow both descriptions to proceed until they reach control points (in this case after one clock cycle), and then prove that in all cases the new states and outputs correspond.

Verification through symbolic simulation is done as described above. There are, however, several problems about symbolic simulation. The most important of them is that in order to verify the equivalence of two descriptions by symbolic simulation, the intended correspondence between the states of the two machine descriptions must be provided by users. Therefore, symbolic simulation is more suited for interactive use. Moreover, symbolic simulation provides much information when debugging designs. So, several debugging systems using symbolic simulation are proposed and developed [6]. They are connected to graphical input/output tools, and present significant supports.

Inductive Assertion

Verification techniques for software programs are proposed and developed by many people, and some of them are applied for verification of hardware. The most popular of these are inductive assertion method proposed by Floyd [7] and developed by Hoare [8]. The application of the method to hardware is found in Pitchumani and Stabler's [9]. They verified register transfer level designs described in an AHPL-like language for describing synchronous circuits. The inductive assertion method requires a user to embed assertions in his program (or design) at the start, at the end, and at least one within each loop. So, inductive assertion belongs to 'assertion verification'. Each assertion is a predicate on the program variables, and specifies what the user expects to be true at that point. The first assertion specifies what the program can assume about the initial values of variables. The assertion at the end specifies what the program has accomplished. The combination of these two assertions serves as the formal specification of the program.

Proof of the correctness of the program consists of a set of path proofs. That is, suppose there is a possible execution path P from a control point with assertion Q to another one with assertion R with no other assertion on the path. The two control point need not be distinct, as when the path P is a loop. Q is the 'precondition' and R the 'postcondition' for the path. We can generate a 'verification condition' for this path. The verification condition is what must be true in order that when control follows path P starting from precondition Q, the resulting state will satisfy postcondition R. This is described as follows:

$\{Q\}\ P\ \{R\}$

In order to generate verification conditions, the axioms of the relation between precondition and postcondition for all the statements of the language to be verified are defined. Pitchumani and Stabler defined the axioms for all the statement of the AHPL-language, and verified several designs. A part from the proof of verification conditions, the inductive assertion method require the proof of termination in verifying software. Pitchumani and Stabler, however, introduces the variable 't' for expressing time. If the loop invariant or postcondition for the loop contains assertions about time and places an upper bound on 't', the loop has to terminate. This allows assertions to specify real-time characteristics to be used to prove termination.

Although Pitchumani and Stabler' method verifies register transfer level design including termination, it is not easy to specify loop invariant assertions and it needs some techniques to verify hardware having more than two automata, that is, several modules working in parallel. Also, automatic proofs of verification conditions require powerful theorem prover, which is also used in the verification techniques of the followings.

Theorem Proving

Another approach to hardware verification is to use logic. A logic circuit is very easy to describe precisely using logic. Once we describe specifications in logic, a theorem prover will do the rest of the work.

The first thing we can do using logic is to compare a combinational circuit with its specification; a logical expression. Figure 4.4.2.4 shows two implementations of EXCLUSIVE-OR and each can be successfully compared with the specifications, 'EXCLUSIVE-OR'. Of course, this type of verification belongs to 'equivalence verification'.

A further step was taken by W agner [10], who reported on hardware verification by a proof checker. He specified hardware using register transfer statement like:

if cond then A ← B,

meaning that if 'cond' is satisfied, then the values of variable B is transferred to A. He verified several hardware, including adders and 8-bit multiplier, just like a proof in mathematics, that is, the specification in register transfer statement is regarded as a theorem and the design as a hypothesis. His proof of an 8-bit multiplier with 260 steps is excellent, but the problem is that the designer must construct a verification through the interaction with the proof checker. Some reports on automatic proof have been reported. They do not use proof checkers but use theorem provers for automatic verification. W ojcik accomplishes the automatic verification of fairly simple circuits, such as, adders, shifters, etc [11]. On the other hand, proofs of complex circuits often fall into a loop, and do not terminate.

Gordon made use of hierarchy of hardware designs, and accomplished the automatic verification of fairly large hardware; multiplier and a simple computer [12,13]. Hardware instinctively has hierarchy, such as, mask level, transistor level, gate level, module level, system level, etc. They have different abstraction levels of structure and behavior. Gordon developed a theorem prover for the verification of two different levels, called 'LCF_LSM', on LCF [14]. LCF is a proof checker for Logics for Computable Functions, and has a programming language in which we can program proof strategies. LSM, Logics of Sequential M achines, is a formal system which extends the logical calculus embedded in LCF with terms based on the behavior expressions of M ilner's Calculus of Communicating Systems (CCS) [15]. So, the details of LCF_LSM is presented in section 4.4.3. Here instead, we present another verification system, called 'verify' and developed by Barrow, which uses very similar verification strategy [16]. Barrow assumes that design descriptions of these levels are all provided, and his system, call 'verify', verifies the whole system by

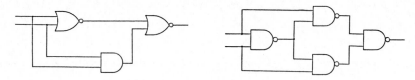

Figure 4.4.2.4 Two Implementation of Exclusive-OR

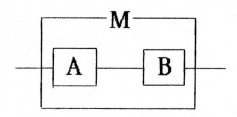

Figure 4.4.2.5 A Hierarchical Module

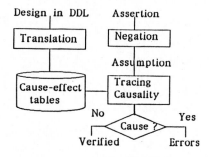

Figure 4.4.2.6 DDL Verifier

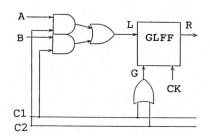

Figure 4.4.2.7 A Circuit

R	(0:0)	1 BIT REGISTER
RANGE	SOURCE	TRANSFER CONDITION
(0)	A	C1
	B	C2

Figure 4.4.2.8 Cause-effect Table for figure 4.4.2.7

showing the equivalence of designs between the two levels. For example, as shown in figure 4.4.2.5, a 4 bit adder is composed of full adders, a full adder is composed of half adders, a half adder is composed of primitive gates, and a primitive gate is composed of transistors. When verifying a 4 bit adder, 'verify' first tries to verify one level below: a full adder. If the full adder is already verified, 'verify' tries to verify the 4 bit adder, assuming that the full adders used are correct. Otherwise, 'verify' tries to verify one lower level designs: a half adder. In this way, 'verify' tries to verify lower level designs until they are already verified or primitive ones whose correctness is guaranteed. 'verify' is implemented in Prolog [18]. The above verification strategy is described in Prolog as follows.

```
verify(Module):- verified(Module).
verify(Module):- primitive(Module).
verify(Module):- composite(Module),
                 verify_parts(Module),
                 verify_whole(Module).
```

The above means that if 'Module' is already verified, verification of 'Module' succeeds (the first definition of 'verify'), if 'Module' is a primitive one, verification of 'Module' succeeds (the second definition of 'verify'), and if 'Module' is neither already verified nor primitive one, the parts of 'Module' are verified using 'verify', and then the entire 'Module' is verified (the third definition of 'verify'). 'verify_whole' tries to verify the whole 'Module', assuming that the parts of 'Module' are all correct. Its definition is like:

```
verify_whole(Module):- spec_behavior(Module,SB),
                       implied_behavior(Module,IB),
                       equiv_behavior(SB,IB).
```

where 'spec_behavior' accepts 'Module' and outputs the specification of 'Module' as 'SB', 'implied_behavior' calculates the behavior that is implied from parts of 'Module' as 'IB', and 'equiv_behavior' checks on the equivalence of two behaviors 'SB' and 'IB'.

To see more details of verification, consider the verification of a design shown in figure 4.4.2.5. Module 'M' is composed of two sub-modules 'A' and 'B'. 'verify' accepts both behavioral and structural description of the design. The behavioral descriptions are like:

```
A.out= f(A.in),
B.out= g(B.in),
```

and the structural descriptions are like:

M .data= A .in,

B.in= A .out,

M .ans= B.out.

Specification to module 'M', that is, 'SB' in the above, may be

M .ans= h(A .in).

From the above, the implied behavior 'IB' is

M .ans= g(f(A .in)).

So, we must verify

h(A .in)= g(f(A .in)),

which is done by 'equiv_behavior'. 'verify' has a theorem prover to show the above. The theorem prover has transformation mechanisms, such as, evaluation, simplification, canonicalization, cancellation, substitution, etc., to verify large types of expressions. Of course, if 'verify' can not verify designs automatically, it requests help to users in order to be a practical tool.

Barrow verified an 8 bit multiplier, called 'd74(8)'. It has 9 levels of designs, 49 different module types, and 18400 transistors. The verification time is about 10 minutes on DEC 2060 Prolog.

Since behavior descriptions for 'verify' are simple state equations, complex timing relations can not easily described. However, it is a pioneer system that verified fairly large systems as a whole.

F. Hanna et al. are developping a verification system using higher order logic [17]. Higher order logic enable us to describe and reason about hardware compactly and smoothly. Also, the concepts in temporal logic (described in the next section) are easily described in higher order logic. Since this approach is very promissing, it is hoped that much progress is done in the near future.

DDL Verifier

DDL Verifier is a verification tool for DDL developed by Maruyama et al. [19,20], and its basic configuration is shown in figure 4.4.2.6. Inputs for the verifier are a hardware design described in DDL and an assertion; a logical formula that is a part of the functional specifications of hardware. So DDL verifier belongs to 'assertion verification'. The DDL description is transformed into cause-effect table by the DDL Translator [3]. Causality is the relation between a cause and its effect. In figure 4.4.2.7, for example, the condition C1 loads register R from A. The value of A in that cycle determines R's output signal value in the next cycle. Tracing causality means, roughly speaking, going from the effect, R's output signal value in one cycle, to its cause, the value of the source A and the loading condition C1 in the previous cycle. For terminals, tracing causality does not mean going back to the previous cycle, but going from the output signal of a combinational circuit back to its input signals. Cause-effect tables provide information for tracing causality. Figure 4.4.2.8 is the cause-effect table translated form figure 4.4.2.7. We can see that the necessary and sufficient condition in the previous cycle for R is

(A & C1) | (B & C2) | (R & ~ (C1 | C2))

We can see that the value of register R is one if and only if the expression above was true in the previous cycle. The DDL Verifier accepts a formula in propositional calculus and some already defined types of temporal logic. The verifier assumes that the assertion is not true and repeats tracing causality to find a feasible cause for the assumption. If no feasible cause exists, that is, the cause is always false, there is no counter example of the assertion, and the assertion is verified. Otherwise, the feasible cause shows why the assertion is not true for the design. In order to check whether the formula obtained is always false, analysis of logical formula is carried out. The basic axioms are as follows:

$A \& \sim A \equiv 0,$

$B \neq B \equiv 0,$

$(B = v1) \& (B = v2) \equiv 0 \ (if \ v1 \neq v2),$

$st1 \& st2 \equiv 0,$

where st1 and st2 are two different states in one automaton. The last axiom is proper to DDL.

It is possible to give assertions in several types of temporal logic expressions [20]. Temporal logic can describe timing relationships among variables, and will be explained in section 4.4.3. Because DDL verifier only handles some fixed types of temporal logic, it can prove only a limited characteristics of hardwares. Complex timing problems, such as interfaces between several modules, however, can be verified by DDL Verifier, though it is very difficult for men. The verifier only investigates causality, which means only the small part of the design is referred even with a very large system. This enable the verifier to prove an assertion for very large systems within a practical time.

4.4.3 Verification with Parallel Execution Models

Formal verification requires some model that has well mathematical backgrounds, and hardwares instinctively have parallelisms. In this section, several parallel execution models that are suited for expressing properties of hardware systems and their application to hardware verification are presented. They are temporal logic [21], CCS [15], and Circal [22].

Temporal Logic

While traditional logic uses such operators as ~ , &, |, → , etc., temporal logic introduces additional operators for dealing with temporal sequences. Temporal operators are defined not on continuous time but on discrete time. So there must be the next state for each state.

While an expression of traditional predicate calculus is assumed to specify properties of the system state at a given time, called the 'present' time, an expression of temporal logic is assumed to specify properties of all possible execution sequences that may evolve from the present system state.

Many temporal logics are proposed differing from each other slightly. Among those, several ones are used for logic design assistance. They are Linear Time Temporal Logic (LTTL) [21], Interval Temporal Logic (ITL) [23], and Branching Time Temporal Logic (BTTL) [24], These and their application to logic design verification are explained in the followings.

Linear Time Temporal Logic (LTTL)

Linear Time Temporal Logic (LTTL) have four temporal operators: (always), ∇ (sometime), ○ (next), and U (until) [21]. The first three are unary operators and the last is a binary operator. Intuitively each operator has the following meanings:

P (with no temporal operators): P is true at present,
○P: P is true in the next state (in sequential circuits, next clock),
□P: P is true at present and all the future times,
∇ P: P is true at least on a time at present and in the future,
P U Q: P is true on all the times until Q becomes true.

Using the above operators, various timing relationships among variables, including hardware properties, can be described. First of all, properties that is satisfied all the times, e.g., 'A is connected to B' is described like:

$\Box(A \leftrightarrow B)$,

because expressions in temporal logic with no temporal operators designate the property of the present time. On the other hand, the property P held in some future time, which is not directly expressed in the traditional logic, is described in temporal logic as follows:

∇ P.

Temporal precedence is described using U operator. For example, 'A precedes B' is expressed as follows:

$\sim (\sim A \ U \ B)$.

Hardware designers usually use timing diagrams to make clear the timing relations among variables. Temporal logic can express such relations with ease and precise expressions as shown below.

For example, the handshaking sequences, shown in the figure 4.4.3.1, is described as follows.

$\Box(\text{Call} \to \nabla \text{ Hear})$,
$\Box(\sim \text{Hear} \to \sim \text{Hear U Call})$,
$\Box(\sim \text{Call} \to \nabla \sim \text{Hear})$,
$\Box(\text{Hear} \to \text{Hear U} \sim \text{Call})$,
$\Box(\text{Hear} \to \nabla \sim \text{Call})$,
$\Box(\text{Call} \to \text{Call U Hear})$,
$\Box(\sim \text{Hear} \to \nabla \text{ Call})$,
$\Box(\sim \text{Call} \to \sim \text{Call U} \sim \text{Hear})$

(expressions marked off by a comma mean the conjunction of those)

Call is a request signal from a calling module to a called module and Hear is a response. The timing diagram says that if Hear is low, then Call rises; if Call rises, Hear rises; if Hear rises, Call falls; and if Call falls, Hear falls.

As shown in the above, various timing relationships can be described in a precise and concise way.

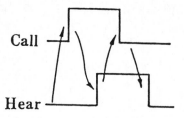

Call

Hear

Figure 4.4.3.1 Timing Diagram for
Handshake Sequences

$\sim p \wedge q$ ① $\sim p \wedge \sim q$
         p
$p \wedge q$ ② ← ③ ↺ $\sim p \wedge \sim q$
         $p \wedge \sim q$

Figure 4.4.3.2 State Diagram Translated
from □(p-> q)&□(q-> q U p)

Call   Call∧Hear

① ② ↺ ⟹

~Call∧~Hear ~Call

Present state	Next state	Condition
1	1	~Call∧~Hear
1	2	Call
2	2	Call∧Hear
2	1	~Call

Call   Hear
design ([O,O], 1, 1).
design ([1,_], 1, 2).
design ([1, 1], 2, 2).
design ([O,_], 2, 1).

Figure 4.4.3.3 Conversion Example of Design to State Diagram

□(Call→◇Hear) ⟶ Call ∧ □ ~Hear
              Negation

                    ↓

Call   Hear   Call ∧ ~Hear
negation_spec ([1,O], a, b).
negation_spec ([_,O], b, b). ← ⓐ   ⓑ ↺

                              ~Hear

Figure 4.4.3.4 Conversion Example of Negation of Specification to State Diagram

A verification method using LTTL is reported in [25], and is explained in the followings.

Any temporal logic expression can be translated into state diagrams by using the temporal operators' expansion rules, and satisfiability of a temporal logic expression corresponds to the existence of an infinite state transition path of that state diagram. For example, a temporal logic expression:

$$\Box(P \to \bigcirc Q) \& \Box(\sim Q \to (\bigcirc \sim Q \ U \ P))$$

translates into the state diagram shown in figure 4.4.3.2. This state diagram has an infinite state transition path like: 1,2,1,2,..., and is therefore satisfiable.

As described in the beginning of section 4.4, verification is to check the next formula:

$$D \to S,$$

provided that S be a temporal logic expression corresponding to a specification and D be a temporal logic expression corresponding to a design. Using proof by contradiction, the premise is first negated and the following expression is obtained:

$$D \ \& \sim S.$$

So, we must verify that this expression can never be satisfied, that is, it is identically false.

As described before, any temporal logic expression can be translated into state diagrams, and satisfiability of a temporal logic expression corresponds to the existence of an infinite state transition path in the state diagram. In order to verify designs using proof by contradiction, we must make sure that the conjunction of a design and the negation of a specification is unsatisfiable. This is done by checking whether there is an infinite state transition path in the state diagram translated from the temporal logic expression.

The verification system uses Prolog [18] as the implementing language, because it has an automatic backtracking mechanism and a powerful pattern matching facility. The automatic backtracking mechanism makes it very easy to check for the nonexistence of an infinite state transition path for all cases, and the powerful pattern matching facility makes programming much easier. In the verification process, both designs and negations of specifications are first converted into state diagram descriptions in Prolog. Then, the verifier checks whether there is an infinite state transition path for the two state diagrams. In other words, it searches for global state diagram for the design and the negation of the specification.

Figure 4.4.3.3 shows an example of the conversion from hardware descriptions to Prolog. (a) is an example of a design of state diagram. Call and Hear are variables or terminals, and 1 and 2 are state names. The state diagram is converted into a relation table between the present and the next state. The table is easily described in Prolog like (c). 'design' is a clause name, and the first argument expresses the values of the two variables Call and Hear, where 0 means inactive and 1 means active. The next two arguments correspond to the present state and the next state.

Like designs, temporal logic expressions are easily converted into Prolog. In the verification process, negations of specifications are converted into Prolog. For example, consider the conversion of the following temporal logic expression (shown in figure 4.4.3.4).

$\Box(Call \rightarrow \nabla Hear)$

First the expression is negated, and becomes the expression figure 4.4.3.4 (b). It is translated into a state diagram by using the temporal operators expansion rules and the result is (c). This is converted into these Prolog statements like (d). This Prolog description expresses the negation of a specification.

In order to execute verification, we must construct Prolog descriptions for the conjunction of design and the negation of specification. This means that we need the global state diagram in Prolog. Since the global state diagram must satisfy the state diagrams for both design and negation of specification, it is expressed as the conjunction of the two state diagram descriptions in Prolog. In Prolog, conjunctions are expressed by connecting descriptions with a comma. So the global state diagram for 'design' and 'negation_spec' in figure ....,is expressed as figure ... 'global' also has three arguments: present values for terminals, present internal states, and next internal states.

To summarize, verification is done in the following way. First design and negation of specification are converted into state diagrams descriptions in Prolog. Then, the global state diagram is made. This global state diagram expresses the relationship 'D $\wedge$S', so it must be checked that this state diagram has no infinite state transition paths. Since we handle only finite state diagram, the existence of an infinite state transition path is equivalent to the state transition path falling into a loop. So, we must check whether any state transition path of the global state diagram falls into a loop. Figure 4.4.3.5 is a flowchart of the verification algorithm. The algorithm checks whether any state transition path of the global state diagram falls into a loop. It is very simple; two step are repeated:

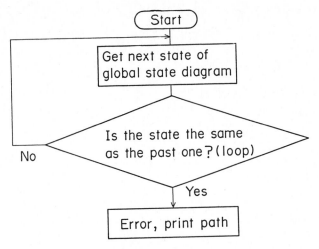

Figure 4.4.3.5 Flowchart of Verification

```
Verify([PI,P2],H):-
% state-transition global state diagram
   design(T,PI,NI),
   negation_spec(T,P2,N2),
% loop-check
   if (member([PI,P2],H))
      then (print([[PI,P2][H])) %error
% if not loop, continue state transitions
      else (verify([NI,N2],[[PI,P2][H])).
```

Figure 4.4.3.6 Prolog Program for figure 4.4.3.5

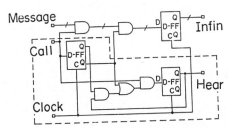

Figure 4.4.3.7 Receiver of a Data Transfer System

getting the next state of the global state diagram, and checking whether whether it is the same state as the past one in the state transition path being examined. If a loop exists, then the state transition path is printed out as a counter example. If there is no next state in the global state diagram, the algorithm terminates. Since we handle only finite state diagrams, the algorithm must terminate.

Figure 4.4.3.6 is the verifier in Prolog, based on the above ideas. The program checks all state transitions of the global state diagram for a loop. 'verify' is a name of the verifier. 'negation_spec' is a Prolog description for the negation of a specification, and 'design' is a Prolog description for a design. 'T' contains the values of all terminals in list format. and 'P1' and 'P2' contain the present states of 'negation_spec' and 'design', respectively. 'N1' and 'N2' contain the next states. 'H' has all states that appear in the state transition being examined, and this corresponds to the loop check. If the predicate 'member' succeeds, that implies the existence of a loop, and the state transition path is printed out as a counter example. If 'member' fails, the verifier continues the examination. The key point of this program is that if 'negation_spec' or 'design' fails, meaning the current state transition path does not satisfy D &^S, the other paths are automatically checked by Prolog's automatic backtracking mechanism. Thus an exhaustive search is easily accomplished.

Because the verifier checks all cases by backtracking, the time required for verification grows exponentially with the scale of the design. So, we need some way of increasing the efficiency of verification, to keep the verification time for large systems sufficiently small. There are three principal methods proposed in [26]; 'memorizing state transitions already treated,' 'utilizing external specifications', and 'extracting the necessary part.' The most important method for increasing efficiency is 'extracting necessary part'. A specifications S has the form of

$$S = S1 \& S2 \& ... \& Sn,$$

where each Si is a fairly simple temporal logic expression for i:= 1 to n. The verifier checks each Si in order. The important point is that only a small of the design is needed for verifying each Si. If we extract the necessary part from the design in advance, the number of cases the cases to be examined by the ' verifier is reduced drastically. For example, consider a circuit for the receiver of a simple data transfer system in figure 4.4.3.7. W hen verifying the specification:

$$\Box(C \, all \rightarrow \nabla \, Hear),$$

only the area inside the dotted lines influence the specification. The extraction algorithm starts with the external output terminals appearing in a specification, in this case 'Hear'. It traces back to inputs, along the networks to be verified, until an external input or a module that was already visited is reached. By this algorithm, we can extract the part of the logic network that influences the specification, which is needed for verification. Since specifications are expressed by the conjunction of a number of simple expressions, the size of the extracted results does not increase with the size of the design, so this technique enables us to verify large systems.

The automatic verification system consists of 2000 lines of code, and is written in C and Prolog [26]. This verification technique is applied to several examples [3-6], including large ones, and the performance was evaluated. The summary of the results are as follows. 'memorizing state transition already treated' speeds up verification by 5 to 100 times. 'utilizing external specifications' speeds up verification by 4 to 100 times. Using these two methods, it takes only a few minutes to verify hardware with less than 100 gates on a main frame computers. 'extracting the necessary part' of the circuit enables us to reduce the circuits to less than 100 gates. Therefore large-scale hardware can be verified.

In the above specifications are described in propositional logic, so only control parts can be easily supported. As for data paths, first order logic (or, higher order logic) is required to specify in a compact and easy-to-understand way. A logic programming language based on first order temporal logic, called *Tokio*, and its interpreter are developed. Several hardware description examples and the results of their simulations are reported in [27,28].

3.1.2 Interval Temporal Logic

Using LTTL, we can describe lots of properties of systems. Describing sequentialities, however, is not easy to express in LTTL, though it is possible.

Interval Temporal Logic or ITL is a temporal logic proposed by Moszkowski [23]. It is based upon 'interval's, which are successive times (or states) of finite length, and has the chop operator ';' (read semicolon) to divide an interval into two subintervals; the former and the latter. Using ';', we can easily express sequentialities in ITL, although it is tedious or not easy to describe such properties in LTTL. For example, as shown in figure 4.4.3.8, 'first P will be true and then Q will be true' is expressed like:

(P ; Q)

ITL has also ○ (next) operator and can express all the operators of LTTL using these.

For example, ∇ operator is expressed in ITL as follows.

$\nabla f \equiv (\text{true} ; f)$

Some operators useful to describe properties of hardware are defined in terms of ; and $\bigcirc$. The beginning and ending time of an interval, beg and fin, are defined as abbreviations of the followings.

beg(P)$\equiv$((empty&P) ; true),
fin(P)$\equiv$(true ; (empty&P)),
where empty=$\bigcirc$fail

In the above, empty means an interval with 0 length, that is, an interval having only one state. Interval length is also defined as:

$\text{len}(n) \equiv \bigcirc^n \text{empty}.$

Using the above operators, a pulse whose width is n is described as:

$\square(0=0) ; \text{len}(1) ; \square(0=1)\&\text{len}(n) ; \text{len}(1) ; \square(0=0).$

Register transfer statements in HDLs or assignment statements in programming languages are described with beg and fin in first order ITL.

$A \leftarrow B \ \square \ (\text{beg}(B=c) \rightarrow \text{fin}(A=c))$

According to ';' operator, ITL can express all the properties expressed in LTTL much easier, especially in describing algorithms, which requires expressions for not only parallelisms but also sequentialities. Many hardware description examples are found in [29]. No formal verification tool has been implemented on computers yet. However, a programming language based on ITL, called *tempura*, is proposed in [30],

Branching Time Temporal Logic (BTTL)

Another important temporal logic is Branching Time Temporal Logic, or BTTL [24]. BTTL handles the expression for several concurrent processes by introducing operators dealing with the next time for each process. The specification techniques using BTTL are very similar to those using LTTL.

A very efficient decision procedure is developed as a Model Checker by Clarke et al. [31]. The Model Checker uses efficient graph-traversal algorithms to check an expression in time linear in the size of the graph and in the length of the expression, in contrast with

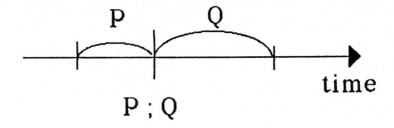

Figure 4.4.3.8 Dividing an interval by ;

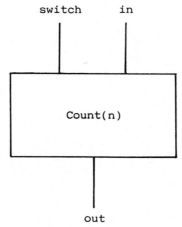

out

Figure 4.4.3.9 A Counter

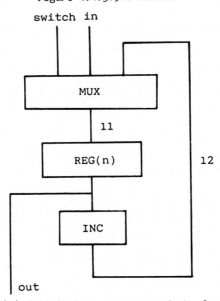

out

Figure 4.4.3.10 An Implementation of the Counter

the fact that using LTTL a specification is verified in exponential order time with the number of the temporal operators embedded in the specification [26]. However, the size of state graph grows exponentially with the size of the circuit to be verified, and the Model Checker must traverse all state transitions, whereas the method for LTTL have only to check a subset of state transitions. Therefore, we can not declare which method is superior to the other.

The Model Checker is written in C and runs on a VAX 11/780 under UNIX. A verification system is also developed using the Model Checker as a central component [32]. The verification algorithm works in two steps: in the first step, it builds a labeled state-transition graph; and in the second step, it determines the truth of a temporal logic expression with respect to the state-transition graph. Two techniques that automatically generate the state-transition graph from HDL sources. The first involves extracting the state graph directly from the circuit by simulation. The second obtains the state graph by compilation from an HDL source.

Several verification examples are reported in [32]. The experimental results show that state graphs with several hundred states can be checked for correctness within several minutes.

CCS

Another parallel execution model is CCS (Calculus of Communicating Systems), which is developed by R. Milner [15]. The basic ideas of CCS are message passing without buffering, and observation equivalence. In CCS a behavior is defined to have a capability of communicating external environment. A behavior has its name and ports to communicate other behaviors. Ports are dynamically connected to other ports which have the corresponding names. For example, a port whose name is π can communicate its complementary port, that is, a port whose name is $\bar{\pi}$. Behaviors are defined by explicitly specifying a sequence of allowed communications as follows:

(1) NIL: a behavior that does nothing,
(2) B+ B': a behavior that does the same as the one that is nondeterministically selected from the two behaviors B and B',
(3) $\pi x_1 ... x_n.B$: a behavior that receives n values from the port π and assigns those values to x_1, ..., x_n and then does the same as B,
   $\bar{\pi} E_1 ... E_n.B$: a behavior that outpus to the port $\bar{\pi}$ n evaluated values from the expressions E_1, ..., E_n and then does the same as B.

σ.B: a behavior that does internal communications which are not observable from the outside and then does the same as B. For example, a behavior that inputs two values from the ports σ_1 and σ_2 and outputs the sum of those values to the port $\bar{\pi}$ is described as:

$$\sigma_1 x.\sigma_2 y.\bar{\pi}(x+y).NIL + \sigma_1 y.\sigma_2 x.\bar{\pi}(x+y).NIL$$

(4) A|B: a behavior that does like the concurrent execution of A and B.

(5) A \π: a behavior that does the same as A except that it does not communicate through the port π.

(6) A $[a_1/b_1, ..., a_m/b_m]$: a behavior whose port names, $b_1, ..., b_m$, are replaced by $a_1, ..., a_m$.

(7) if E then A else B: conditional statement.

Any behaviors that are constructed from the above rules and that do not have any free variables correspond to concurrent programs.

Reasonings in CCS are based on the idea of 'observation equivalence'. The fact that p and q are observation equivalent means that all the observable sequences that can be realized by p and q are equal. CCS supplies axiomatic system for observation equivalence.

M. Gordon made a verification system, called LCF-LSM [12], that is based upon LCF (Logics for Computable Functions) [14] and CCS. Here LCF-LSM is briefly explained by a very simple example. Consider a counter shown in figure 4.4.3.9. This counter is described as follows:

Count(n) = = dev{switch, in, out}.{out= n}; Count(switch -> in |n+ 1).

The argument of 'Count', n, show an internal state of 'Count', and switch, in, out are ports. {out= n} means that the value of the internal state is outputted to the port 'out'. The description after ';' shows the next state or the next behavior. 'switch -> in |n+ 1' is a conditional statement, and means that if switch is true, the next state is in, and if not, the next state is n+ 1 (the state is incremented).

Now suppose that the counter is implemented by multiplexer, register, and incrementer, as shown in figure 4.4.3.10. Each component, i.e., multiplexer, register, and incrementer, is described as follows:

MUX = = dev{switch, in, l2, l1}.{l1= (switch -> in |i2)}; MUX.
REG(n) = = dev{l1, out}.{out= n}; REG(l1).
INC = = dev{out, l2}.{l2= (out+ 1)}; INC.

In order to verify that figure 4.4.3.9 and figure 4.4.3.10 are equal, we must check those are observation equivalent. The description correspond to figure 4.4.3.9 is Count(n), and the one correspond to figure 4.4.3.10 is

MUX | REG (n) | INC.

The observation equivalence of figure 4.4.3.9 and figure 4.4.3.10 means that the two are logically equal. LCF_LSM system supplies us the various tools to show the equivalence. In this case, we can show the equivalence by using hide operation, that is, removing internal lines.

Gordon verifies a rather large example, a simple micro programmed computer [13], and now tries to use LCF_LSM in an industrial environment.

Circal

G. Milne developed another calculus called 'Circal' [22]. He introduces another operator, 'dot operator', in order that complex devices can be modeled hierarchically with the behavior of a composite device being established from the behavior of the parts using the dot operator. The dot operator, together with an abstraction operator, allows us to remove information and describe a device at a more abstract level.

For example, consider a hardware module composed of four sub-modules. Also suppose each sub-module, A, B, C, D, has a behavior in Circal as:

A < = (a d) A 1,
B < = (a b) B1,
C < = (b c) C1,
D < = (c d) D 1,

where a, b, c and d are ports for communications like CCS. The first line of the above means that A first executes communications through the ports a and simultaneously and then behaves just like A 1. The Circal description for the whole module composed of the above four modules is constructed by using the dot operator '·'.

A·B·C·D

By the properties of the dot operator [], the above expression is reduced to the followings:

A·B·C·D = (a b c d) A 1·B1·C1·D 1.

The above means that the whole module first executes communications through the ports a, b, c and d and behaves just like A 1•B1•C 1•D 1.

Using the same techniques, we can easily get the descriptions in Circal for RS flip-flip from the description for two N or gates, etc.

Several verification examples are found in [22]. Supporting tools for Circal are also developed [33].

W e have shown various verification techniques in the above. A lmost all of them are now developing or used only for experiments. However, we believe the time comes when formal verification techniques can be applied to industrial use.

References

[1] M.A. Breuer and A.D. Friedman, A. Isoupovicz, "A Survey of the State of the Art of Design Automation", Computer, Vol.14, No.10, pp.58-75, October 1981.

[2] J.R. Duley and D.L. Dietmeyer, "A Digital System Design Language (DDL)", IEEE Trans. on Computer, Vol.C-17, No.9, pp.850-861, September 1968.

[3] N. Kawato, T. Satito, F. Maruyama and T. Uehara, "Design and Verification of Large-Scale Computers by Using DDL", ACM IEEE 16th DA Conference, June 1979.

[4] W.C. Carter, W.H. Joyner Jr. and D. Brand, "Symbolic Simulation for Correct Machine Design", ACM IEEE 16th DA Conference, June 1979.

[5] W.M. VanCleemput, "A Hierarchical Language for the Structural Description of Digital Systems", ACM IEEE 14th DA Conference, June 1977.

[6] T. Saito, T. Uehara and N. Kawato, "A CAD System for Logic Design based on Frames and Deamons", ACM IEEE 18th DA Conference, June 1980.

[7] R.W. Floyd, "Assigning Meanings to Programs", Proc. of Symposium in Applied Mathematics, Vol.19, Mathematical Aspect of Computer Science, American Mathematical Society, New York, 1967.

[8] C.A.R. Hoare, "An Axiomatic Basis for Computer Programming", Communications of ACM, Vol.12, No.10, pp.576-580, October 1969.

[9] V. Pitchuman and E.P. Stabler, "A Formal Method for Computer Design Verification", ACM IEEE 19th DA Conference, June 1982.

[10] T.J. Wangner, "Hardware Verification", Dept. of Computer Science, Stanford Univ., Report STAN-CS-77-632, 1977.

[11] A.S. Wojcik, "Formal Design Verification of Digital Systems", ACM IEEE 20th DA Conference, June 1983.

[12] M. Gordon, "LCF_LSM: A system for specifying and verifying hardware", Univ. of Cambridge Computer Laboratory Technical Report No.41, 1983.

[13] M. Gordon, "Proving a Computer Correct with LCF_LSM Hardware Verification System", Univ. of Cambridge Computer Laboratory Technical Report No.42, 1983.

[14] M. Gordon, R. Milner and C. Wadsworth, "Edinburgh LCF", Lecture Notes in Computer Science 78, Springer-Verlag, New York, 1979.

[15] R. Milner, "A Calculus of Communicating Systems", Lecture Notes in Computer Science 92, Springer-Verlag, 1980.

[16] H.G. Barrow, "VERIFY: A Program for Proving Correctness of Digital Hardware Designs", Artificial Intelligence, Vol.24, Norh-Holland, 1984.

[17] F.K. Hanna and N. Daeche, "Specification and Verification Using-Higher Order Logic", IFIP 7th Computer Hardware Description Languages and their Applications, August 1985.

[18] W.F. Clocksin and C.S. Mellish, "Programming in Prolog", Springer-Verlag, New York, 1981.

[19] F. Maruyama and M. Fujita, "Hardware Verification", IEEE Computer Magazine, Vol.18 No.2, February 1985.

[20] T. Uehara, T. Saito, F. Maruyama, N. Kawato: "DDL Verifier and Temporal Logic", IFIP 6th Computer Hardware Description Languages and their Applications, May 1983.

[21] Z. Manna and A. Pnueli, "Verification of Concurrent Programs, Part 1: The Temporal Framework", Dept. of Computer Science, Stanford Univ. Report STAN-CS-81-836, June 1981.

[22] G.J. Milne, "Circal: a calculus for circuit description", INTEGRATION Vol.1 No.2 and 3, North-Holland, 1983.

[23] B. Moszkowski: "A Temporal Logic for Multi-Level Reasoning about Hardware", IFIP 6th Computer Hardware Description Languages and their Applications, Pittsburgh, May 1983.

[24] E.M. Clarke and E.A. Emerson, "Design and Synthesis of Synchronization Skeletons Using Branching-Time Temporal Logic", Proc. of Logics of Programs, New York, May 1981.

[25] M. Fujita, H. Tanaka and T. Moto-oka, "Logic Design Assistance with Temporal Logic", IFIP 7th Computer Hardware Description Languages and their Applications, Tokyo, August 1985.

[26] M. Fujita, "Logic Design Assistance with Temporal Logic", PhD Dissertation, Information Engineering, University of Tokyo, 1984.

[27] T. Aoyagi, M. Fujita and T. Moto-oka, "Tokio: A Logic Programming Language based on Temporal Logic - a natural extension of Prolog", Prof. Logic Programming Conference '85, Lecture Notes in Computer Science, Springer-Verlag, 1986.

[28] S. Kohno, T. Aoyagi, M. Fujita and H. Tanaka, "Implementing Tokio", Prof. Logic Programming Conference '85, Lecture Notes in Computer Science, Springer-Verlag, 1986.

[29] B.C. Moszkowski: 'Reasoning about Digital Circuit', Rep. No.STAN-CS-83-970 Dept. of C.S. Stanford Univ. July 1983.

[30] B.C. Moszkowski, Z. Manna: 'Reasoning in Interval Temporal Logic', Rep. No.STAN-CS-83-969 Dept. of C.S. Stanford Univ. July 1983.

[31] E. Clake and B. Mishra, "Automatic Verification of Finite State Concurrent System Using Temporal Logic Specifications: A Practical Approach", 10th ACM Symposium on Principles of Programming Languages, Texas, January 1983.

[32] M. Browne, E. Clarke, D. Dill and B. Mishra, "Automatic Verification of Sequential Circuits Using Temporal Logic", Dept. of Computer Science, Carnegie-Mellon Univ. Report CMU-CS-85-100, 1985.

[33] G.J. Milne, "A Model for Hardware Description and Verification", ACM IEEE 21st DA Conference, New Mexico, 1984.

4.5. RT LANGUAGES IN GOAL-ORIENTED CAD ALGORITHMS

Andrea WODTKO

Department of Computer Science, University of Kaiserslautern
Postfach 3049, D-6750 Kaiserslautern, F.R.G.

This chapter briefly describes examples of design automation systems based on HDL source input. This chapter stresses goal-oriented algorithmic approaches of automated design procedures using AI methods, rather than tool-box-oriented CAD methods which interactively augment human design activities. Mainly two application areas are covered by this paper: automated synthesis and optimization, as well as automatic test pattern generation. This chapter tries to demonstrate the importance of modern software engineering methods for the design of complex digital hardware systems, the importance of methods using goal-oriented AI algorithms in symbiosis with modern hardware description languages on hardware development.

1. Introduction

Although hundreds of Hardware Description Languages (HDLs) have been proposed in the last 30 years, and, although quite a few of it have been implemented, only a very few approaches really promise an efficient support of the VLSI and digital system design process. However, increasing demand due to increasing VLSI circuit complexity more recently stimulated a number of research teams to focus on architectural CAD applications of HDLs rather than on language design issues. That's why recently not only compilers and simulators, but also a few integrated CAD systems based on hardware description languages have been implemented. An example of a CAD tool set following this trend is the *CVT CAT Environment* ([59], *CAT* stands for *Computer-Aided Testing*).

This section of the book, however, gives an overview almost only on such HDL-based systems, where at least subsystems of it follow an algorithmic design automation approach. Since the term *CAD* associates the augmentation of human skills by using of CAD *tools interactively,* we prefer to use the term of *goal-oriented algorithms,* instead. First the role of hardware description languages (HDLs), especially that of RT-level languages, will be explained briefly.   Furtheron concepts of lower description levels (such as e. g. at gate level) and its impact on research in the area of RT-level systems will be shown. Finally the state of art in RT-level-oriented systems and future directions of research will be discussed.

1.1 HDLs In The Design Process

Originally HDLs have been designed to support a formalization of otherwise rather informal hardware descriptions ('narrative descriptions' mixed with more or less informal diagram and schematics notations), such as e. g. found in catalogues of hardware vendors, and sometimes in text books on computer structures and principles of computer hardware design. That's why these languages have been used for more concise hardware descriptions. Another motivation has been the fact, that HDL descriptions are parsable, so that HDLs have been used for decades as input languages to RT level and microprogram level hardware simulation systems.

Currently HDLs and their compilers are applied to support many aspects of hardware design, such as for example a notational vehicle for specification process or as an input medium to algorithms and CAD tools which analyze, verify, or synthesize hardware. Also the source notations used as input to silicon compilers (see chapter by M. Glesner within this book) are more or less HDLs, or, show much similarity to HDLs. Many software packages have been implemented to cope with the complexity of the processes of hardware design and test design. It is the intention of this chapter, to give an overview on HDL-based algorithms having been implemented to handle the problems of hardware design and analysis. This section is mainly restricted to systems using an automation approach via goal-oriented algorithms rather than tools mainly being used more or less interactively by a human user. For an example of how hardware description languages may guide a designer look [55].

1.2 What means *goal-oriented* ?

Because of increasing complexity of VLSI chip design and other forms of digital hardware systems design in many cases a much higher increase of designer productivity is needed, than that given by the CAD approach. The term CAD (Computer-Aided Design) expresses, that the CAD tool box approach just augments human skills. This approach is only a means to cope with increasing complexity of designs. It is an aid for the designer to keep up with rapidly growing requirements. However, it did not really relieve the burden from the designer's shoulders. That's because in this tool-oriented approach the goal-oriented ingredients of the design process have still to be carried out by the human designer.

A new direction of research has been opened up, which is exploring the use of artificial intelligence techniques in digital system design. Instead of *tools* more and more goal-oriented algorithms and other concepts are used, which very often are more sophisticated than tool implementations. So the use of modern software engineering methods, instead of using conventional programming techniques, are highly advisable, since these techniques encourage designers and programmers to focus much more on pre-specifications, before starting implementation. The main aspects of a problem solving process, which conflict with conventional programming techniques, are :

- search processes within large amounts of data are to be carried out
- more than one criterium should be the basis of optimization
- there may exist no optimal solution, but several suboptimal solutions
- each decision is influenced by many different factors
- implications by particular solutions may be recognized in very late phases of the design process
- many alternative solutions must not be excluded too early (see above)

So an almost infinitely wide multi-dimensional space of possible design solutions is opened up, so that it is by far impossible to fully explore it. That's why intelligent search strategies are needed to substantially reduce the amount of search steps required to find a good solution. So to achieve a higher degree of automation within the design process we have to implement algorithms, which allow dynamic ordering and reordering of program steps depending on the state of computation, and which are capable to independantly decide the selection of alternatives.

We should use rule-directed or *rule-driven* search algorithms, which automatically subdivide the general problem into several subproblems to be solved much easier. A rule determines a set of states of computation, and for each of these states an action associated to it, that can be executed to reach another state from the current one. In one state more than one rule may be applicable and heuristics may define a hierarchy that determines which rule should be used under given circumstances. So without rewriting parts of the program we can influence the behaviour of the program by modification of the set of rules, such as e. g. by adding new rules, or, by changing the weights of rules.

Algorithms using these techniques are oriented toward the identification of subgoals on the way to find a final solution. That's why we here talk about *goal-oriented* algorithms. A second way to achieve additional efficiency is to give the algorithm some capability to remember former experience, so that by adaptation the algorithm improves its performance with respect to optimal subgoal selection. To store and handle this knowledge you need a so-called *knowledge base,* which essentially consists of a data base and a knowledge acquisition modul to enlarge and modify this data base and its structure. Two main advantages of using knowledge based systems are:

- performance improvement can be achieved incrementally

Better performance can be achieved by more and more detailed knowledge. The knowledge is represented by rules, which are independent of each other. The improvement of knowledge is achieved by simply adding new rules or defining more detailed rules.

- better understanding of the systems actions

Most rules have the format ' if <situation> then <action>' , so that it is quite simple to implement a small reasoning module displaying the rules, which lead to a particular solution. If the rules are described in a self-documenting mnemonic format, the user can easily comprehend the computers argumentation.

It is not the author's intention to give a tutorial on AI techniques, but rather to illustrate its applications to hardware synthesis and test generation. Specialized information about the structure of these algorithms and the organisation of such systems can is found in the extensive supply of artifical intelligence literature [1]-[3], and many others, where keywords of interest are mainly: *problem solving, heuristics, expert systems, knowledge representation,* and others. More details will be given throughout this text, whenever it seems to be useful for understanding a particular application.

2. Recent Results in goal-oriented methods at lower levels

Recently also some goal-oriented algorithms have been implemented to solve the problems of circuit extraction, logic synthesis and test generation, but most of them use input sources at levels lower than that of register transfer descriptions. Most heuristics-driven search algorithms known from AI literature can be directly applied to routing and placement. Design rules and other rules at lower levels, used as fundamentals of optimization algorithms, using parameters, such as e.g. the length of wires, minimum distances of material layers, scaling of pull-ups/pull-downs, can be easily described in detail. However, the formulation of such rules for an optimal design at RT level, is a much more difficult task, because of much higher complexity. Even splitting-up the problem still leaves behind relatively complex subproblems to be solved. That´s why the development and application of those algorithms started at low level descriptions and is currently is still in an early phase of research at RT-level.

At logical level (gate level) and even at lower levels, many goal-oriented optimization approaches have been implemented, such as for example, some programs for PLA optimization by PLA folding, and programs for PLA test pattern generation [4]-[6]. Also synthesis systems based on standard cell placement and routing have been implemented at this level (TALIB [7]). Test pattern generation for logical circuits can also be done automatically. Here a fundamental algorithm is the *D -Algorithm*. This algorithm, or, versions of it, have been implemented within a number of systems.

This D-Algorithm is suitable for programming in a goal-oriented manner, because it is a formal specification of a path sensitization method and can be divided into several subgoals. The symbol 'D' stands for **Discrepancy** and is assigned to each fault sensitive line within a circuit, e.g. the output line of the 'and' gate in fig. 1.

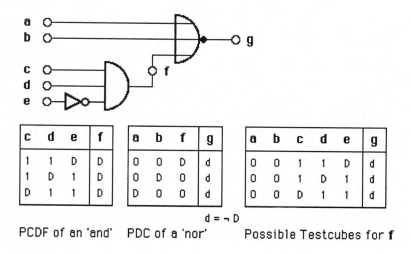

Fig. 1 : Example of the D-Algorithm Tables

The Primitive D-Cube of Failure (PDCF) of a gate defines the input values to sensitize the gate to the fault condition at its output. In case of the three input 'and' two inputs must be '1' to sensitize the third input, which means that each change at this input will be recognized as a change of the output, too.

Finally the 'D' at the gate's output must be forwarded to a primary output. This step is normally called *Propagation* or *D-drive* and the cube defining the other input values for propagation is called Propagation D-Cube (PDC). The 'nor' gate in fig. 1 has a PDC of $0^a 0^b D^f$ and propagates the negated D.

The D-Algorithm starts with selecting a fault and generating the PCDF. Next the D-drive process finds a PDC, which must match with the pattern derived already by the PCDF. Then the D-drive goes on propagating the next PDC until a primary output is reached.

A final consistency operation mostly called justification computes the values of the primary inputs according to the inputs of the PCDF and the PDCs. So, if the D-drive through one node of the justification of one input is interpreted as a sub-goal of the complete test pattern generation task, and a multiple fanout net provides different paths, the above described techniques are needed to handle greater networks within adequate time.

In literature you find many descriptions of its implementation, also of its version, called the *subscripted D-Algorithm* [8]-[14]. In Levendel's and Menon's paper [11] a starting point of using a generalisation of this algorithm adapted to high level language constructs is discussed.

Further AI applications in the field are found in analysis programs for testability evaluation, such as for critical path analysis, and for controlability and observability analysis, and also for gate level verification (Critical Path Tracing and Fault Simulation [15],[16], PODEM-X [17],[18], SCOAP [19], VERENA [20]). All these systems mentioned here use lower level input sources, such as gate level descriptions or comparable formalisms.

3. Applications at RT-level

At RT level relatively a few programs or software systems are found, which have been implemented. The reasons for this have been mentioned in the previous section. At RT level two difficult tasks have to be solved by implementing efficient algorithms for digital hardware design: the synthesis problem itself, as well as the automatic generation of test patterns, both based on derivation from HDL description sources. Figure 2 illustrates these two main research areas where the hardware synthesis procedure is subdivided into five subtasks. Each of these subtasks is a separate application area of goal-oriented algorithms.

In the following chapters, some systems will be discussed and illustrated. Starting with synthesis systems the *CMU-DA*, the Carnegie-Mellon-University Design Assistant, is discussed first. This system has been developed during the last 15 years and has been improved by a knowledge base to increase the capability to handle highly sophisticated design problems. It supports the design of hardware by correcting manual designs, proposing design alternatives with various degrees of

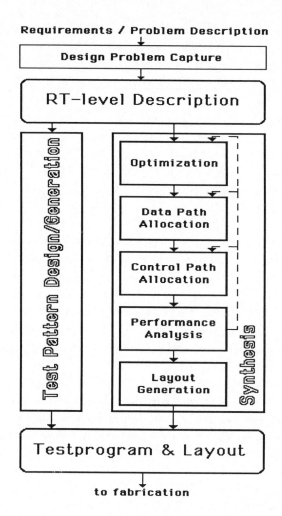

Fig. 2 : Application areas of goal-oriented algorithms at Register-Transfer Level

parallelism, etc. It will be described in the next paragraph. The *REDESIGN* from Rutgers University system assists a designer during improvement steps. REDESIGN is a knowledge based system which gives advice to the designer by checking the consistency of design refinements. Next the *MIMOLA* facilities (developed at the Universities of Kaiserslautern and Kiel) for design and optimization of hardware for microprogrammed architectures will be discussed. The synthesis system *CADDY* developed at the Karlsruhe University combines many algorithms, that can also be found in several other systems, and links them together using one single language, DSL, as input format. So CADDY's importance is not mainly the presentation of new and better algorithms but the unification and combination of existing ones by using one general description language.

With respect to the field of automatic test pattern generation only very few approaches at RT-level have been published. The systems described in literature are mostly based on extensions of the *D-algorithm* or the *subscripted D algorithm,* applications of which can be found at lower levels (also see above).

It has already been mentioned before, that Levendel and Menon [11] describe the application of the D-algorithm to high level language constructs, however, we do not have any indication of an implementation of this proposal. Breuer and Friedman [10] also give a rather theoretical description of a possible application of a generalized version of the D-algorithm at levels higher than gate level, however, an implemented system using these results is not known to us. The language they use is described rather informally and the authors give no hints, whether the methods they introduced are integrated into an existing hardware design environment.

A system of this class, which has been implemented is *SCIRTSS* [47]-[51]. It has been developed at the University of Arizona. The input language for the hardware description is AHPL [47]. It not only uses a modified D-algorithm, but also another heuristics-driven tree search algorithm. Finally the *KARL* system [52]-[60] will be introduced, with the focus its tool for automatic testpattern generation, called *KARATE,* which is actually under development at the Kaiserslautern University. The hardware is described either in KARL-III, a textual description language, or ABL, the graphic companion language [52],[53].

3.1 Synthesis Systems

The main goals of automated synthesis are the improvement of the process of hardware design by accelerating the design iteration step, and by a guarantee of consistency to the refinements in a top-down design. Especially the design is a very difficult creative process, where the skills and experience of a single designer affect the result in many ways. An important problem is the formal definition of design rules at a high level, which characterize the quality of each design decision possible at any stage of the design process. This is a problem because of the great variety of alternatives of a very wide design space, and, since the advantages and disadvantages in consequence of earlier decisions may become obvious in much later phases of the design process. So you may find quite different approaches in the implementation of synthesizers to solve or at least reduce the above problem of hardware design. Shiva [21] subdivides experimental synthesis systems into the following classes :

- design style-dependent, RT level- to -logic synthesizers
- global design space exploration systems for design optimization
- interactive design aids exploring local design space
- logic array synthesizers
- silicon compilers
- systems attempting to follow the human design processes

In general you may have the following criteria to evaluate the quality of hardware synthesis systems :

- interactive use (tool) versus automatic synthesis

- technology-dependent versus technology-independent input language
- down to logic level versus down to layout
- generator part: random logic versus regular structures (PLA, ILA, etc.)

The languages as input sources to those synthesis systems should support descriptions, that are rather independant from a special hardware especially from a special technology. An example for this is the declaration of variables, they should be declared with a defined path width but not bound to special registers. This second step, mostly called hardware allocation, should be already task of the synthesis systems.

The control of the allocation process normally is influenced by giving constraints, e.g. timing or area constraints of the single modules. The first class of constraints will enable a highly parallel design tending to invent hardware operators for each expression of the language description, the later will create a more sequential design, using only few operative hardware elements with multiplexed inputs. The constraints should also be described by the RT language source of the synthesis system. This leads to language structures, dividing a language and the descriptions into a behavioural information part and a structural informaton part. An example of such a language is the MIMOLA language (see chapter 3.1.3).

Another approach used in knowledge based systems is the use of goals describing the quality criteria the final design should have. These goals control the application of the rules. This way the language itself need not have features describing the area used or the time needed by a certain hardware module. The REDESIGN system (chapter 3.1.2) is an example of such goals and rules.

3.1.1 The CMU-DA System

The acronym *CMU-DA* stands for *C* arnegie *M* ellon *U* niversity -*D* esign *A* ssistant. Its task is to support the design of hardware in many ways. So, for instance, the CMU-DA capabilities are:

- correction of manual designs,
- multi-level representations
- Proposal of design alternatives, e. g. varied degree of parallelism

Fig. 3 shows the global flow diagram of the CMU-DA design process. The input language is ISPS, a behavioral description language, which has been developed at Carnegie-Mellon University in the seventies [25],[26]. The CMU-DA converts the ISPS source into an internal representation, the *value trace,* which is an important data format, a source to other processes within the system, such as:

- plotting and displaying of the value trace
- applying metrics for analyzing the value trace
- transforming of the value trace to improve the design
- partitioning of the design
- allocating control steps depending on parallel-to-serial ratios
- allocating data paths to get multilevel representations

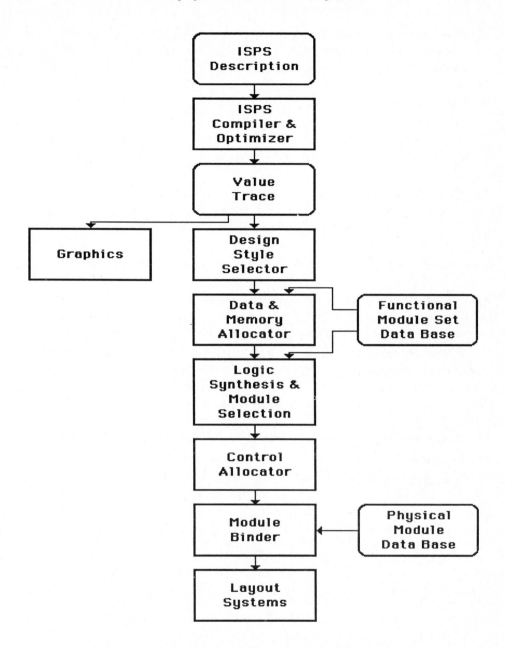

Fig. 3 : The Carnegie-Mellon-University Design-Assistant (CMU-DA)

During the last years an improved system called *DAA* or *Design Automation Assistant* has been developed. This system is a knowledge based system which consists of three main modules, namely a work space memory, a rule memory and a rule interpreter. DAA makes use of goal-oriented algorithms during partitioning, and also inside its program called EMUCS. The function of the EMUCS program will be explained later in this section. Most well known partitioning procedures, e.g. [23, 24], couldn't be used, because they assume a description with fixed hardware elements. The algorithm used in the DAA must do the partitioning before any hardware allocation has taken place.

Main criteria of the partitioning algorithm described by Michael and McFarland [30] are

- data and function, that can efficiently share the same hardware should be within the same partition
- interconnections between single partitions are minimized.

To achieve an effective sharing of hardware, similar objects have to be allocated close together. That's why a metric to define similarity between objects has to be found. Michael and Farland decided to implement such a metric according to the so-called *Jaccord* measure with two different similarities to be computed, one of it is called *operator proximity* whereas the other one is called *register proximity*. The algorithm implementation is integrated into the *VTDRIVE* program, a part of the CMU-DA system.

Another application of goal-oriented algorithms and heuristics can be found in the EMUCS program, which attempts to find a *minimum cost* implementation of the hardware represented by the value trace body. Within this program hardware design rules are used to control the binding of value trace elements to real hardware. Cost tables reflect the feasibility of those bindings. Potential conflicts are solved by looking not only for the least but also for the second least costly elements. Changing the design may be done by changing the cost table database. It is beyond the scope of this paper to go much more into detail here. The interested reader will find more details about the system in [25]-[33].

3.1.2 The REDESIGN System

REDESIGN from Rutgers University is an interactive design tool to support a designer during the functional design of digital circuits. The CRITTER system, a subsystem of REDESIGN, can be used to criticize an existing chip design. The internal data representation distinguishes two structures:

- modules, which consist either of single components or of clusters of modules
- datapaths, which consist of wires or groups of wires.

The history of data values on datapaths is represented by a data structure called *data streams*. The operations of modules are performed by *functions*.

REDESIGN is a knowledge based system and the stored knowledge is represented by a set of *implementation rules of design tactics*. The rules are of the following structure:

        'IF <situation> THEN <design_action>'

according to the definition of the introducing chapters. REDESIGN knows two modes of reasoning

- causal reasoning and
- reasoning about purpose.

Causal reasoning gives an answer to the question: 'what are the consequences of a certain input or output condition ?' The question is answered by examination of data streams or parts of it. Reasoning about purpose answers the question: 'what is the purpose of module M ?' This question is answered by examination of the *design plan*. This plan is a data structure, which shows how circuit specifications are decomposed and implemented by the circuit design. It shows the conflicts and subgoals as well, which arise during design. In this reasoning mode the original design process is essentially viewed as a planning problem, with subgoals derived both from the decomposition of parent goals and from conflicts between other subgoals.

An example of a rule describing a design decision representing control knowledge is shown below :

        IF the goal is to minimize the delay of an adder

            THEN prefer a CARRY-LOOKAHEAD implementation

                over a RIPPLE-CARRY implementation

The current REDESIGN system has many limitations, e.g. limitations in the size the circuits which can be handled. Only board-level circuits built from standard TTL MSI parts have been used as example circuits, the work is followed up by other projects. *VEXED,* an interactive intelligent consultant is actually under development. This system is intended to be able to show alternative designs of a given functional description. It also is intended to keep the structure of the rules as used in REDESIGN, but its main purpose will be the design, not the redesign; the example of the rule given above is part of VEXED's knowledge. Another intention of the actual research is the extension of VEXED to be able to acquire new rules by observing the users' design steps.

Also the CRITTER subsystem is intended to be the starting point of the development of a chip debugging aid, which determines whether failures are due to design or manufacturing faults. Additionally it will attempt to localize the reason of the failure. But, as mentioned before, these are subject of actual and future research.

3.1.3 The MIMOLA Design System

The acronym *MIMOLA* stands for *M*_achine *I*_ndependent *M*_icr*O*_programming *LA*_nguage. The MIMOLA design system is used to generate microprogrammable hardware structures. It has been published first in 1976 [38]. Meanwhile a second version of the system [36] has been developed, including an improved hardware allocator and new features, for example for generation of diagnostics and detection of parallelism, which have not been available within in the first version of the system. Together with the new precompiler you can even use modular PASCAL-like program notations as an input source to the system, instead of the BASIC-like source language of the old version of MIMOLA. The MIMOLA-guided design process is illustrated by fig. 4.

The MIMOLA design system starts with a behavioral description of how the future hardware is supposed to behave. From this source it synthesizes a processor structure. The processor is described by a RT level program notation derived by replacing the high level elements of the functional

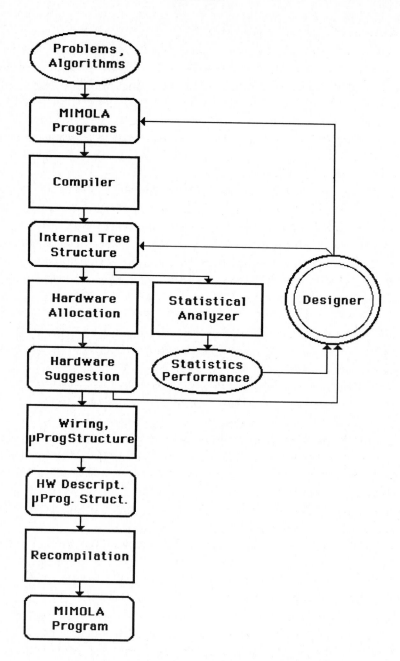

Fig. 4 : Synthesis within the MIMOLA Design System

description by equivalent RT level elements. The goal is to find an optimal hardware structure for a given set of descriptions under given constraints. These constraints may be: execution speed, hardware cost, space needed, and power consumption, as well as the amount of program and microprogram storage needed. They are given in the structural part of a MIMOLA description listing all available hardware modules of the target architecture and their actions. Operators, e.g. can be specified together with *cost* and *time* requirements, storage elements with *cost* and *size* parameters. The behavioural and procedural part of the description consists of so called *elementary statement blocks* (ESBs), that are executable in parallel.

Starting with a MIMOLA program containing a set of available hardware modules and an alogrithm description, that should be implemented on a synthesized architecture, first the syntactic correctness is explicitly in his high level description using ESBs, statements, that might be executed in parallel, are detected and put into parallel blocks. An example of replacement rules to increase the number of statements, that may be executed in parallel, is the non-decomposition of conditional statements. They are not replaced by conditional jumps and unconditional assignments in order to increase parallelism. The RT level description stored in the data base after this run contains a union of all hardware resources needed to execute each microstatement of the high level description within one step. The realization of this hardware would be too expensive, because many hardware elements allocated by this first run are rarely used by most of the microstatements.

Now the designer has to set up certain restrictions to optimize the usage of the resources. The statistical analyzer tells him, how often a hardware element is used and the designer can start with deleting those high-cost low-utilized elements by declaring hardware constraints within his description or changing the *cost* parameter of some hardware modules in the structural part. The optimizing criterium is the factor *time* x *cost*.

At a new run more resources will be requested than are available. The compiler now tries to solve these problems by sequentialization, introduction of storage cells for intermediate results, or by restructuring. All these actions increase the number of steps needed to execute some of the microstatements, of course, but they also increase the throughput of the described and allocated hardware. Since the MIMOLA system is an interactive aid, unsolvable situations will be reported to the designer. The statistical analyzer shows the improvement in design after each iteration step.
The final hardware description may be recompiled to get a MIMOLA program again, which is more readable than the internal tree representation.

3.1.4 The CADDY Synthesis System

CADDY (*CA*rlsruhe *D*igital *D*esign *SY*stem) is based on the *D*igital System *S*pecification *L*anguage *DSL* and it has been developed at the University of Karlsruhe ([40] - [44]).
The compiler transforms the DSL description into the internal flow graph, since the synthesis algorithms are based on data flow analysis. This data flow representation at first provides maximum parallelism, that means, that the control part is very small and it is only needed for data dependent branches and loops with either a data dependent number of iterations or iterations, that depend on former iteration results of the same loop. The area is only minimized under given timing requirements. These timing constraints and the area used by the actual unit are given within the DSL description,

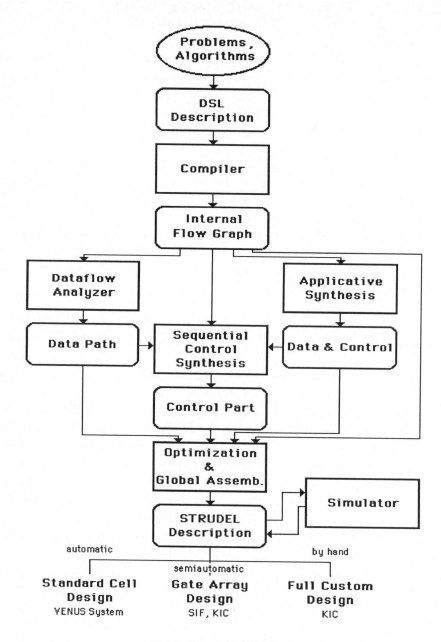

Fig. 5 : The CADDY System

marked by the keywords *AREA* and *FREQUENCY* .

Also the technology used lateron and design style can be specified. Currently available for the DSL description are TTL, CMOS, nMOS, pMOS, ECL and I^2L technologies as well as one-phase-clocked (*PHASE1*), two-phase-clocked (*PHASE2*), *MICROPROGRAM* , *BUS* structured or *PIPELINE* design styles (compare [41]).

Starting with this highly parallel description the optimizing algorithms work on the flow graph representation. Per options the system is told to minimize the processing speed of the hardware and the area consumption locally and globally. Special options may tell the system to reduce area instead of increasing processing speed.

These two tasks may be divided into subproblems by optimizing either the data path or the control path or combining both doing an overall optimization.

The algorithms at RT level are mainly performed by local transformations and based on pattern matching mechanisms. Rules describe the kind of transformations like the balancing or debalancing of arithmetic expressions to make use of the commutativity or the logic expansion of operations with constant operators, e.g. "((a + b) + c) + d" may be decomposed to " (a + b) + (c + d)" or " a + const " may be decomposed to a constant increment with a .

Other RT level optimizations are the so called data path foldings, which mean either that the same hardware operator module is used for several operations by multiplexing the inputs of the operator or that two different operators are joined to form a more universal component. The synthesis can also be controlled by inserting a delay into the components description, so that the system trying to optimize does not take the components with the greater delay time when searching a high speed implementation. The delay or cycle definition is part of the DSL description, all other control mechanisms are system's option. At the moment there are about 20 transformation rules at RT level, but they cannot be modified by the user up to now. This will be part of further research.

The output after the optimization and assembling steps is a structural hardware description in its own description language called STRUDEL (*STRU* cture *DE* scription *L* anguage). This is an input language for a simulator or several layout design programs like VENUS (Standard Cell Design, full automatic process), SIF/KIC (Gate Array Design, semiautomatic process) or KIC (Full Custom Design, done by hand). This format can be also translated into the LOGE system input format [44] doing logic level optimization.

3.2 Test generation Systems

The second large problem in hardware design occuring at RT-level is the testpattern generation. It is not sufficient to describe and produce the hardware, but also to make sure, that it fulfills the specification. Up to now testing and test design is the task of highly skilled specialists with a feeling for fault occurences. Like a 'devil's advocat' they postulate a series of defects which may be present and then develop test sequences to detect these particular defects.

'Capturing all possible faults' is always an assertion which is based upon an underlaid fault model. But even in a minimal fault model, which assumes only stuck-at-0 and stuck-at-1 faults, you cannot generate all combinations of testpattern to detect those faults one after another, because this would spend too much time. You have to find minimal testpattern sets that cover the list of possible stuck-at faults. This restriction causes the loss in error location. As we consider go/no-go tests used during fabrication of VLSI chips this is not relevant, because in this case no error correction can take place. There are several strategies in use to find out, whether a chip is faulty or not :

• random testing;

This is one of the simpliest concepts and is based on a random number generator. The generator produces input vectors, that are applied in parallel to the chip under test and to a simulator, which simulates the correct description or to another chip, that is correct. Still the question of fault coverage cannot be answered, so that this method can only be used, if no specific test confidence is required.

• algebraic techniques;

these methods are based on boolean expression manipulation. This is not useful at RT level, since there are too many boolean equations to be handled at gate level, and, since a RT algebra has not yet been elaborated.

• trace directed methods;

they seem to achieve the best results at RT-level; basing on the generation of sensitive paths through networks by backtracking mechanisms, the amount of computer storage is nearly constant and proportional to the circuit size; complexity increases by the number of choices in the graph search alogrithms. More detailed information about the single methods can be found in [10],[11],[45],[46].

Automatic test pattern generation is needed to cope with the complexity of large VLSI circuits, no matter what method is used. The problem is, that most tests generated automatically either need too much computation time, or, don't achieve enough fault coverage. Again we need a system with a creative capabilities and to get efficient minimized test pattern sequences. Again search and opitimization problems have to be solved. Again there is more than one problem to be solved by the same system. And again we cannot foresee the consequences of a step choosen at a certain stage of the design process.

3.2.1 The SCIRTSS system

The HDL integrated into *SCIRTSS* (*S*equential *CIR*cuit *T*est *S*earch *S*ystem) is *AHPL* (*A H*ardware *P*rogramming *L*anguage), an APL dialect [47],[48]. The input to SCIRTSS is a circuit interconnect list, compiled from an AHPL source, and a set of faults, which have to be detected by the generated tests. The system has been built to generate tests for sequential circuits. This problem is treated as a problem to be refined in a problem reduction graph (PR graph), which provides a useful basis for guiding a search. Based on the single permanent fault assumption the system recognizes a fault either by the occurrence of wrong outputs or by the occurrance of different states between a correct machine and the faulty machine.

A modified D-algorithm observes the behavior of the machine for only one clock period. The results are value vectors for the primary inputs and a *current state*. Then a heuristic tree search algorithm starts to find a transition sequence from the initial state to that current state, so that the test input stimuli can be applied to the machine when in the initial state. If the fault is detected by wrong primary outputs, the next fault in the set of faults can be treated. Otherwise the fault can be detected by a wrong next state. In this case a propagation algorithm tries to find a sequence, which causes a faulty primary output after a finite number of clock cycles.

The goal of both search algorithms, the sensitization, as well as the propagation, is to reduce search cost by finding a minimal sequence of inputs to reach the desired state. Their strategy is an adaptation of that described in Nilsson's book [1]. The problem reduction graph used for sensitization contains three kinds of nodes:

• nodes for registers loaded with a specific value (re:val);

• nodes for values assigned to inputs (in:val);

• nodes for control states (CS,k);

For termination of the PR graph generation and guidance of heuristic search it is necessary to set up an algorithm to compute the probability of satisfying each problem node before the search starts. Basic measurements for the distance to the goal are :

• the number of still unsatisfied nodes and

• the probability of not solving the problem graph at the given state;

For example (in:val) nodes get the probability 1 to indicate that they can be assigned as the user likes it. Also it is useful to include knowledge from earlier searches in the actual probability values. In this way the system gains useful knowledge from circuit features, that have been visited earlier, and use it within future runs.

In spite of all these efforts the resulting test sequences would be very long, if they are computed and applied one after another for each fault of the fault set. An identifying step is added to gather all faults, which have already been tested by another sequence, so that they can be eliminated from the set of untested faults.

3.2.2 The KARATE experimental system

KARATE is part of the *KARL* system [54], which is based on the hardware description language KARL-3 (see also [54]). The system consists of a compiler, a library manager and a simulator. Actually under development is an additional interactive graphic editor using the *ABL* architectural diagram language [52],[53] and an automatic functional test generator. (The abbreviations KARL, or ABL, respectively, stand for *KA* iserslautern *R* egister-transfer *L* anguage, or, for: *A B* lock diagram

_L_anguage, whereas KARATE means _KAR_ l _A_utomatic _T_est _E_xtraction. Figure 4 illustrates the structure of the KARL/KARATE system and its application.

The KARL compiler checks syntactic correctness of the RT-level description program written in KARL. The output of the compiler is an intermediate form, the _Register Transfer Code (RTcode or RTC)_. This form is accepted by the simulator as well as by the test generation system described here, and other KARL-related design tools (e.g. VERENA [20]). The test generation system includes a C test-based library of test patterns for each KARL primitive (C-test: e. g. see [58]). The user also has the chance to insert his own test pattern for special primitives, changing the underlying fault model this way. The primitives themselves are considered to be bit-slice-partitioned, so that path width adaptation is very simple.

From a KARL source KARATE derives a pseudo-structural model of the circuit under test (CUT). This model is a net with nodes, that model KARL functions. A test for a CUT is derived from a combination of node tests. To find a test for a current node under test (NUT) a _sensitization_ has to be found by patterns forwarded from _primary inputs_ (of the CUT) to the NUT, and, a _propagation_ has to be found, an observation path to forward the NUT's responses to primary outputs (of the CUT). KARATE includes a tree search algorithm module knowing two search modes, a backtracking mode, and a forwarding mode to find both, the sensitization, or, the propagation, respectively. Since scan path design strategies, such as e. g. using Level Sensitive Scan Path Design (LSSD), allow the break-up of feed back loops in data paths, KARATE terminates backtracking or forwarding, when reaching primary inputs, primary outputs or registers.

The algorithm used in the KARATE sensitization and propagation subsystem is substantially different from the D-Algorithm. The D-Algorithm just computes _input values_ needed to evoke particular responses subcircuit under test. The KARATE algorithm, however, distinguishes two kinds of values, namely the _control values_ and the _data values_ and computes the control values, which enable the trace of input data values from primary inputs to the node under test (NUT). On this data path both values, the zero as well as the one, can reach the NUT, if the circuit is testable. So KARATE needs to compute such path only once for a NUT, independant of the number of test vectors conveyed by it.

In order to search a way back to the primary inputs the inputs of two and more input operators are subdivided into paths, which carry the common data vector, a control path, and unused paths as has been told before. The choice, which path carries the data vector, is driven by priorities associated to each primitive as a node of the actual net. Reaching a dead end the algorithm returns to the previous node, where this time another combination of data path and control path is tried.

Searching for a propagation is quite similar. Reaching a KARL primitive while carrying a value vector the algorithm determines the value of the other inputs of the node to enable the forwarding of this value vector and starts the backtracking routine for each of these values. To minimize the number of tests

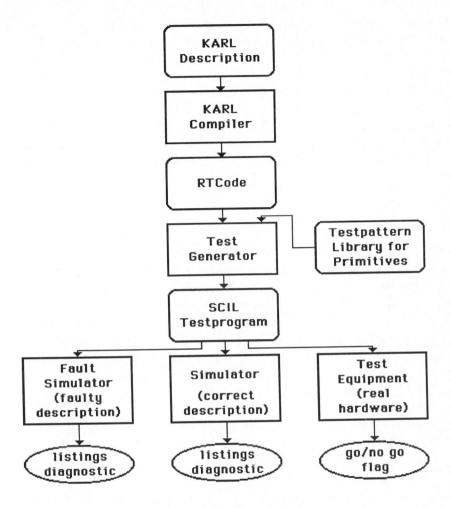

Fig. 6 : The KARL/KARATE System

new backtracking and propagation vectors are computed, whenever a conflict with previous vectors occur. Otherwise the tests can be executed in parallel. An example of the strategies can be found in [57].

The output of KARATE is a simulation program which can be applied to a fault simulator [60] to check the quality of the primtives' test pattern as well as to a automated test equipment. These applications are currently under development at Kaiserslautern. *SCIL*, the $\underline{S}$ imulator $\underline{C}$ ommand and $\underline{I}$/O $\underline{L}$ anguage [56] recently has been extended to include test equipment instruction features.

4. Conclusions

In the past few years also in the area of design sciences trends toward a change in implementation techniques have become more evident. First these techniques have been applied to solve the problems occuring at lower levels of the design process. But meanwhile the problems to develop methods based on higher levels like register transfer or higher functional levels are objects of research. Most of the systems having been developed and implemented up to now are prototypes and starting points to further research. Hardly a fully automated system can be found, but nevertheless very good support for the designers is provided to achieve better results.

The author intended to give a brief overview on the advances made on goal oriented circuit design automation techniques based on register transfer level methodologies. Only a few aspects out of a very large large area of research could be covered within the scope of this chapter. The following list of literature references, may help the interested reader to find more details and more application examples on techniques having been discussed here.

5. Acknowledgments

The author wants to thank the editor, Prof. Hartenstein, for his encouraging advisement and proposals on this section. She also acknowledges Dr. Rosenstiel from Karlsruhe University for his informations concerning the description of the CADDY system and Mr. Franz Schreiner from Kaiserslautern University for his support and demonstration of the MIMOLA facilities. Last but not least she thanks the members of her study group at Kaiserslautern University for the fruitful discussions.

6. Literature

/Artificial Intelligence/

[1] Nilsson, Nils J., Problem solving methods in artifical intelligence, Mc-GrawHill, New York, 1971.

[2] Barr, Avron, Feigenbaum, Edward (eds.), The handbook of artifical intelligence, Volume I - III; HeurisTech Press, Stanford, California, and William Kaufmann Inc., Los Altos, California, 1981-1982.

[3] Pearl, Judea, Heuristics - Intelligent search strategies for computer problem solving; Addison-Wesley Publishing Company, 1984.

/Gate Level Applications/

[4] Eichelberger, E. B., Lindbloom, E., A heuristic test pattern generator for programmable logic arrays, IBM J. R&D, Vol.24, No. 1, Jan. 1980, p. 15-22.

[5] Grass, W., A Depth-First Branch-and-Bound algorithm for optimal PLA folding, Proc. of the 19th Design Automation Conference, 1982, p. 133-140.

[6] Lewandowski, J. L., Liu C. L., A Branch-and-Bound algorithm for optimal PLA folding, Proc. 21th Design Automation Conference, 1984, p. 426-431.

/TALIB/

[7] Kim, Jin H., McDermott, John and Siewiorek, Daniel P., Exploiting domain knowledge in IC cell layout, IEEE Design and Test, (August 1984), 52-64.

/D-Algorithm and its applications/

[8] Benmehrez, C., McDonald, J. F., Measured performance of a programmed implementation of the subscripted D-algorithm, Proceedings of the 20th Design Automation Conference, 1983, p. 308-315.

[9] Benmehrez, C., McDonald, J. F., The subscripted D-algorithm - ATPG with multiple independent control paths, IEEE ATPG Workshop, San Francisco, March 1983, p. 71-79.

[10] Breuer, Melvin A., Friedman, Arthur D., Functional Level Primitives in Test Generation, IEEE Trans. on Computers, C-29, No. 3, March 1980, p. 223-235.

[11] Levendel, Ytzhak, Menon, Premachandran R., Test generation algorithms for computer hardware description languages, IEEE Trans. C-31, No. 7, July 1982, p. 577-588.

[12] Roth, J. Paul, Bouricius, Willard G., Schneider, Peter S.,Programmed algorithms to compute tests to detect and distinguish between failures in logic circuits, IEEE Transactions on Electronic Computers, Oct. 1967, p. 567-580.

[13] Svaenes, Dag, Aas, Einar J., Test generation through logic programming, Integration, the VLSI journal, no. 2, North-Holland Publ. Co., 1984, p. 49-67.

[14] Vaughn, Glen D., CDALGO - A test pattern generation program, Proc. of the 13th Design Automation Conf., 1976, p. 186-193.

/Critical Path Tracing & Fault Simulation/

[15] Abramovici, M, Menon, P. R., Miller D. T., Critical Path Tracing - An alternative to fault simulation, Proc. of the 20th Design Automation Conf., 1983, p. 214-220.

[16] Hong, Se June, Fault simulation strategy for combinational logic networks, Proc. 8th Int'l Symp. on Fault- Tolerant Computing, June 1978, p. 96-99.

/PODEM-X, SCOAP, VERENA/

[17] Goel, Prabhakar, Rosales, Barry C., PODEM-X: An automatic test generation system for VLSI logic structures, Proceedings of the 18th Design Automation Conference, 1981, p. 260-268.

[18] Goel, Prabhakar, An implicit enumeration algorithm to generate tests for combinatorial logic circuits, 9th Ann. Int'l Conf. on Fault-Tolerant Computing, FTCS, 1979, p. 145-151.

[19] Goldstein, Lawrence H., Thigpen, Evelyn L., SCOAP: Sandia controllability/ observability analysis program, Proc.17th Design Automation Conf., 1980, p. 190-196.

[20] Grass, W., Schielow, N., VERENA: A program for automatic verifications of the refinement of a register transfer description into a logic description, IFIP International Symposium on CHDL '85, Tokio, Japan, 1985.

/Hardware Synthesis/

[21] Shiva, Sajjan G., Automatic Hardware Synthesis, Proceedings of the IEEE, Vol.71, No.1, January 1983, p. 76-87.

[22] Sudo, Tsuneta, Ohtsuki, Tatsuo, Goto, Satoshi, CAD Systems for VLSI in Japan, Proceedings of the IEEE, Vol. 71, No. 1, January 1983, p. 129-143.

/Partitioning Algorithms/

[23] Kernighan, B.W. and Lin, S., An efficient heuristic procedure for partitioning graphs, Bell System Technology Journal, 49 (2), 1970, p. 291-308.

334

A. Wodtko

[24] Payne, T. S., vanCleemput, W. M., Automated Partitioning of hierarchically specified digital systems, Proc.19th Design Automation Conf., 1982, p. 182-192.

/CMU-DA , DAA/

[25] Barbacci, Mario R., Siewiorek, Daniel P., Applications of an ISP Compiler in a design automation laboratory, Proc.12th Design Aut. Conf.,1975, p. 69-75.

[26] Barbacci, Mario R., Instruction Set Processor Specifications (ISPS) : The notation and its applications, IEEE Transactions on Computers, Vol.c-30, No. 1, January 1981.

[27] Granacki, John J., Parker, Alice C., The effect of register-transfer design tradeoffs on chip area and performance, Proceedings of the 20th Design Automation Conference, 1983, 419-424.

[28] Hitchcock III, Charles Y.,Thomas, Donald E., A Method of automatic data path synthesis, Proceedings of the 20th Design Automation Conference, 1983, 484-489.

[29] Kowalski, T. J., Thomas, D. E., The VLSI design automation assistant: Prototype System, Proceedings of the 20th Design Automation Conference, 1983, p. 479-483.

[30] Michael, C., McFarland S. J., Computer-aided partitioning of behavioral hardware descriptions, Proceedings of the 20th Design Automation Conference, 1983, p. 472-478.

[31] Thomas, Donald E. et al, Automatic Data Path Synthesis, IEEE Computer, December 1983, p. 59-70.

[32] Tseng, Chia-Jeng, Siewiorek, Daniel P., FACET: A procedure for the automated synthesis of digital systems, Proceedings of the 20th Design Automation Conference, 1983, p. 490-496.

[33] Walker, Robert A., Thomas, D. E., Behavioral level transformation in the CMU-DA system, Proceedings of the 20th Design Automation Conference, 1983, p. 788-789.

/REDESIGN/

[34] Kelly, V., The CRITTER system : Automated critiquing of digital hardware designs, Technical Report WP-13, Rutgers AI/VLSI Project, November 1983, also : Proceedings of the 21th Design Automation Conference, 1984.

[35] Steinberg, Louis I., Mitchell, Tom M., A knowledge based approach to VLSI CAD - the REDESIGN system, Proceedings of the 21th Design Automation Conference, 1984, p. 412-418.

/MIMOLA/

[36] Marwedel, Peter, The MIMOLA design system: Tools for the design of digital processors, Proceedings of the 21th Design Automation Conference, 1984, p. 587-593.

[37] Marwedel, Peter, The MIMOLA design system: Detailed description of the software system, Proceedings of the 16th Design Automation Conference, 1979, p. 59-63.

[38] Zimmermann, G., Eine Methode zum Entwurf von Digitalrechnern mit der Programmiersprache MIMOLA, in: Neuhold, E. J. (edt), GI - 6. Jahrestagung, Informatik-Fachberichte, Vol. 5, Springer- Verlag, Heidelberg, 1976, (in German).

[39] Zimmermann, G., The MIMOLA design system: A computer-aided digital processor design method; Proceedings of the 16th Design Automation Conference, 1979, p. 53-58.

/ CADDY/

[40] Camposano, R. and Rosenstiel, W., A Design Environment for the synthesis of integrated circuits, Preprints of the EUROMICRO Symposium 1985, Brussels, 1985, p. 211-215.

[41] Camposano, R. and Weber, R., DSL - Eine Sprache zur Spezifikation digitaler Schaltungen, Internal Report 24 /84, Informatics Dept., Karlsruhe Univeristy, 1984 (in German).

[42] Rosentiel, W., Optimizations in high level synthesis, Internal Report, Informatics Dept., Karlsruhe University, 1985.

[43] Rosenstiel,W.and Camposano, R., Synthesizing Circuits from Behavioural Level Specifications, Proc. of the 7th Int'l Conference on Computer Hardware Desription Languages (CHDL) 1985, Tokio, Japan, 1985, p. 391-403.

[44] Keller, H., Erzeugung der LOGE Funktionstabellen aus DSL Programmen, MS Thesis, Informatics Dept., Karlsruhe University, 1986 (in German).

/Testing/

[45] Akers, Sheldon B., Test Generation Techniques, Computer, March 1980, p. 9-15.

[46] Putzolu, GR., Roth, J.P., A heuristic algorithm for the testing of asynchronous circuits, IEEE Transactions of Computers, Vol. C-29, June 1971, p. 639-646.

/SCIRTSS/

[47] Hill, F.J., Introducing AHPL , IEEE Comouter, Vol. 7, No. 12, 1974, p. 28-30.

[48] Barbacci, Mario R., A comparison of register transfer languages for describing computers and digital systems, IEEE Transactions on Computer, Vol. C-24, 1975, p.137-150.

[49] Hill, F. J., Huey, B. M., SCIRTSS: A search system for sequential circuit tests sequences, IEEE Transactions on Computers, Vol. C-26, May 1977, p. 490-502.

[50] Huey, B. M., Guiding sensitization searches using problem reduction graphs, Proceedings of the 15th Design Automation Conference, 1978, 312-320.

[51] Torku, K. E., Huey, B. M., Petri net based search directing heuristics or test generation, Proceedings of the 20th Design Automation Conference, 1983, p. 323-330.

/KARL, KARATE/

[52] Girardi, G., ABL editor: User Manual (draft), CVT report, CSELT, Torino, Italy, 1984.

[53] Girardi, G., Hartenstein, R. W. and Welters, U., ABLED - Rt level schematics editor and simulation interface, International EUROMICRO Symposium, Brussels, Belgium, 1985.

[54] Hartenstein, Reiner W., Lemmert, Karin, Wodtko, Andrea, KARL-3 Reference Manual, CVT-Report, Kaiserslautern University, Kaiserslautern, March 1986.

[55] Hartenstein, Reiner W., Lemmert, K. , The hardware description language KARL-III : its integration into a CAD tool box for VLSI, Internal Report, Kaiserslautern University, Kaiserslautern, 1986.

[56] Hartenstein, Reiner W., Mavridis, Angelos, SCIL-3 Language Specification, Kaiserslautern University, Kaiserslautern, 1985.

[57] Hartenstein, Reiner W., Wodtko, Andrea, Automatic generation of functional testpatterns from RT language source, Interner Bericht, Kaiserslautern University, Kaiserslautern, March 1985, also: Proceedings of the international EUROMICRO Symposium, Brussels, September 1985.

[58] Liell, Peter, Test pattern generation for data paths using iterative arrays of cells; Ph.D. dissertation, Kaiserslautern University, Kaiserslautern, 1983, (in German).

[59] Morpurgo, S., Hunger, A., Melgara, M. and Segre, C., RTL Validation of VLSI : an integrated set of tools for KARL, IFIP International Symposium on CHDL, Tokio, Japan, 1985.

[60] N.N., CVT fault generator and simulator level on KARL description nets, CVT report, Ing. OLIVETTI S.p.A., Ivrea, Italy, 1985.

HARDWARE DESCRIPTION LANGUAGES
R.W. Hartenstein (Editor)
© Elsevier Science Publishers B.V. (North-Holland), 1987

4.6. FAULT SIMULATORS AT FUNCTIONAL LEVEL

Marcello MELGARA

CSELT – Centro Studi e Laboratori Telecomunicazioni
Via G. Reiss Romoli, 274, 10148 - Torino, Italy

The fault simulation procedures aim to evaluate the capability of an input sequence to detect and locate faults within
a circuit. The increasing complexity of VLSI devices demands
for higher level of description and validation to reduce computational problems.
At first, this chapter will analyse general problems related
to the fault simulation at different levels. Then, a description and a classification of fault models, with particular
attention to functional models, will be carried out. Finally,
the most common fault simulation techniques, together with
some practical implementation will be presented.

4.6.1. INTRODUCTION TO THE FAULT SIMULATION

The use of HDLs is not confined to the simulation of a digital circuit in normal
operation condition, in order to verify the correctness of the design. Sometimes
it might be advisable to analyse the circuit behaviour under abnormal situations,
due either to physical failures or to particular boundary conditions.

This different approach of the problem is called "Fault Simulation". Given the
description of a fault-free network, the fault simulation process consists of the
following steps:

- define a set of faults according to chosen fault hypothesis;
- modify the circuit model by injecting a fault selected in the fault set;
- apply the same input pattern to the correct and to the faulty circuit description;
- compare the outputs looking for differences that can reveal the presence of the
  fault.

The fault hypotheses, as we will see in the next sub-section, can spread from the
physical failures (short between two adjacent wires, polysilicon or metal line
cut), to the anomalous behaviour of functional blocks.

The main goals of fault simulation are:

- to evaluate the capability of an input sequence to detect and locate faults
  within the circuit;
- to analyse the response of the circuit under the effect of a given set of
  faults.

The general problem encountered in fault free simulation, related with the great
amount of time required to deal with very large circuits, becomes dramatic when a
fault simulation is tried. The computational complexity of the fault simulation
has been estimated to be proportional to the number of the circuit elements to
the power of two [Wil81]. It is clear that, as the network dimension increases,
the time needed to perform fault simulation becomes unacceptable.

In the VLSI world, in which the circuits are composed of tens of thousands of
gates, a logical level approach, i.e. describing the network in term of elementary logical functions such as AND, OR, inverters, cannot be faced with normal
general purpose computers. The only way of getting results in a reasonable time
is to adopt ad hoc hardware simulation engines. These machines, organised as
multiprocessors, can evaluate in parallel more than one element at a time,
achieving very high through-put. This solution is not always acceptable since the
hardware is very expensive. Furthermore, a simulation engine is often a hardwired implementation of a particular algorithm. If a more effective algorithm is
found, it could be very difficult to transform the hardware in order to adopt the
new solution.

A more feasible approach consists in changing the level of circuit description,
trying to reduce the total amount of the primitive elements employed, saving computation time. Functional cells, like registers, multiplexers, ROM, RAM and so
on, can be modeled with a single or few code lines, in spite of many logical
gates. Though this solution doesn't affect the accuracy of a fault free simulation, it can create problems to fault simulation. Since few nodes are reachable
in the circuit description, it might be difficult to inject faults into a primitive: particular care must be given to the fault hypothesis assumptions.

A fault simulation procedure may be used not only to verify if a generated test
pattern can detect a given set of faults, but also to build a diagnostic dictionary, allowing the identification of the fault which caused the observed
error. When fault simulation is performed to compute the fault coverage, it is
advisable to delete, from the fault list, any fault, as soon as it has been
detected, in order to reduce the overall number of faults to be simulated. This
approach, called SOFE (Stop On First Error), is generally very time effective,
since the first patterns have a high probability of covering a large amount of
faults.

However, SOFE techniques may generate some ambiguity if the final aim of the
fault simulation procedure is to provide information about fault location. In
this case, it is better to simulate all the faults with all input patterns. A
table is built, putting in the rows the test patterns and in the columns the
faults, eventually saving, for each pattern, the effect of each fault on the pri-

mary outputs. When the test patterns are applied to the circuit in a given sequence, it is easy to identify the fault that caused the errors by intersecting the rows of the table corresponding to the used input set. It is evident that this technique makes the simulation process very heavy or even unapplicable to VLSI devices, due both to the high device count and to the relevant number of test patterns needed to test such circuits.

Several fault simulation algorithms have been developed [But81][Lev81]. Among these, the most popular are: deductive simulation [Men78], concurrent simulation [Hen80], parallel simulation [Ses65], parallel value list simulation [Moo83]. These methods, except deductive fault simulation, have been applied to several commercial fault simulation programmes. In the following sections the simulation algorithms will be described, analysing a practical implementation for each approach.

4.6.2. FAULT MODELS

The fault simulation process aims to analyze the behaviour of the circuit containing failures. The choice of the kind of failures to inject is both bounded with the adopted description level and the required accuracy.

A first classification of errors tends to distinguish between logical faults and parametric faults.

A logical fault causes the transformation of the logical function of an element into another function. logical faults are usually caused both by design errors (the expected function was synthesized wrongly by the designer) and by physical failures (open bonds, open interconnections, bulk shorts, shorts due to scratches, shorts through dielectric, pin shorts, cracks, and so on).

A parametric fault generally alters the magnitude of a circuit parameter, such as the delay introduced by an element, its response speed, its switching thresholds, currents and voltages. Parametric faults are usually due to modifications of the integration process or to environmental conditions, such as temperature, humidity, power supply fluctuations.

It is very unlikely that parametric faults can be mapped into logical fault models. Although parametric faults are rarely considered in fault simulation process, they represent a real design problem. It is therefore useful to devote a small talk about this subject.

Parametric fault effects are usually studied performing a fault free simulation, with a careful delay analysis. As mentioned before, parametric faults usually generate changes in element delays and driving capabilities. Using a thorough electrical simulation it is possible to compute the range in which the parameters can vary: for each element minimum, typical, maximum delay values are defined. A double delay simulation is performed, in order to verify the overall circuit timing.

Let us consider for example a NOT gate with minimum and maximum delay dmin and dmax respectively. suppose that at time t the input signal switches from the logical value 1 to the logical value 0. If single delay simulation is performed, the output will take the logical value 1 at time t+d, where d is the element nominal delay. On the contrary, in double delay simulation, at time t+dmin the output enters an undefined state till t+dmax when it switches to the final logical value.

Using double delay simulation it is possible to check if all signals can be correctly propagated through the circuit in worst case condition, verifying the absence of races, glitches or skew that can compromise the circuit behaviour. Double delay simulation is also called "worst case simulation".

Some problems may rise if double delay simulation is performed using classical four level logic, in which values 0, 1, X (undefined) and Z (high impedance) are assumed. The time interval during which the signal is in an ambiguous state (it is impossible to identify the switching edge) is different from the undefined state X (0 and 1 have the same probability, this is peculiar condition for uninitialised registers). It is dangerous to assign during the transition period the X value, since mistaken undefined states can be propagated through the circuit, modifying the contents of memory elements.

For those reasons, transition states should be modeled with ad hoc levels, switching to eight, nine or sixteen value simulation, including either a subset or all possible transition among the logical values (i.e. 0->1, 0->Z, 1->0, 1->Z, etc.). This approach, from a theoretical point of view, can be applied to any simulation level, supposing that all basical values are implemented and a delay is concentrated onto the element output.

In the following sections, only logical fault modeling and simulation will be considered, leaving the analysis of parametric fault problems to other bibliography more concerned with simulation problems.

4.6.2.1. Structural fault models

Logical circuits are usually described as networks of primitive elements. Each element implements a particular function, receiving data from its predecessors and providing the computed outputs to its successors. The function implemented by each element may vary according to the level of description. In the past, the most common level was the logical level, in which basic boolean functions such as bitwise AND, OR, NAND, NOR, NOT, etc., were considered.

The most diffused fault model is derived from the analysis of the interconnection network among the elements; it is based on the consideration that a physical fault inhibits the switching capability of a signal, fixing its value either at logical one or at logical zero.

Those faults are called "stuck-at" faults. Given an elementary logical function, it is possible to define both a stuck-at-0 (s-a-0) or a stuck-at-1 (s-a-1) on each input and output of the primitive. The total amount of stuck-at faults for a logical primitive with n input signals and one output signal is $2*(n+1)$.

In a VLSI circuit, composed of thousands of logical gates, the number of possible faults connected to each primitive input and output is too high to allow any kind of fault simulation. A possible partial solution to this problem is to try to reduce the amount of distinct faults to be simulated. This procedure, called "fault collapsing", is based on logical properties of faults [Bre76].

Given two distinct faults, f1 and f2, they are "equivalent" if and only if there doesn't exist a test pattern that can distinguish f1 from f2. Let us consider for example a two input AND gate; Single stuck-at-0's on its input wires are equivalent since the only pattern we can apply to detect them is A=1 and B=1. If two faults are equivalent, it is sufficient to consider only one of them, since if a test pattern detects it, surely it will detect the equivalent one.

Another rule for fault collapsing is based on the concept that a fault f1 may be covered by the test pattern generated for fault f2, although f1 and f2 are not equivalent. Given two faults f1 and f2, f1 is said to "dominate" f2 if the set of test T which detects f2 detects also f1. This means that it is sufficient to generate tests for f2, the dominated fault, since the dominant fault is surely detected by them. Note that if also the set of test for f1 detects f2, it means that f1 and f2 are equivalent.

The application of the equivalence and dominance rules allows the reduction of the total fault count to be considered for elementary logical gate to the ones listed below:

AND, NAND: each input stuck-at-1, any input stuck-at-0;

OR, NOR: each input stuck-at-0, any input stuck-at-1.

This result may be extended to complex combinational circuits. It can be shown that, in a fanout free combinational circuit, any test set which detects all stuck-at faults on the primary inputs detects also all internal stuck-at faults. If some fanout point is included into the circuit, the test set must also cover the stuck-at faults on the branches of the fanout points.

It must be noticed that the fault collapsing procedure may distort the fault coverage concept. Suppose, for example, there are faults f1,..,f10. The

collapsing procedure has reduced the faults from f6 to f10. Suppose then there
exists a test that covers the faults from f1 to f5. The apparent coverage,
defined as the ratio of covered fault versus the total number of faults, is 83.3
%; unluckily f6 represents five faults: the real probability that the test can
detect a fault is 50 %. To avoid this problem a figure must be associated to each
fault giving its weight, i.e. the number of faults it represents.

Other kinds of faults can be easily modeled on a structural circuit description.
It may happen for example that a wire is cut, leaving the input of a gate not
controlled, in an high impedance state. According to the device technology, this
input can be seen as a logical 0, 1, or better as a high impedance.

Another classical failure condition is the short between two adjacent wire. The
circuit behaves as if a new AND or OR gate, according to the device technology,
has been added between the wires.

However, recent studies have shown that stuck-at, open and short circuit faults
are unsuitable for MOS VLSI circuits, because the effect of many physical
failures in complex MOS gates cannot be represented by the stuck-at fault model.
Failures can change the logical function of a gate and also change into sequential, a combinational circuit and vice versa [Gal80,Cou81,Ban83]. This calls for
the use of lower level circuit and fault models in order to achieve meaningful
results from simulation.

This assertion is particularly true for CMOS technology. Let us consider an elementary CMOS NOR gate (figure 4.6.1 a,b). The cut of the line creates a dynamic
memory node, whose value is related to the previous input sequence. Some proposals have been made [Red84] to model the CMOS latch-up effect as a consequence of
open faults. The output node of the CMOS gate (figure 4.6.1c) is substituted by a
Set/Reset flip-flop; the NMOS and the PMOS planes are converted into the equivalent gate level representation, inverting the input signals of the PMOS plane;
the NMOS and PMOS network output wires are connected to the reset and to the set
inputs of the flip-flop, respectively. The cut fault considered in figure 4.6.1b
can be modeled as a stuck-at-0 on the OR gate input.

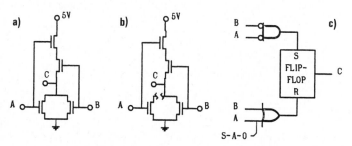

Figure 4.6.1 - CMOS NOR gate; a. fault free circuit; b. faulty circuit;
            c. SET-RESET model.

Although this methodology solves the problem of modeling open faults for CMOS gates, it must be noticed that, in the very simple example shown, a single NOR gate has been substituted by an OR, a NOR and a flip-flop. It is clear that this procedure, if applied to a complete circuit in a flat way, increases many times the degree of the problem complexity. A fault simulator oriented to this kind of modeling must be able to handle the automatic explosion of a single gate at a time, in order to keep down the total number of elements.

Generally speaking, the complexity of a VLSI circuit is too high to perform a complete fault simulation at a low description level in a reasonable time. The only way of solving this problem is to try to rise the circuit description from the gate level to the register transfer level, introducing more complex functional blocks which can substitute many elementary logical gates, reducing the amount of primitives to be processed.

However, since the functional primitives implement more complex functions, the previously defined stuck-at fault models, already uneffective for logical gates, cannot cope with the problem of describing wrong internal behaviours of the new elements. Hence, the adoption of higher level functional primitives demands for the definition of more powerful fault models.

4.6.2.2. Functional fault models

The choice of describing an integrated circuit at an high level, such as register transfer (RT) -level using languages like Karl III [Lem84], allows the reduction of the number of primitives and to adopt the array notation to represent the typical parallel structures of RT-level HDL. An RTL usually includes primitives implementing registers, multiplexers, arithmetic functions, RAM, ROM, decoders, encoders, buses, etc.. Those primitives can substitute tens, hundreds, even thousands (in the case of RAM and ROM) of logical gates.

A drawback in the adoption of RT and functional primitives derives from their inherent complexity. It is in fact impossible to limit the fault analysis only to stuck-ats, opens and shorts of the interconnection lines, without considering internal faults.

As far as simple combinational circuits are concerned, it can be assumed that the internal faults are modeled by changes of the primitive truth table. On the contrary, as far as complex and sequential primitives are concerned, it would be advisable to change the procedure that describes their function. Furthermore, if the circuit has been described in a structured way, as the interconnection of complex functional blocks, each one composed by elementary primitives, it might be interesting to deal with the faulty behaviour of such large device sub-units.

The reasons for analysing the possible faulty behaviour of complex, but well defined functional blocks are, on the one hand, the attempt to reduce the problem dimension and, on the other hand, the idea that faults can be seen as differences from normal behaviour of the block. In the literature a good number of fault models based upon behavioural consideration for general structures, composed by registers, busses, RAM, Programmable logic Arrays [Tha79] can be found.

Let us consider, as an example a simple structure composed by two registers, A and B, connected to a bus. The behavioural faults related with the operation "transfer the content of A on the bus" may be:

(1) neither A nor B can be transferred on the bus, that is kept in an high impedance state;

(2) the content of B, in spite of the content of A, is transferred on the bus;

(3) both the contents of A and B are transferred together on the bus: the final data will be either the logical AND or the logical OR, according to the technology, of A and B contents.

These fault models can be used both for test generation and for fault simulation procedures. It is however clear that this kind of fault hypotheses covers only the functionality of the blocks, leaving uncovered many physical failure possibilities. Furthermore, it is a very hard job to try to redefine by hand both the truth tables and the functional descriptions of complex units, taking into account all reasonable faulty behaviour.

A solution to this problem consists of using low level circuit and fault models to analyze the faulty behaviour of each circuit block. Fault simulation can then be executed at the RT-level by substituting the faulty blocks in the circuit with automatically built functional description of their behaviour in presence of the simulated physical fault. This approach allows one to work at RT-level with transistor level resolution.

Some tools were developed to perform the aforementioned circuit analysis. In the sequel, CVT-FERT [Mel84], a transistor level fault simulator, will be thoroughly described.

CVT-FERT: functional fault model generator

CVT-FERT is a transistor level fault simulator. It takes as input an electrical description of a circuit macrocell and a list of physical failures, both extracted from the circuit layout, and gives as output the flow table of each faulty version of the macrocell. CVT-FERT can handle dynamic and static memory conditions, as well as oscillations, hazards and races. The faulty truth tables computed by CVT-FERT may be used as a functional fault model in a RT-level fault simulator.

The block scheme of CVT-FERT is given in figure 4.6.2. The tool is composed of seven main blocks: the command interpreter, the fault reduction block, the network analyzer, the fault injector, the state variable identification block, the flow table generator and the network simulator.

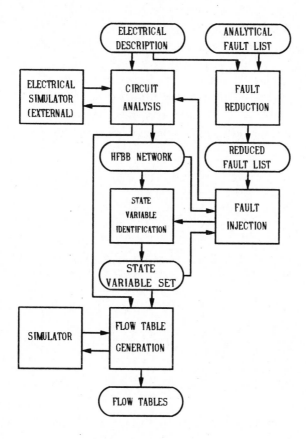

Figure 4.6.2 - CVT-FERT block scheme.

The programme inputs are:
- a description of the circuit at the transistor level, including the parasitic elements (i.e. ground capacitances, coupling capacitances, resistors...);
- a list of analytical (i.e. physical) faults.

Both the description and the list are extracted from the circuit layout. The fault list is currently built according to the level-1 analytical fault hypotheses [Cou81] listed below:
- single polysilicon, diffusion or metal line cut;
- single short between two adjacent diffusion or metal lines;
- single missing device or contact.

A user-defined list of analytical faults can however be input when required.

Before the simulation process begins, the fault list is reduced by collapsing the equivalent faults.

An analytical fault simulator must deal with some non-trivial problems. The introduction of an analytical fault in a circuit may cause a change in its logical function, and maybe more. Changes in a sequential function may be due to changes in the circuit that cause the appearance of hazards, or to changes in the circuit delays that make the network sensitive to pre-existent hazard conditions. Memory points in the circuit may also appear or disappear, and the behaviour of the new memory points may still depend on the propagation delays in the network. The simulator must be able to handle each of these cases in order to work out the actual circuit behaviour.

The whole problem could have been faced with a careful electrical simulation of the circuit for each fault. Unfortunately the electrical simulation run time makes this solution unfeasible. In order to cut simulation time and to meet the requirement related to the sequential circuit analysis a structured top-down approach has been followed.

The circuit is modeled as a network of interconnected hazard-free blocks. Each block is made up of an input set, an output set, a transition delay set and an input-output functions set. The potential memory nodes of the network are identified and associated with the circuit state variables.

The concept of potential memory is related with the topology of the electrical circuit. A potential memory is a circuit configuration that, by itself, could operate as a memory. A potential memory, however, acts as a real memory only if the rest of the circuit generates the proper commands.

The circuit flow table is then determined by simulation. Firstly, the set of the steady states (and of the oscillation conditions) corresponding to each input combination is determined. A series of simulation runs is performed in which the closed loops in the network are opened, the input vector is applied to the network primary inputs, one of the possible present states is applied to the loop inputs and the corresponding future state is read from the loop outputs. The procedure described above is repeated for each input vector. Secondly, the network loops are closed again and the circuit transition table is computed. This time only those input patterns for which more than one state is allowed are considered. A simulation run is performed to check every possible transition from the distance-1 input patterns (and from every allowed state for those patterns) to the chosen pattern. At the beginning of each run the potential memory nodes are forced (if necessary) to their initial values, corresponding to the current state; when the simulator has reached a steady state, the control of these nodes ceases, the input transition is performed and the network evolves freely to its future state. This accounts for the effect of hazards and races when evaluating the network behaviour.

The collected information is finally compacted and the macrocell flow table is built.

The operations described above are performed on the fault free circuit and on every faulty version of it. When a fault is injected into the circuit, the network blocks of interest are opportunely modified. Then the new circuit state variables are identified and the simulation cycle is performed.

Every time a new flow table has been computed, it is compared against the formerly computed ones, in order to check for the existence of non-trivial fault equivalence relationships.

The programme produces as its final output the flow table describing the behaviour of the fault free circuit and a set of faulty flow tables. Each table is associated to the list of equivalent faults that generates it.

The network analysis procedure

The network analysis procedure takes as input an electrical description of the macrocell and gives as output the circuit equivalent representation used by the simulator.

An example of this equivalent representation is shown in figure 4.6.3. The circuit is modeled as a network of hazard free basic blocks (HFBB).

Each basic block is associated with an electrically continuous portion of the circuit, i.e. a circuit part such that there is a low resistance path between any

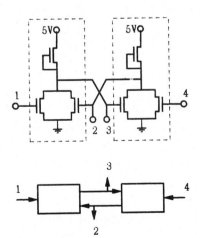

Figure 4.6.3 - Hazard free basic block identification and connection.

two points of it when all the MOS in it are switched on. A HFBB is described by
the set of the block inputs and outputs, the block input-output functions and a
list of input-output delay parameters.

Electrically continuous blocks in a MOS circuit are identified and characterised
by a trace algorithm. Blocks are expanded following the MOS drain-source paths in
the circuit until a ground node or a power node is reached. If a drain-source
path belongs to a block, the corresponding gate node is an input of that block.
The block outputs are the nodes in the block that are connected to a gate. When a
HFBB is identified, its output functions and delay parameter list are computed
performing an exhaustive electrical simulation. The trace algorithm computes the
input sequence for the circuit block, translates the HFBB description into the
electrical simulator language, analyzes the simulation results in order to
extract the requested information.

The network analysis procedure also identifies and handles the potential dynamic
memory nodes in the circuit.

Definition: a node in a MOS circuit is a **potential dynamic memory node (PDMN)** if
and only if

(1) at least one gate is connected to the node, and

(2) the local topology of the node allows it to be put in a high-impedance state,
    and

(3) when all MOS devices are switched on, resistive paths to ground and power
    exist.

A PDMN is modeled as both input and output of the block to which it belongs and
the simulator treats it as a static memory. This causes no error in the open-loop
identification of the network steady states, while the data decay effect due to
leakage currents is easily modeled when the flow table is computed. In the present version the evaluation of the function and of the delay parameters is performed by an accurate electrical simulation of all blocks in parallel, in order
to reduce computation time.

The MOS gate connected to the output is substituted by an equivalent capacitor.
Every capacitor in the HFBB is kept in the model. Particular care is given to
capacitors connecting two different blocks; they are considered as part of both
HFBB and their connection to the outside of the block is considered to be a block
input.

While in a NMOS logic only those block outputs that are not connected to a pull-
up device are PDMNs, in a CMOS logic every HFBB output is a PDMN. Since PDMNs are
treated as inputs in an exhaustive simulation process, their number has to be
kept as small as possible.

The state variable identification procedure

A circuit is sequential if the knowledge of its input values is not always sufficient to completely determine its state. In this case, the problem arises of determining the state variables of the circuit - that is, the smallest possible set of network nodes such that the knowledge of their state and of the input vector assures the knowledge of the whole network state, whatever the input vector is. In the circuit equivalent model, the state variables set can be shown to be equal to the potential memory nodes set defined below.

Definition: a **potential memory nodes set (PMNS)** is a set S of network nodes such that
1) for every loop in the network, at least one node in the loop belongs to S, and
2) the cardinality of S is minimum.

Many different PMNSs may exist for a given network. Since they are all equivalent, any of them may be used as the state variable set of the network (figure 4.6.4).

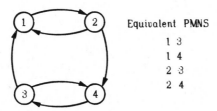

Equivalent PMNS
1 3
1 4
2 3
2 4

Figure 4.6.4 - Example of equivalent Potential Memory Node Sets.

Determining the PMNS of a network is a classical graph theory problem. A new algorithm, based on knowledge algebra and formal equations, has been developed to solve it. A description of the procedure adopted and further details on knowledge concepts, theorems, graph reduction rules and algorithms can be found in [Mel84,1].

The algorithm is based on the knowledge propagation through the graph. The state of a node is known if either all the inputs of that node are specified or it is externally forced.

The PMNS search is performed by a recursive procedure of graph simplification and choice of a node. Simplification rules are given below.

Graph reduction rules

Rule 1: A node whose state is known is deleted from the graph together with all the edges to and from it.

Rule 2: The state of a node with no input is supposed to be known. Primary inputs are modeled as nodes without inputs and hence they can be deleted from the graph.

Rule 3: A node with no outputs is not included in any PMNS because the knowledge of its state gives no further information, and it is deleted from the graph. Primary outputs are modeled as nodes without outputs.

Rule 4: A node that is a direct predecessor of itself is included in every PMNS, and deleted from the graph, since its state is fully known if and only if it is externally controlled.

Rule 5: A node with only one input (output) is deleted from the graph and all its outputs (inputs) are transferred to its predecessor (successor), since if it is included in a PMNS, another PMNS surely exists in which the node is replaced by its predecessor (successor).

Rule 6: An edge not included in any prime loop is deleted from the graph, since it doesn't increase the information.

The flow table generation procedure

The flow table generation procedure takes as input the HFBB network describing the circuit and gives as output the circuit flow table.

Given a network described in terms of HFBBs, all previously characterised by electrical simulation, open-loop zero-delay simulations are performed in order to identify the steady states corresponding to each input vector. Every loop in the circuit is cut, and each input (output) of a cut loop is associated with a current (future) state variable. Signals are propagated through the network by assigning to each HFBB output the asymptotic value implied by its inputs. To minimise the computation time, the propagation order through the HFBBs is precom-puted by the state variable identification routine. A circuit state is identified as steady if the future state vector is equal to the current state vector, i.e. if the output digital value of every loop is equal to its input digital value.

When an input vector corresponds to more than one steady state or to a possible oscillation condition, the actual circuit behaviour has to be determined by a closed loop simulation. In fact, this is the only way of taking into account the effect of hazards in the feedback loops. As mentioned in section II, a simulation run is performed to check every possible transition from every distance-1 input vector. The simulation results are then interpreted by the procedure to build the flow table in its final form.

The determination of the circuit steady states and the transition analysis are performed using a dedicated simulator, described in the next section. The deve-lopment of such a simulator was necessary to optimise the trade-off between result precision and computation time.

The simulator

The CVT-FERT simulator is based on a first-order electrical model. Each HFBB output is associated with a table in which asymptotic voltage values are listed against the block input logical values. For each input of a given HFBB a commutation threshold is given to describe the correspondence between logical values and electrical levels. The current state of a connection within the network is modeled by a parametric representation of the connection-ground voltage. The current state value ranges between 0 and 1000. The chosen model can take into account the transition delays due to charge storage.

The simulator works on an event-driven base; when a logical transition is detected, i.e. the value on a node reaches the computed threshold of an interconnected block input, the eventual transition of the block output is scheduled. The output node will tend linearly to the asymptotic value, with a slope obtained from the previous HFBB thorough analysis. The input slope affects the block response delay only imposing an output signal temporal shift equal to the period necessary to reach the threshold. The input signal slope doesn't influence the output voltage slope. The threshold and the slope are chosen in order to optimise such an approximation. It must be observed that the HFBB definition surely excludes the possibility that the output voltage could begin moving after the input one reached the final value. Such a behaviour can however be obtained by cascading several HFBB.

The problem of determining if a circuit state is or is not a steady one in electrical simulation is overcome by the event-driven nature of our simulator. A state is seen as steady when no further transition is scheduled in the event table.

Dynamic memories can cause trouble when the steadiness of a state has to be detected, since their decay time depends on the quantity of charge stored in them. If decay causes no transition, or if the first transition caused by decay is scheduled far in the future, it can be ignored. On the contrary, if decay causes a short-term transition, it has to be taken into account. For this reason, a time threshold has been defined in the event table to cut out meaningless solutions. The threshold value is supplied by the user as a programme input.

The fault reduction procedure

Before performing the fault simulation, a first reduction of the fault list is executed. All the equivalent analytical faults are collapsed to a single analytical fault. This fault reduction procedure is local to each block and is based on topological properties easy to be identified. No global fault reduction is performed at this stage, because it would take too long. Anyway, a functional fault

reduction is performed at the end of the programme, identifying all the faults
that produce the same flow table.

The reduction rules are as follows.

- All the cuts and all the MOS stuck-at-open on a path between two branch points
  are equivalent (figure 4.6.5a).
- All the shorts between MOS source and drain and all the MOS stuck-at-on in a
  single level OR structure are equivalent (figure 4.6.5b).
- A short between a power line and a line driving an enhancement MOS gate is
  equivalent to a NMOS stuck-at-on or to a PMOS stuck-at-open.
- A short between a ground line and a line driving an enhancement MOS gate is
  equivalent to a PMOS stuck-at-on or to a NMOS stuck-at-open.

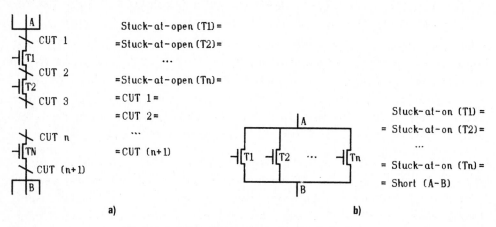

Figure 4.6.5 - Equivalent physical faults.

The fault injection block

The fault injection block picks faults from the reduced fault list and modifies
the fault free descriptions of the circuit according to the chosen fault.

Since the effect of many faults involves only a small portion of the circuit, it
is pointless rebuilding the whole circuit equivalent model and recomputing the
state variable set from the start every time a new fault is considered. The fault
injection procedure performs the following tasks:

- identification of the circuit blocks modified by the fault;
- computation of the new table of block description;
- modification of the circuit equivalent network by constructing and inserting
  the descriptors of the modified blocks;
- computation of the new state variable set, starting from the previous PMNS.

The modified network is then fed into the flow table generator to compute the
minimised flow table of the faulty cell.

Fault collapsing is performed at the end of the procedure, comparing the new flow table against all the already computed ones to check if some other faults exist that generate the same behaviour.

Applications and concluding remarks

FERT has been developed under a research project sponsored by European Economic Community, the CVT project (Cad Vlsi for Telecommunication), whose aim is to build an integrated CAD environment for integrated circuit design and validation. The whole system, as it will be deeply described in chapter 5, is hinged on a common Data Base, that ensures design consistence, and a User Interface. CVT-FERT belongs to a group of RT-level tools, composed of a simulator, a fault simulator, a testability analysis programme. In such an environment CVT-FERT represents the link towards the topological and electrical implementation of the circuit.

The flow tables are computed only once - i.e. when a circuit block is designed -and can be used in every RT-level fault simulation that contains that block. This is particularly useful in gate array and standard cells design environments. Furthermore, if a circuit block is redesigned, the RT-level fault model of the whole circuit can be easily updated changing the flow tables of that block in the fault simulator input.

The use of CVT-FERT is not however limited to the RTL fault simulation environment. CVT-FERT is the connection between the easy to manage, but poorly detailed functional description level and the richly detailed, but hard to manage transistor level. This makes it a useful analysis and verification tool both in circuit design and test. For example, when designing concurrently self-testing circuits and checkers, CVT-FERT can be used to verify if they have characteristics as strongly fault secureness, strong code disjointness or the totally self checking property with respect to a given class of faults.

4.6.3. FAULT SIMULATION TECHNIQUES

The goal of fault simulation is, given a circuit described at a certain level using HDL, to verify if a computed input sequence can detect and locate faults belonging to a given set. The simpliest way of performing fault simulation is:
- simulate the fault free machine;
- for each fault in the fault set:
  - inject the fault;
  - simulate the faulty machine;
  - compare the obtained results looking for differences.

The complexity of the problem rises from the high number of faults to be considered, since the simulation process must be repeated as many times as the number of faults. In order to reduce the computational complexity of the problem, several algorithms have been developed. In the following sections, the most common fault simulation techniques will be described, together with some commercial implementations.

4.6.3.1. Parallel fault simulation

Parallel fault simulation exploits the computer word parallelism to treat many faults at a time. Let us first examine, for simplicity, two value simulation.

Given a multi-bit computer word, for example a 32 bits word, whose bits are named from 0 to 31. Suppose one assigns to bit 0 the good machine and to the other bits (1,..,31) 31 faulty machines. At primary input nodes, the input pattern is assigned to the word. Suppose now this computer word is propagated through the circuit, performing on it all bitwise computations requested by the primitives encountered. When a faulty wire is encountered, the corresponding bit in the word is forced to the faulty value, using precomputed masks. For example, in order to inject a stuck-at-0 in the i-th bit, an AND operation is performed between the word and a mask containing all ones, except the i-th bit at 0; on the contrary, to inject a stuck-at-1, an OR operation is performed with a mask of all zeroes except the i-th bit

When the activity in the circuit, due to a primary input change, is concluded, primary output are considered: bits from 1 to 31 are compared against the first one. Faulty machines, corresponding to the bits whose value is different from the first one, are detected at the primary outputs.

Let us consider the example in figure 4.6.6. A four bit computer word is used; the following assignment is assumed:
- bit 0: the fault free machine,
- bit 1: a1 (wire a stuck-at-1),
- bit 2: d0 (wire d stuck-at-0),
- bit 4: e0 (wire e stuck-at-0).

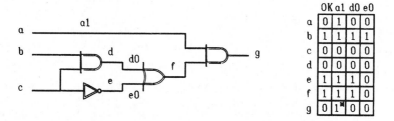

Figure 4.6.6 - Example of parallel fault simulation procedure.

Let us apply the pattern a=0, b=1, c=0, and propagate the four bit word through the circuit. At node a, the fault a1 generates a change in bit 1, since this faults implies a fixed value 1 in that node. This condition is obtained by ORing the word with "0100": the other bits aren't influenced by the fault. At nodes b and c no modification is observed. At node d, being bit 2 already at 0, d0 doesn't effect the value (i.e. ANDing the word with "1101" nothing changes). On the contrary, at node e, bit 3 is forced to 0 by e0. The e0 effect is propagated to node f, but it is masked at node g by the value a=0. At node g, the primary output, the only difference from bit 0 is observed in bit 1: only a1 is detected.

The estimated computational complexity of parallel fault simulation is proportional to the cube of the number of gates. However, in [But81] is referred that, if SOFE technique is adopted, the complexity will be reduced to the 2.5 power rule.

Parallel fault simulation allows simultaneous handling of many faults. Given an n bits computer word, if we exclude the time necessary for masks building, output comparisons and house keeping procedures, the theoretical net gain of parallel simulation is (n-1), since that is the number of faults simulated during each step, in almost the same time as one single fault. If m faults are included in the fault set, the procedure must be repeated m/(n-1) times. The number of faults considered simultaneously may be greater than the computer word if powerful string operations are provided. When a certain number of faults has been detected, the bit assignment is modified, in order to compact the information, reducing string length.

This figure can also be improved if several "independent" faults are simulated in the same bit position. Two faults f1 and f2, defined on wires a1 and a2, are said to be independent if the set of wires S1, interested by f1 effect, has no intersection with the set of wires S2, affected by the propagation of f2. In other words, f1 and f2 never affect the same signal lines, but there are distinct path from the fault site to the primary outputs. A reasonable upper limit to the number of independent paths is the number of primary outputs p. Hence, the minimum number of steps to be performed is m/((n-1)*p). However, the procedure of identifying independent faults may be so complex that the time saved in simulation is lost.

Parallel simulation is easily extended to any number of logical values, by considering more than one word at time. The logical values are coded using one bit of the different words, assigning the same bit position to the same machine. Let us consider for example three value logic. Two words are needed; in the sequel, we will identify the first word as X', and the second ones with X''. A usual coding scheme for three logic value is:

```
X' X''  value
0   0    unknown
0   1    1
1   0    0
1   1    unused
```

Suppose, now, one propagates the two coded words through elementary logical gates. The propagation rules to be applied are listed below, where A', A'', B', B'', C' and C'' represent the first and the second word of A, B, C, respectively.

```
AND: C   = A AND B
     C'  = A' OR B'
     C'' = A'' AND B''
OR:  C   = A OR B
     C'  = A' AND B'
     C'' = A'' OR B''
NOT: C   = NOT A
     C'  = A''
     C'' = A'
```

If an eight value logic is chosen, three words are needed. In this case, more complex function must be defined, such as C'=f(A',A'',A''',B',B'',B'''), and so on.

Parallel fault simulation can also be extended to functional primitives. In that case, the function describing the relation among the different words of input and output signals are more complicated than those defined for logical gates. If two value simulation is used, the function may be expressed with the switching algebra. Some problems may rise if high level functional primitives deal with array variables, rather than single bit signals. In that case, when such a function is encountered, the parallel word must be unpacked, the bits rearranged to create arrays to be evaluated by the primitive; afterwards, the results are packed again in the parallel words.

To cope with fault hypotheses that are different from classical stuck-at faults, the functions used for primitive evaluation and for mask construction may be changed. For example, a delay fault may be handled by modifying the scheduling procedure that should update the "faulty bit" when a change in the "fault free bit" is observed. Other faults, like internal wrong behaviour of high level primitives, require that more complex function must be defined.

In the sequel, Hilo2 fault simulator will be shortly presented.

Hilo2 fault simulator

Hilo2 [Hil84] is a simulation environment developed by Cirrus Computer Inc. Brunel (GB), sold by GenRad, Milpitas (Ca,USA).

Hilo simulator is based on a multi level description language that includes primitives like bi-directional transfer gates, elementary logical gates, buffers, tristates, together with high level constructs, allowing the description of the circuit at a behavioural level of abstraction.

These levels can be combined in the same circuit model, allowing the user to define different description granularity. This approach can be very useful both to reduce the simulation time, exploding only some blocks to be deeply analysed during the actual simulation session, and to perform a top-down design development, proceeding step by step in the refinment process.

An interesting feature of high level language is the concept of "event" that can condition the execution of statements. An event, more than a simple logical variable, is a flag that can be set by a process and tested by another to implement a sort of synchronisation.

Hilo2 description language supports array definition at all levels, allowing a dramatic reduction of the description development time and increasing the listing readability. However, the array notation doesn't modify the simulation time, since the compiler expands the arrays into single bit wires. By the way, since the simulation algorithm is an event driven one, it is more convenient to deal with single bit wires, since, a change on a wire doesn't schedule the modification of the whole array, but only of the affected successors.

The simulation algorithm is based on a four level logic (0, 1, high impedance, undefined). Particular care is given in Hilo2 to delay modeling. It is possible to specify, for each output wire, absolute rise and fall time, or marginal rise and fall time, defined as the slope of the changing edge, related with the capacitance to be controlled by the output. The input capacitance may also be defined. Those parameters allow the modelling the effects of big loads, due to long connection lines, big buffers or large number of controlled gates, that heavily modify the delay introduced by a gate. Furthermore, for each parameter, minimum, typical and maximum value can be specified, in order to perform worst case simulations.

Hilo2 simulation environment includes also a fault simulator and a test pattern generator, both working at logical level.

The test pattern generator is based on a critical path algorithm [Bre76]. It only deals with combinational circuits described as logical gate networks. If some asynchronous feedback loop is included in the circuit, it must be cut using ad hoc primitives that avoid signal propagation within a test generation step.

The fault simulator is based on a parallel algorithm. One hundred and fifty nine faults are treated simultaneously. The fault models supported are stuck-at-0, stuck-at-1, wired-and, wired-or faults.

The fault simulator has shown to work very well also on complex circuits. For example the fault simulation of a 14x10 bits parallel multiplier plus input and output register, including about 2000 gates and 5000 faults, with 20 pattern took about one hour on an Apollo 660.

However, parallel fault simulation techniques, as already indicated, has a complexity that increases as the 2.5 power of the number of gates. Cirrus Computer, in the new version of Hilo, named Hilo3, implemented a new algorithm, called "Parallel list fault simulation". The algorithm, together with its application in Hilo3 will be described in a following session.

4.6.3.2. Deductive fault simulation

Deductive fault simulation was introduced by Amstrong [Ams72] and Godoy [God71]. As far as we know, there is no commercial implementation of the deductive simulation. However, this technique is shortly described for its theoretical importance. In deductive simulation technique, a fault list is associated with each gate output. The evaluation of a gate is scheduled if there is a change in the fault free machine or in the fault list of the fanin nets. A change in the fault list is called a list event. The content of the fault list, at the gate output, is deduced from the predecessors state and fault list.

Let us define $Li = \{f1,..,fk\}$ as the set of faults at the gate input i. If two value logic is considered, the rules for fault list calculation for elementary logical functions are:
- if all inputs are at non-controlling value (0 for OR, 1 for AND):
  Output fault list = (U Li │all gate inputs) U (output stuck-at controlling value);
- if some inputs are at a controlling value (1 for OR, 0 for AND):
  Output fault list = (∩ Li │controlling inputs) ∩ NOT (U Li │non controlling inputs) U (output stuck-at non controlling value);
where (Li │inputs) means for all inputs, and NOT (L) means set of faults not in L. Figure 4.6.7 shows an example of fault list calculation.

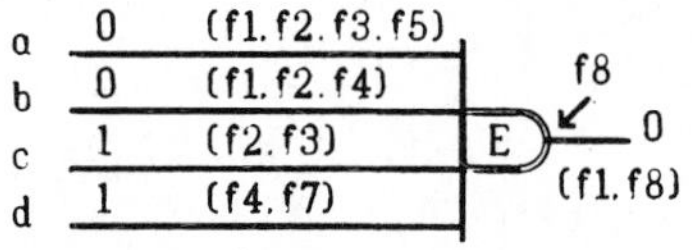

Figure 4.6.7 - Example of deductive fault simulation procedure.

Let us consider the four input AND gate. Note that wires a and b are at controlling value: the second rule applies. The intersection of wire a and b fault set gives {f1,f2}, while the union of c and d fault set gives {f2,f3,f4,f7}. The intersection between the first and the complement of the second fault set obtained is {f1}, since f1 is in the first, but not in the second set. Finally, the union with the non controlling output value gives {f1,f8}, that is the final fault set of gate E.

If a three value logic is considered, the fault list calculation becomes more complex. One way of handling such problem is to keep separate fault lists for the two states different from the good machine: this solution doubles the gate processing time. Another possible solution is to ignore the the faults associated with good machines in the X state. If the good machine state is either 0 or 1, and the faulty state is X, let the fault be denoted by f*.

In order to apply the aforementioned rules, the set union and the intersection must be defined as in figure 4.6.8

A	B	$A \cup B$	$A \cap B$	$A \cap \overline{B}$	$B \cap \overline{A}$
f^x	0	f^x	0	f^x	0
f^x	f	f	f^x	0	f^x
f^x	f	f^x	f^x	f^x	f^x

Figure 4.6.8 - Rules for list operations with faults at X state.

Ignoring faulty machines, when the good machine is X, it may provoke some troubles. Some tested faults may not be marked as detected, and glitches like 0-X-0 may occur and set off oscillations in the sequential elements, in the presence of faults. This behaviour increases the simulation time. A possible solution is to add to the list faults even if the good machine enters the X state, from a 0 or 1 previous state. This technique prevents glitch generation when the good machine changes its state. Another way of escaping from this problem is to use special latch models to handle fault list evaluation in the sequential blocks [Ams72].

When deductive simulation is applied to very large circuits, the storage requirement, especially for the first test patterns, may grow exaggerately. In that case, the fault set must be split into subsets, to be simulated in following steps.

The deductive fault simulation technique has been extended to the functional level of description by Menon [Men78], defining new fault set propagation rules both for combinational functional blocks and for memory elements.

The computational complexity of deductive simulation is estimated to be in the

magnitude order of the square of the number of gates. Compared against parallel
fault simulation, there is an improvement in the performances. In any case, the
problems related to the difficulties of calculating the fault lists and of scheduling events in nominal delay simulation make deductive fault simulation less
interesting than other techniques, such as concurrent simulation.

4.6.3.3. Concurrent fault simulation

Concurrent fault simulation was introduced by Ulrich and Baker in 1974 [Ulr74].
This technique combines the features of parallel and deductive simulations. In
parallel simulation, fully functional and faulty circuit are simulating together,
computing the response of all primitives, although the logical values are equal;
in deductive simulation only the fault free machine is simulated, while faulty
machine behaviour is deduced from the good machine state. In concurrent fault
simulation, a fault is simulated only when it causes some wire to get a different
value from the corresponding ones in the good machine. This method, with respect
to the parallel ones, allows the reduction in the total number of faults to be
simulated for each input pattern.

Let us consider the NOR gate in figure 4.6.9. The input signal A has a stuck-at-1
fault (s-a-1). Since there is a difference in the logical state of good and
faulty machines, the gate will be simulated. If A switches to the logical state
1, the fault will have no influence, its state is identical to the good machine;
hence its simulation is stopped.

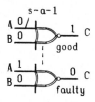

Figure 4.6.9 - Condition for gate calculation in concurrent fault simulation.

Given a network composed of primitives, a fault list is associated with each element, as in deductive simulation. An entry in the fault list contains the fault
index and the input-output value of the gate induced by the fault. A faulty
machine is said to be ACTIVE at a given node if the logical state of that node
differs from that of the good machine. Gate calculation are scheduled for significant events of the good machines and for activated faults. Separate entries in
the work stack are used for good and faulty machine.

Gate evaluation in a general concurrent simulator can be organised as a three
step procedure. We assume that the simulator will be an event driven simulator,
in which the calculation of a gate successor is conditioned by the presence of a

significant event, i.e. by a change in the previous logical value. Supposing that the calculation of a gate was scheduled by a previous event.

```
begin
step 1: if the gate was scheduled by an event in the good machine
        then begin
                    evaluate the new good machine output value;
                    if there is a change in the output value
                    then schedule the successors of the gate for next good machine
                    calculation step;
        end;

step 2: if any faulty machine is active in the considered node
        then begin
                    if the fault is in the list
                    then update the state in the list entry for the faulty machine;
                    else add a new entry to the list;
                    compute the output state;
                    schedule the gate successors if the fault is still active;
        end;

step 3: if good machine output changes creating significant events at the site of
        not yet activated faults
        then schedule those faults for calculation at successor sites;
        remove any fault entry no longer active (i.e. with all input and output
        states equal to the good machine ones);
end.
```

Let us consider the example of figure 4.6.10. Gates are identified by capital letters, wires by small ones; a small letter followed by a number (0 or 1) represent a stuck-at-0 or a stuck-at-1 of the wire. Assume that $a=1$, $b=1$, it follows that $c=1$: fault $a0$, $b0$ and $c0$ are active at gate A. But the value of c, under the effect of $a0$ and $b0$ is equal to the good machine ones: $a0$ and $b0$ are deleted from the entry list, only $c0$ is scheduled in gate B list. The output of gate B, being $c=1$ and $d=1$, is $f=0$. Faults $d0$ and $f1$ are added to the list and scheduled for calculation at gate C. We assume now $e=1$, which forces $g=0$. Faults $c0$, $d0$ and $f1$, scheduled for calculation, generate no active machine at gate C output: only faults $e0$ and $g1$ are detected at the primary output.

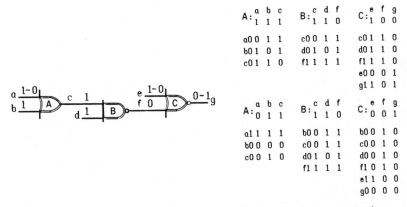

```
A: a b c      B: c d f      C: e f g
   1 1 1         1 1 0         1 0 0

   a0 0 1 1      c0 0 1 1      c0 1 1 0
   b0 1 0 1      d0 1 0 1      d0 1 1 0
   c0 1 1 0      f1 1 1 1      f1 1 1 0
                               e0 0 0 1
                               g1 1 0 1


A: a b c      B: c d f      C: e f g
   0 1 1         1 1 0         0 0 1

   a1 1 1 1      b0 0 1 1      b0 0 1 0
   b0 0 0 0      c0 0 1 1      c0 0 1 0
   c0 0 1 0      d0 1 0 1      d0 0 1 0
                 f1 1 1 1      f1 0 1 0
                               e1 1 0 0
                               g0 0 0 0
```

Figure 4.6.10 - Example of concurrent fault simulation procedure.

At this point a and e are changed to 0. Faults a0 and e0 are no loger active: they are dropped. Since c=1 doesn't change, no activity is scheduled on gate B. Since b0 is now active, at A and B output, it is added to gate B and C entry list. Since e changed, g is recomputed: its new value cause all faults to be active, the old and the new ones are all observed.

From the example we can see that, any time there is a change in the circuit, and a faulty machine is scheduled for calculation, the successor gates fault list must be scanned to verify if the fault is already present.

Since the fault list in a concurrent fault simulator contains entries for each active machine, for any circuit node, while deductive simulators preview only entries to the difference of the gate output state, concurrent fault technique requires more storage space. Furthermore, owing to the separation of scheduling between good machine and faulty values, the work lists are longer.

The problem of the large memory requirement can be solved by splitting the fault set and performing many simulation steps with sub-sets of faults. The number of faults to be considered during each step is related to the memory space. The figure may be defined as an upper limit of the active fault number, in spite of a fixed number of faults per step, since the memory requirement is directly related with the activity in the circuit. This solution is also preferable because it allows the tuning of the simulation system to the computer configuration, rather than to each different circuit to be simulated.

Gate calculation is faster than in deductive simulation . The computational complexity of concurrent simulation is estimated to be proportional to the square of the number of gates in the network.

However, it has been shown that concurrent simulation works better with "short and fat" circuits (i.e. circuits in which many short, possibly distinct paths may be identified between primary inputs and outputs): for such circuits, the entry list is not too long and the searching time in the list is reduced.

Concurrent simulation is maybe the best one for dealing with multivalued logic and particular delay conditions, since the procedures used in the faulty circuit evaluation are the same as those developed for the fault free ones.

This peculiarity of concurrent simulation has allowed its easy extension to functional models which are more complex than elementary logical gates [Hen80][Abr77]. No general theoretical problem is encountered during this extension. As far as fault lists are concerned, all primitives can be seen as black boxes, since all decisions about the state of a given fault are taken after having executed the primitive evaluation. However, the great advantage of concurrent simulation, that of analysing only active faults may be partially lost if this technique is applied to relatively complex functional blocks. Let us consider for example an n-bit, two way multiplexer. If the new data to be propagated affects only bit i, it will be uneffective to recompute and check all the n bits, looking for differences: in gate level simulation, only the active line would have been considered. This disadvantage decreases if the simulator, on the one hand, can treat in an effective way primitives modeling a high number of equivalent gates and, on the other hand, it keeps the capability of manipulating bit information of gate level primitives, injecting faults only on the active single-bit wires.

Some commercial application of concurrent simulation technique are CADAT, LASAR, OFSKA. In the sequel OFSKA, a RT-level concurrent simulator, will be analysed in detail.

CVT fault simulator based on Karl III

The RT-language adopted by CVT project (see 4.6.2.2) was Karl III [Lem84], developed by the University of Kaiserslautern.

The term Karl III is used to identify a simulation environment composed by a compiler, a simulator, an handler of compiled network libraries and various utility programmes.

Karl III simulator works on a four values logic (0, 1, undefined, floating). A network is represented in Karl by a cell, characterised by its inputs, its output and the logic it contains. The logic consists of primitives and other lower level cells. There are about sixty primitives, ranging in complexity from simple gates to registers, multiplexers, RAMs, ROMs. The compiler produces the Register Transfer Code (RTC) that is fed to the simulator. The simulator makes use of a Simulation Commands Language (SCIL) that allows the application of stimuli to the network inputs and to observe the logical state of various circuit nodes.

The fault simulator [Mor85], called OFSKA (Olivetti Fault Simulator based on KArl), designed by Olivetti's researchers within CVT project, is one the tools developed around Karl III for test pattern generation and validation [Mor85,1]. The fault simulator has been built as a shell around the Karl III simulator. The concurrent algorithm was chosen since it seemed the only one that met the requirements of working speed and compatibility to minimise the development time.

OFSKA takes as input the RTC of the fault free network, an RTC-like description of faults to be injected, libraries of fault free and faulty cells. The fault set, generated by a user friendly fault generator, includes the following class of faults:

- stuck-at-0, stuck-at-1, stuck-at-undefined, stuck-at-floating;

- wired-and bridge, wired-or bridge, X-bridge (if values are different, the undefined state is entered;
- truth table changes;
- function changes.

Table and function change models can be automatically computed, starting from the
circuit layout, using CVT-FERT, the functional fault model generator described in
4.6.2.2. Fault coverage figures are computed taking into account the real fault
weight, that may be different from one, due to collapsing procedures.

The simulator makes it possible to simulate many single faults in parallel: to
the introduction of a single fault there is no corresponding effective alteration
to the network, but only recording of the fault.

The fault simulator commands have been added to standard SCIL. They allows the
definition of primary outputs, fault manipulation and examination. The assignment
commands act simultaneously on fully functional network and on the faulty networks, to guarantee that the test pattern applied is identical for all the networks. Particular attention was given to the fact that a simulation session may
last for a considerable time, and so it has to be su ended and taken up again
later: it is possible to save and restore the whole state of the simulator,
stopping the simulation at any time.

All the faults are loaded at the beginning of the session. If no fault is loaded,
the fault simulator behaves as a normal logical simulator. In presence of faults,
however, it also simulates the faulty nets, and updates their "state". The state
is a variable associated to each fault; its content can be seen at any moment
during the simulation: it tells the designer whether the fault has already been
observed, controlled or neither. Figure 4.6.11 shows the finite state machine
graph of fault states. A short description of each state follows.

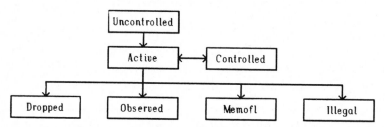

Figure 4.6.11 - The states of faults in OFSKA.

UNCONTROLLED: it is, normally, the initial state of all faults.
ACTIVE: the state of the faulty machine differs from that of the fault free
              one in at least one node. The nets are actually simulated only
              while in this state, since it is the only one in which they behave
              differently from the fault free one.

CONTROLLED: the states of the faulty and of the fault free net are identical, but the fault has been already active at least once.

OBSERVED: this is a terminal state. The fault has been observed at the outputs. The corresponding faulty machine is deleted, and the time of observation and the name(s) of the output(s) on which the fault was observed are saved to be displayed in the final reports.

X-OBSERVED: rather than a state, this is a flag condition. When a difference is detected between the logical levels on the primary outputs of a faulty net and those of the fault free one, the corresponding fault is observed. But it may happen that the only difference is an "unknown" on one of the outputs of either net and a solid value on the corresponding output of the other net: in this case the fault is not marked as observed; it is flagged only as "maybe observed", and its simulation is continued, hoping to get a "solid" detection. In the printed reports, an X-OBSERVED fault is denoted by an "X" prefixed to its state.

MEMOFL: this is also a terminal state. If all the available memory has been occupied, while more is still needed to simulate all the faults, the "exceeding" ones are put in this state, and henceforth ignored. Those faults shall constitute the input set of a subsequent simulation session; this process may need to be iterated a number of times if the memory is saturated again.

DROPPED: another terminal state: it is entered if the user decides to stop the simulation of a specific fault; this might be the case for one or more X-OBSERVED faults.

ILLEGAL: another terminal state. It is entered if a fault causes some condition that the simulator cannot handle (typically, an "unknown" value on a control input of a primitive, or a feedback loop with no delay, and the like).

Every time the computing command STEP is given, all the networks (both faulty ad fully functional) are simulated in parallel. Figure 4.6.12 shows a flow chart of STEP commands.

There are three areas of memory on which STEP works continuously: **main, old, new.** **Main** is the main work area, in which all the calculations are performed; **old** and **new** contain the initial and final states respectively of the simulation step of the fully functional network, and are of fundamental importance in the operation of reconstruction and comparison of the states of the faulty machines.

Thus at first the fully functional network is simulated, saving its initial and final states in **old** and **new** respectively. Then, one at a time, the faults are simulated. Not all of them, only those in the ACTIVE state and those which,

```
Procedure STEP;
begin
    save status in OLD;
    compute new status;
    save status in NEW;
    while not end of faults do
    begin
        next fault;
        restore status from OLD;
        inject fault;
        update the status of the nodes following the indications
         contained in the faulty machine description;
        compute new status;
        compare the status with NEW and generate
         new faulty machine description;
    end;
    restore status from NEW;
end STEP.
```

Figure 4.6.12 - Procedure implementing the STEP command in OFSKA.

though in the UNCONTROLLED or CONTROLLED states, if inserted into the circuit, generate a different activity from that of the fully functional network at the fault location: they are potentially able to pass into the ACTIVE state. Simulating only these faults permits considerable CPU time saving without negatively affecting the result obtained.

The simulation of each fault may be subdivided into three very distinct phases: the reconstruction of the associated faulty machine, the actual calculation, and the comparison with the faulty free machine. The reconstruction is prompted by consideration of memory economy: since it is presumed that the effects of a fault will have repercussions on a reduced percentage of primitives, it is preferable to record on the faulty machine only the state of those primitives, and to take the value of the remaining ones directly from the fully functional machine. The injection of the fault also forms part of the reconstruction phase. At the end of these operations the state of the faulty machine has been reconstructed in the main work area exactly as it was at the end of the preceding step.

There follows the simulation of the faulty machine, at the end of the which the new state of the faulty machine is in the main work area. Finally a primitive-by-primitive comparison is made between the new state of the faulty machine and of the fully functional ones. In the machines associated with the fault is recorded the state of all primitives which show differences in the course of the comparison. If a difference is found at a primary output, the fault is marked as observed, the associated faulty machine is deleted and the fault is no longer simulated. If the effect of a fault disappears, the fault is put in the CONTROLLED state, its simulation stops until a value at the fault site may prime new potential activity.

When simulation of all faults is complete, the new state of the fully functional machine is reinstalled in the main work area, and control passes to the following command.

The fault simulation programme has been developed in Standard Pascal on VAX/780 under VMS 4.2 operating system. It is also running on VAX under Berkeley Unix and on Apollo workstation under Aegis 9.0 operating system. The OFSKA fault simulator belongs to the test pattern generation and validation environment built around Karl III and connected to the CVT design data base COSMIC. The other tools (figure 4.6.13), described in [Mor85,1], are:

- OFGKA: fault generator;
- OTAKA: testability analyser;
- FERT: functional fault model generator;
- TIGER: test pattern generator and validator.

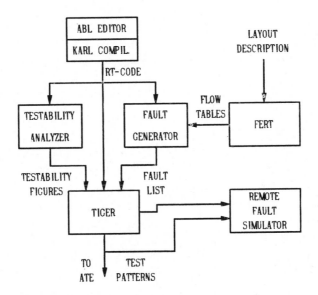

Figure 4.6.13 - Test pattern generation and validation tools in the Karl III environment.

TIGER has been built as a further shell around the OFSKA fault simulator. A designer, after having described and compiled his network using the ABL graphic editor [Gir85] and the Karl III compiler, can enter the user friendly, interactive environment of TIGER, to perform fault free simulation, test pattern generation and fault simulation, using an extended set of SCIL commands. The advantage of this approach is to provide the user with a single environment, saving him the trouble of learning different languages, and allowing him to switch easily from the validation of the fault free design to the analysis of faulty conditions.

4.6.3.4. Parallel value list fault simulation

The parallel value (PV) list approach [Moo83] is a combination of the parallel technique and a list structure, similar to the deductive and the concurrent technique. The faults are collected into groups of a size equal to the computer word size. Each fault is identified by the group number and the bit position in the word. Each wire has a fault free value and a fault list of equal sized cells, containing the value of a group. Each group is present if at least one bit in it has a value different from the fault free.

The parallel value list technique requires less memory than concurrent technique, since values are coded in parallel words, and dynamic creation and deletion of the cells is fast. It has been shown that parallel fault simulation is more suited for highly sequential circuits, while concurrent fault simulation works better with "short and fat" circuits. PV list is a trade off between those characteristic; if faults are chosen to make up a group of "near equivalent" faults, the faulty activity produced will lead to shorter lists and a higher concentration of the faulty effects in the cell. When faults are detected, they are deleted and the remaining faults are compressed to reduce the group number.

The PV lists are propagated, following two methods, according to the circuit activity. If the fault free value changes, involving many changes in the faulty values, the evaluation process is similar to the deductive techniques, where set union and intersection operation are applied to the fault lists. If the fault free value doesn't change and only a few faulty values are affected, the evaluation procedure ignores the inactive groups; the fault effects are propagated using an approach similar to the concurrent technique that evaluate and schedule the active faulty elements.

Let us consider the example in figure 4.6.14, derived from [Moo83]. Suppose that the computer word is eight bit (0,..,7). The following notation is adopted to represent the groups and values. Given the string a=0-G7(p2=1,p3=X), it means the fault free value of the wire a is 0; in group 7, bits 2 value is 1 the bit 3 value is X. Suppose to switch the value of wire a to 1-G1(p7=0)-G3(p3=0). The evaluation of the NAND gate A involves the set union of the two PV lists 1-G1-G3 and 1-G3-G7-G11. The groups 1,7,11 will simply be copied and inverted, but group 3 will be processed. The obtained PV list 0 - G1(p7=1) - G3(p0=p2=X,p3=1) -G7(p5=X,p7=1) - G11(p6=1) is put into a event cell and it is scheduled wire c switching and wire c PV list assignment. The evaluation of gate B involves the intersection of the two PV lists 0-G1-G3-G7-G11 and 0-G7: only group 7 will require Boolean equation processing, the other groups being ignored. The result of this evaluation is 1-G7(p5=X). At this point, because the fault free value in node e is not changing, the resultant PV list is converted to a different type of list where there is no fault free value and the cells in the list indicate active groups only. The new PV list scheduled on wire e is identified by *-G7(p5=X). When the value changes on node e, during the evaluation of gate C only the active group 7 will be considered; groups 28 and 29 will be ignored. The result of the evaluation is *-G7(p2=p3=0,p5=X), which will be scheduled on the output wire g and propagated until the activity in the group dies.

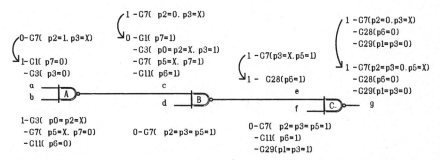

Figure 4.6.14 - Example of parallel value list simulation procedure.

The PV list technique has been implemented, by Cirrus Computer in Hitest and in Hilo3. In the sequel, some notes about Hilo3 are reported.

Hilo3 fault simulator

Hilo3 simulation environment derives directly from Hilo2 one. The same HDL is adopted, to guarantee a full transportability between the two environment. The main differences between the two packages are the number of logical values and the fault simulation algorithm. Hilo3 is a 15 value simulator. The five basic values are: strong 0(0), weak 0(L), high impedance (Z), weak 1 (H), strong 1 (1); furthermore, the following transition are defined: P(1-H), R(H-Z), F(L-Z), N(0-L), T(1-Z), B(0-Z), W(H-L), U(1-L), D(0-H) and X(0-1). All these levels allow an accurate MOS circuit modeling. These values are applied both to fault free and to fault simulation, at switch, gate and RT level of description.

In Hilo2, the fault simulator was based on a parallel technique, while in Hilo3 a PV list technique is adopted. In Hilo3 fault set are first of all collapsed, looking for fault equivalence not only at the primitive boundary, as Hilo2 did, but also analysing the network. A classical example of simple global fault collapsing may be applied to a chain of n inverters. If the single inverter is considered, its collapsed fault set includes a s-a-0 and a s-a-1; it follows that the total number of faults to be injected into the chain is 2*n. However, if a global fault collapsing is performed, since there are no fan-out points in the chain, the real collapsed fault set includes only a s-a-1 and a s-a-0.

The collapsed fault set is then subdivided into groups, and list of groups are built. It is possible to select the number of faults to be considered during each simulation pass, splitting the complete fault simulation procedure in subsequent sessions. This parameter must be chosen in order to optimise the memory size and the simulation time. If too large a set of faults is considered, the virtual memory size will explode, loosing computer time for page fault handling. On the other hand, a very small fault subset requires many simulation passes, and many initialisation procedures. A reasonable set would include from 500 to 1000 faults.

Other improvements are the possibility to perform fault simulation in an incremental way: the procedure may be stopped at any time, saving the overall status, and restarted later. This approach also protects the simulation result from dangerous system failures, avoiding the loss of partial information already computed. Percentual sub-sets of faults may also be considered to reduce the overall simulation time, obtaining statistical fault coverage. In Hilo3, fault simulation has been extended to the behavioural level. Fault hypotheses added for this abstraction level deal with the "event". A behavioural fault may be seen as an event inhibition or forcing. Using this approach, faults like those presented in [Tha79] are easily modeled.

Some benchmarks performed by Cirrus Computer show that Hilo3, with PV list simulation, is from two to seven times faster than Hilo2, with parallel fault simulations. However, there are some unlucky circuits for which the two simulation times are almost the same. PV list simulation technique, implemented in Hilo3, represents a theoretical enhancement, with respect to the parallel technique, as for simulation time, and to the concurrent ones, as far as memory requirements are concerned, although PV list still requires some fine tuning, maybe, in group building procedure.

4.6.4. CONCLUDING REMARKS

The problems and the reasons related to the fault simulation of digital integrated system have been discussed in this chapter. The fault simulation is a very important step during the design of a VLSI device, since it allows both the analysis the circuit behaviour in critical and faulty conditions, and to evaluate the coverage offered by a test pattern. If the first reason may acquire importance for particular devices for which a fault secureness is compulsory, as it has in the aerospace, telecommunication and military application, the second reason involves commercial considerations.

The cost of an integrated device is related also to the quality degree assured by the supplier: the higher the coverage of the test sequence applied during the device validation, the lower the residual defect level of a given stock, the higher the probability that the boards, built using these components, will work.

However, the fault simulation procedures, due to their inherent complexity and to the amount of computation time required, are rarely performed, especially in the case of VLSI devices. This class of circuits, composed of hundreds of thousands of transistor, are hardly handled by fault free simulators. The only reasonable way of dealing with them is to perform their validation at an high description level, like register transfer level, reducing the total number of elements to be considered. Some algorithms have been presented, together with their commercial implementation. In our opinion, concurrent and parallel value list technique are the most suitable for functional fault simulation.

However, as previously underlined, an abstraction level that is too high may create a discrepancy between the description and the physical circuit, although the behaviours are the same. This separation requires a careful choice of the fault models to be considered. New functional models, that take into account topological information, transferring it to the higher level of description, represent the only reasonable link that can assure both acceptable fault simulation time and reliable simulation results.

ACKNOWLEDGEMENT

A particular acknowledgement must be given to Mr. Maurizio Paolini (CSELT), for his contribution to the section about CVT-FERT, to Mr. Cesare Segre and Stefano Morpurgo (Olivetti), for their co-operation in the writing out of section on CVT-OFSKA, and to Declan Farrelly for his friendly help.

REFERENCES

[Abr77]: M. Abramovici, M.A.Breuer, K. Kumar, Concurrent fault simulation and functional level modeling, 14th Design Autom. Conf., (1977) 128-137.

[Ams72]: D.B. Amstrong, A deductive method of simulating faults in logic circuits, IEEE Trans. Comput., vol. C-21, (1972) 464-471.

[Ban83]: P. Banerjee and J. A. Abraham, "Generating tests for physical failures in MOS logic circuits", IEEE International Test Conference 1983, Digest of Papers, October 1983, 554-559.

[Bre76]: M.A. Breuer, A.D. Friedman, Diagnosis & reliable design of digital systems, (Computer Science Press, Inc., Woodland Hills, Ca-USA, 1976).

[But81]: P.S. Buttorff, Computer aids to testing - An overview, in: P. Antognetti, D.O. Pederson, H. De Man, Computer design aids for VLSI circuits (Sijthoff & Noordhoff, Alphen aan den Rijn, N, 1981).

[Cou81]: B. Courtois, "Failure mechanisms, fault hypotheses and analytical testing of LSI-NMOS (HMOS) circuits", VLSI 81, (1981) 341-350.

[Gal80]: J. Galiay, Y. Crouzet, and M. Vergniault, "Physical versus logical fault models MOS LSI circuits: impact on their testability", IEEE Trans. Comput., vol. C-29, No. 6, June 1980, 527-531.

[Gir85]:   G. Girardi, R. Hartenstein, U. Welters, ABLED: a RT-level schematic
            editor and simulator interface, Euromicro '85, (1985) 193-200.

[God71]:   H.C. Godoy, R.E. Vogelsburg, Single pass error effect determination
            (EED), IBM Tech. Journal. vol. 13, (1971) 3343-3344.

[Hen80]:   L.P.Henckelels, K.M. Brown, C. Lo, Functional level, concurrent fault
            simulation, 1980 IEEE Test Conf., (1980) 479-485.

[Hil84]:   Hilo2 usermanual, (Cirrus Computer, Gen-Rad, Milpitas, Ca-USA, 1984).

[Lem84]:   K. Lemmert et al., Karl III reference manual, CVT report, University
            of Kaiserslautern (1984).

[Lev81]:   Y.H. Levendel, P.R. Menon, Fault simulation methods - Extension and
            comparison, The Bell System technical Journal, vol.60, n. 9 (1981)
            2235-2258.

[Mel84]:   M. Melgara, M. Paolini, R. Roncella, S. Morpurgo, CVT-FERT: Automatic
            Generator of Analytical faults at RT-level from Electrical and
            Topological Descriptions, 1984 IEEE Test Conf., (1984) 250-256.

[Mel84,1]: M. Melgara, M. Paolini, and R. Roncella, An algorithm for asynchronous
            NMOS/CMOS network analysis in a CAD tool for physical fault simula-
            tion, Microprocessing and microprogramming, vol. 14, n. 3,4 (1984)
            117-123.

[Men78]:   P.R. Menon, S.G. Chappell, Deductive fault simulation with functional
            blocks, IEEE Trans. Comput., vol. c-37, n. 8 (1978) 689-695.

[Moo83]:   P.R. Moorby, Fault simulation using parallel value list, ICCAD-83,
            (1983) 101-102.

[Mor85]:   S. Morpurgo, C. Segre, Fault simulation at register transfer level of
            VLSI, Olivetti Res. & Tech. Review, n. 3 (1985) 83-92.

[Mor85,1]: S. Morpurgo, A. Hunger, M. Melgara, C. Segre, RTL test generation and
            validation for VLSI: an integrated set of tools for Karl, CHDL Conf.,
            (1985) 261-271.

[Red84]:   S.M. Reddy, V.D. Agrawal, S.K. Jain, A gate level model for CMOS com-
            binational logic with application of fault detection, 21 Design
            Automation Conference, (1984) 504-509.

[Ses65]:   S. Seshu, On an improved diagnosis program, IEEE Trans. Electron.
            Comput., vol. EC-14 (1965) 76-79.

[Tha79]   S.M. Thatte: "Test generation for microprocessor", report R-842,
            University Of Illinois, May 1979.

[Ulr74]:   E.G. Ulrich, T. Baker, Concurrent simulation of nearly identical digi-
            tal networks, Computer, vol. 7 (1974) 39-44.

[Wil81]:   T.W. Williams, Design for testability, in: P. Antognetti, D.O.
            Pederson, H. De Man, Computer design aids for VLSI circuits (Sijthoff
            & Noordhoff, Alphen aan den Rijn, N, 1981).

5. The HDL Subsystem of an Integrated CAD System

5. THE HDL SUBSYSTEM OF AN INTEGRATED CAD SYSTEM

Guglielmo GIRARDI, Silvano GIORCELLI and Giuseppe GIANDONATO

CSELT – Centro Studi e Laboratori Telecomunicazioni
Via G. Reiss Romoli, 274, 10148 - Torino, Italy

Hardware Description Languages are going to play a more important role in modern CAD systems. In this chapter, we shortly recall the motivations for the use of HDLs and their associated tools in some current integrated CAD systems. Next, the nature and the role of a specific, recently developed HDL subsystem within an integrated CAD system (CVT) is described in detail. Mainly via examples, we illustrate the type of hardware descriptions, mixed structural/behavioural, that can be entered in both textual and graphic form. The latter, provided by a specialized graphic editor, driven by the syntax and reflecting the semantics of the HDL, is enough general to allow the connection with a chip floor plan tool in order to evaluate, at an early stage of the design process, the layout area occupancy. The correctness of the descriptions is checked through simulation at the underlying structural level, while the testing aspects are taken into account by a set of programs working at RT level, but with a transistor level resolution in modeling the functional faults. Finally, we shortly describe an experiment in designing a complex integrated circuit by means of the system reported here.

5.1 INTRODUCTION

The concept of multilevel descriptions of a circuit is now generally agreed in the VLSI area. But even considering only the architectural design phase (here meant as the phase from the initial specification to the net list of predefined cells), which is the concern of our work, the involved levels range from purely functional (i. e. algorithmic) to purely structural (set of interconnected user-defined and primitive blocks); only different and separated HDLs are today commercially available to cover this wide spectrum.

As a consequence the description of a circuit seldom is completely performed using a top-down methodology: often behavioural descriptions go in parallel with structural and logic descriptions, and re-adjustments of the project, when further details are added, start directly from the lower levels, thus generating inconsistencies in the design data base.

Moreover in our experience, describing with the class of Pascal-like behavioural HDLs, which looks friendly to tool designers, in practice, turns out to be a

tricky way to express ideas for hardware designers, who often think of the circuit in terms of interconnected machines working in parallel, even since the initial conception of its functionalities.

Efforts were done and are still under progress to fix a common way to generate
families of homogeneous HDLs, in order to help the user to change level of representation without learning new languages and grammars, and the tool maker to deal
with a set of common implementation concepts.

As a less ambitious goal, to improve the above situation in the context of a
short-term CAD project, the HDL subsystem described in the following sections has
been developed with these features:

- mixed-mode, behavioural and structural, descriptive capabilities;
- semantics of the language constrained (compared e. g. with Pascal-derived
  languages) to realize a one-to-one mapping with hardware primitives and concepts;
- graphical editing facilities allowing a direct translation into the textual
  form.

Besides the traditional purposes of supporting documentation, specification for
the design at lower levels and verification through simulation, the role of the
HDL subsystem in our case is to meet, the new and different requirements connected to modern VLSI oriented integrated CAD systems. Mainly these are:

- to integrate the architectural design activity with the other design tasks, in
  particular those related to layout floor planning and testability analysis;
- to constitute a formal front-end language for successive mechanized steps in
  the design process (e. g. automatic derivation of PLAs and/or micro-code from
  behavioural descriptions, formal verification, silicon compilation, etc...).

In this chapter we will first consider some existing integrated CAD systems,
that, at least partially, anticipate the above-mentioned trends. After, we will
present the HDL subsystem embedded in the integrated CVT system. CVT is a project
partially funded by EEC and stands for CAD system for VLSI circuits in
TELECOMMUNICATIONS and is the common effort of CSELT (Centro Studi E Laboratori
Telecomunicazioni, Torino, Italy), CNET (Centre National d'Etudes des Telecommunications, Grenoble, France) and FI/DBP (ForschungsInstitut der Deutschen Bun-
des Post, beim FTZ, Darmstadt, West Germany) to provide an integrated set of
tools suitable to solve problems in the fields of Telecommunications in which the
three partners are engaged.

Integration in the CVT system is achieved both at the user level, by presenting a
unified view of the tools and making him aware of the current state of the
design, and at the system level, by a common Data Base Management System, pro-

viding a common interface to a variety of tools and ensuring consistency and coherence of data.

5.2 DESCRIPTION OF INTEGRATED CAD SYSTEMS

In this paragraph we will survey some aspects of the CAD systems from INTEL, TEXAS and IBM.

5.2.1 The INTEL system

In the INTEL system ([11], [12]), the design process is structured in the following way:

- at register transfer level an Algol like language called MAINSAIL (derived from SIMULA) is used for the architectural definition phase. The language allows separate compilations, and dynamic linking of blocks;
- a simulator (microSIM), built as a shell around MAINSAIL, is used to verify the descriptions. These can be refined with successive steps (micro-architecture definition). To check whether two descriptions, at different levels of detail, describe the same behaviour, the original functional model (not refined) is simulated with a set of stimuli. Later on, the refined micro-architecture is simulated as well, with the same test patterns, and the results are compared;
- at logic level, a schematics data entry, running on work-station is used to edit the descriptions: refined MAINSAIL textual forms are used as specification of the gate level;
- a MOSSIM schematic simulator is used the check the graphic descriptions. The results are compared with those obtained by micro-SIM and if they agree the two descriptions are coherent;
- to check the entire circuit, mixed-mode simulation is accomplished in two different ways:
  - an RT description of a chip is decomposed in a set of modules. These have usually two descriptions (already proved to be equivalent by the above mechanism) at both architectural and logic level. While the whole circuit is simulated at RT level, some modules are activated at gate level. Hence the two simulators are active at the same time: the compatibility is ensured because micro-SIM is written in MAINSAIL;
  - a complete simulation can be carried out at gate level (MOSSIM), even for large chips, substituting some schematic macro-blocks (like ROMs, RAMs, PLAs and registers) with their functional descriptions automatically generated in MAINSAIL.

The main entry point of the system is the behavioural level: then the descriptions are manually refined and translated into schematics, sheet by sheet.

On the whole the main idea is: to have behavioural and gate level descriptions, to simulate both at its own level, to check the coherency, by comparing the simulation outputs, and finally to simulate the whole chip in a mixed-mode fashion with a top-down or bottom-up approach.

5.2.2 The TEXAS system

In TEXAS ([13]), the designers use a unified HDL to cover the different levels of descriptions (from the system specification to the level in which one finds directly implementable primitives).

Each level has two descriptions: the first for the behaviour and the second for the structure implementing the behaviour. The basic idea is that a structure consists of a list of interconnected behaviours at a lower level of description.

The behaviour is a program in a Pascal-like language, able to represent what the outputs are as functions of inputs and time.

Verifications can be carried out in two different ways:

- first, the behaviour of a level is simulated. Its structure is simulated too, simulating the behaviours of the interconnected blocks. If the results are equal, under the same set of stimuli, the decomposition is validated.
  This decomposition mechanism can be applied to each level of description and separately all modules are tested;
- an alternative method, that allows the simulation of the whole system, without flattening completely the hierarchy, consists of expanding some behaviours in the structures they represent, at different levels.

As one can see there is an analogy with the INTEL system in the philosophy of validating parts of the circuit by compared simulations and verifying the whole chip by mixed-mode simulation.

The system is based on two main concepts:
- a hierarchical decomposition method;
- a unified hardware description language.

Because of that, the HDL is suited to be the reference for the entire design process and to provide the common data base, being multilevel, decomposable and hierarchical. Hence, other design aids can be easily connected: a timing analyzer, a drive/load analyzer, a testability analyzer, a test pattern generator and so on.

5.2.3 The IBM system

At IBM, a number of HDLs have been defined and implemented as part of several design automation systems (for an overview of them see [1]). They range from purely functional to purely structural and are used as input to a transformation process, that generates the real hardware. For example, there are languages oriented (with textual and graphic input) to describe logic gates, to facilitate the design of PLAs, to document different stages of the design process, to describe structures (counters, ALUs, data flows, macro blocks, etc...) and behaviours (sequences, procedures, control flows, etc...) and both. Efforts are also devoted to the VHDL (VHSIC Hardware Design and Description Language, [16]), whose main features are technology independency, hierarchical decomposition, functional, RT (structural and behavioural) and logic description capability, a CONLAN-like model of time, the explicit separation of a design entity into interface and body descriptions.

Here we restrict our overview to some aspects of the DAV system, a new logic design workbench, [14], that is built on various internal experiences of both tool-makers and users.

It is composed by a set of application programs, that can be run independently or together. The integration of the whole system is achieved by a common data base and a common user interface. The data base provides the backup function of the working copies of the design data, directly used by the designers.

The common user interface presents to user an integrated view of the system by an unified hierarchy of menus. In a single design session, for example, the designer can:

- enter his functional description using a graphic design language, which is technology independent, hierarchically structured and provided with generic operators and a powerful bundle notation;
- submit the description to the simulation and timing verification tools, with the results directly displayed on the input diagram on the screen;
- translate the description into a specific technology, manually or by the automatic logic transformation system;
- perform testability constraint analysis, e. g. checking whether LSSD rules are satisfied.

Moreover, in case of manual transformation, two different descriptions of the same combinatorial network can be formally verified for equivalence with powerful algorithms of the kind reported in [17].

5.3 SHORT OVERVIEW OF THE CVT CAD SYSTEM

The CVT system focuses mainly on the architectural level of design and is oriented to (future) tasks of synthesis, in order to create a direct bridge between the architectural definition and layout generation.

In the project, the following kinds of researches are present:

- researches on the integration of a CAD system based on tools not yet commercially available, but using mature and stable concepts;
- researches on advanced methodologies and techniques targeted to the generation
  of new tools;
- feasibility studies to explore the possible evolution of the system towards
  innovative concepts.

Moreover in the field of telecommunications a large number of design methodologies should be supported by a CAD system, because of the different types of circuits to be designed, ranging from digital filters and digital signal processors
to microprocessors and concurrent systems.

As a consequence a wide range of tools is needed and should be offered to the
designers and, what is more important, these different design methodologies
should be incorporated and managed by the interface between the designer and the
variety of tools.

Fig. 5.1 depicts a general outlook of the CVT system, designed to meet the above
requirements.

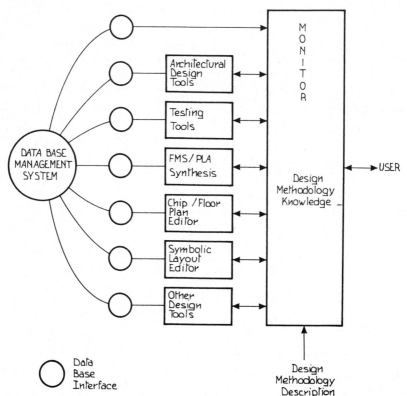

Fig. 5.1 - The CVT system.

The common glue among the different tools consists in:

- a data base management system (DBMS);
- a common user interface (MONITOR).

The DBMS creates a data management environment that can be used by the tools to store and retrieve VLSI design data in an integrated way [3]. The main facilities provided are [4]:

- facilities for data structure description (related to how design objects are built from primitives, to different levels of representations of models, to the temporal evolution of the design process and so on);
- a convenient interface (DBI of fig. 5.1) for:
  - library management (different versions of devices and cells);
  - design hierarchy management;
  - graphic management (geometric descriptions of the cells);
- storage organization on physical supports;
- services for accessing data (creation, deletion and retrieval of design objects);
- maintenance of data coherency versus system crashes;
- ensuring data consistency (simultaneous access to objects shared by different designers within the same project).

The MONITOR ([25], [26]) helps the users to select the appropriate tools, driving him through the 'maze' of representation levels, hierarchical descriptions of the cells and application programs. Altogether it gives the user an easily intelligible description of the current state of the work: which tools generated the descriptions and which programs he/she is allowed to run at a given phase. This is flexibly realized through an extended (with inhibitors) Petri net model, whose mechanization within the MONITOR program, displays on the menu screen the current state of tools (enabled, disabled, ...) and internal design data (operated, out of date, ...), which corresponds to the marking of the net.

Main advantages offered by MONITOR are:

- the couloured graphic representation of the above net allows readable and concise descriptions of the actions that can be performed and of the results of them;
- the user can have at his finger tips the whole set of tools, the links among them and is constantly aware of what he/she can do or is doing;
- possible commands are displayed according to the criterion that the more the command is used the more easily it is reached;
- an appropriate help is provided, being both context and user's skilfulness dependent.

The main operations performed during an interactive session are:

- display of the tools and cells and their relationship;
- invocation of the execution of the selected tools;
- updating the description of the project;
- execution of jobs on different machines in a transparent way.

On the whole the following main characteristics of the CVT system are:

- automation of the design at architectural level;
- bridge between architectural definition and layout generation;
- the system is oriented to automatic layout synthesis (from the architectural level of description);
- incorporation of different design methodologies.

5.4 DESIGN TOOLS AT ARCHITECTURAL LEVEL

The set of tools for designing circuits at architectural level includes:

- KARENE: a language for mixed-mode (structural and behavioural) textual descriptions;
- KARL III: a textual language for structural descriptions [20], [2], [23];
- ABLED: an editor for the structural ABL language, which is the graphical counterpart of KARL [20], [9], [21];
- RT-simulator: a four value simulator at register transfer level [10];
- TEST-TOOLS: a set of tools for generation and validation of test patterns [31], [32], [33];
- ARIANNA: a symbolic chip floor plan editor [29].

These cover a wide range of applications and can be considered as general purpose tools.

Notice that the tools grow around the RT level structural language KARL III and implement design functions connected to well mature concepts.

Besides these, other tools are provided, for more specific aims:

- a generator/optimizer of PLA-based finite state machines ([28], [29], [30]);
- a microprogram compiler;
- a communication protocol verifier;
- specialized silicon compilers for digital filters;
- a tool for synthesis of Petri net based control units.

Moreover, a feasibility study is carried out on formal verification of KARENE/KARL descriptions based on the temporal logic model checking.

Our approach consists in keeping in a homogeneous environment the user, from the initial specification, that corresponds to a set of interconnected high-level behaviours, down to the purely structural description. At each step of this process, a decomposition of a single behaviour into a structure of interconnected behaviours and/or structures or of a single structure into a set of lower level interconnected structures, is accomplished.

The following table summarizes the foreseen usage of the general purpose tools at different representation level.

ARCHITECTURAL DESIGN METHODOLOGY		
Level of description	Tool	Design activity performed
pure behavioural	KARENE	-specification -simulation
behavioural and structural	KARENE	-decomposition -simulation
structural	KARL III ABLED	-decomposition -simulation
structural layout	TEST-TOOLS	-fault model extraction -fault simulation -testability analysis -test pattern generation
symbolic layout	ARIANNA	-layout construction

The data flow among the general purpose tools is depicted in fig. 5.2.

The designer, using a textual editor, can enter behavioural and/or structural descriptions. Structural descriptions can be also entered using the graphic editor ABLED allowing for a more comfortable way to assemble structural objects, that is the way preferred by the majority of hardware designers.

The behavioural descriptions (through a pseudo-hardware generator) and the graphical descriptions (through an ABL-to-KARL translator [22]) are converted into textual KARL form.

The structural descriptions, are assembled using objects taken from a Register Transfer level library. The library is internally stored in both graphical and compiled representations (RTCODE).

The two internal representations express the same information. Therefore the

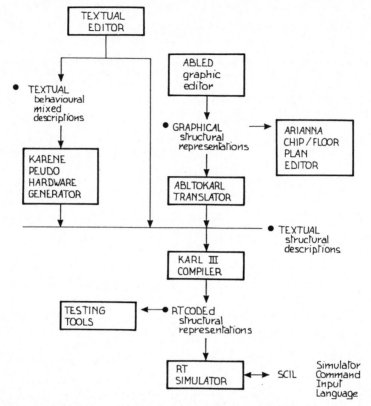

Fig. 5.2 - Data flow among general purpose architectural tools.

other tools, working at the same level, can use the one that is more easy to manage for them. The chip floor plan editor ARIANNA has access to the graphical representations, for its symbolic layout construction, while the testing tools make use of the compiled RTCODE representations for simulation of faulted descriptions and testability analysis.

In the following paragraphs we explain the details of structural, behavioural and mixed-mode descriptive capabilities of our HDL subsystem.

5.4.1 Structural descriptions with KARL III and ABLED

Differing from CAD systems in which graphics facilities are supported only at the lower levels and higher level schematic entries are only an extension of these (e. g. by a macro block capability), in our case, we have an HDL able to express the same concepts with exactly the same semantics, in both graphic and textual forms.

Fig. 5.3 shows two structural descriptions generated using ABLED. They are simple

examples taken from the literature [24] and are the structural descriptions of
the two basic cells used to build a generic one bit slice of an ALU.

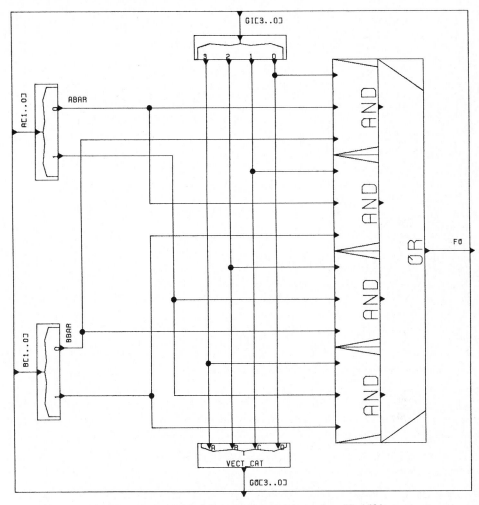

Fig. 5.3.a - Graphic representation of cell 'f1'.

Along the boundary input/output ports are placed. Two bit wide ports (A and B in
'f1' and C and PP in 'carry') convey the corresponding signal and its negated
value.
In cell 'f1', depending on the value of the selection variable GI (4 bit wide),
16 different logic operations on A and B inputs are computed and brought to the
output FO. The control wires (GI) enter the 'front' side of the cell and are connected to the opposite 'back' side, passing through the cell.
The cell 'carry' describes the carry chain circuit: the inputs K and P are used
to kill or to propagate the CIN, carry input signal, to the output COUT. The

signals P and CIN are also rightward propagated.

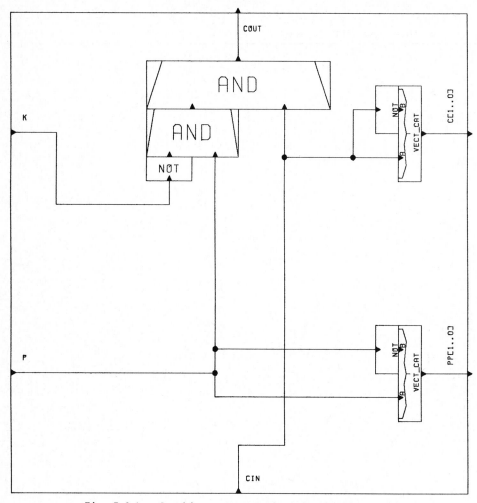

Fig. 5.3.b - Graphic representation of cell 'carry'.

The corresponding KARL III textual forms (output of ABLTOKARL translator) are
given in fig. 5.4.

```
        cell  FL        ( front in  GI[3..0] right out FO back
                          out GO[3..0] left  in  A[1..0]; B[1..0]);
          node      FO ; GO [3..0];
        begin (* of FL *)
         terminal FO .= (((B [1])AND (A [1])AND (GI [3]))OR
                          ((B [0])AND (A [1])AND (GI [2]))OR
                          ((B [1])AND (A [0])AND (GI [1]))OR
                          ((B [0])AND (A [0])AND (GI [0])));
          terminal GO .= ((GI [3]). (GI [2]). (GI [1]). (GI [0])));
          end (* of FL *).
```

Fig. 5.4.a - Textual form of cell 'fl'.

```
cell  CARRY     ( front out COUT right out C[1..0];
                   PP[1..0] back  in  CIN
                   left  in  K; P);
  node       C [1..0]; COUT ; PP [1..0];
begin (* of CARRY *)
  terminal C .= (CIN . (NOT(CIN )));
  terminal COUT .= (((NOT(K ))AND P )AND CIN );
  terminal PP .= (P . (NOT(P )));
end (* of CARRY *).
```

Fig. 5.4.b - Textual form of cell 'carry'.

In Fig. 5.5 the two cells are instantiated and assembled together to form a new cell 'alu_reg_1', which is a complete one bit slice of the ALU, according to the scheme of [24].

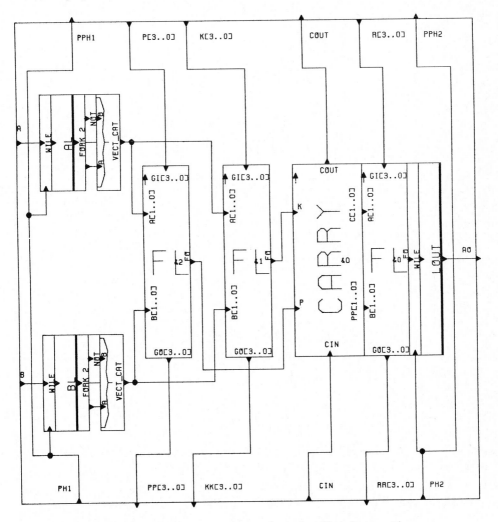

Fig. 5.5 - Graphic representation of cell 'alu_reg_1'.

The two input operands (on left side) are latched (registers AL and BL) with a clock PH1, while the output result (right side) is latched too (register LOUT), but with PH2 before feeding the external circuitry. Input and output ports, placed on 'front' and 'back' sides, are used to propagate the control signals to abutted slices of the same multi-bit ALU.

Notice how ABLED makes it possible to express, beside the usual RT concepts, also some important geometrical information (in this case orthogonality between data and control lines) and how proximity and topological information is specified too (the instance '&0' of the cell 'f1' is left abutted against the instance '&0' of the cell 'carry' and right abutted against the output register LOUT).

The corresponding KARL III textual description follows in fig. 5.6:

```
cell  ALU_REG_1 ( front out COUT in  K[3..0]; P[3..0]
                  out PPH1; PPH2 in  R[3..0]
                  right out AO back  in  CIN out KK[3..0]
                  in  PH1; PH2 out PP[3..0]; RR[3..0]
                  left  in  A; B);

cell  FL      ( front in  GI[3..0] right out FO back
                out GO[3..0] left  in  A[1..0]; B[1..0]);
              external;
cell  CARRY   ( front out COUT right out C[1..0];
                PP[1..0] back  in  CIN
                left  in  K; P); external;
  node FL$2_FO; FL$1_FO; FL$0_FO;
       V_CAT_0[1..0]; V_CAT_1[1..0];

begin (* of ALU_REG_1 *)
  make FL&2 (front in P right out FL$2_FO
             back  out PP left in  V_CAT_0; V_CAT_1);
  make FL&1 (front in K  right out FL&1_.FO
             back  out KK left in  V_CAT_0; V_CAT_1);
  make CARRY&0:FL&0(front out COUT; in R
                    right out FL&0_FO
                    left in FL$1_FO; FL$2_FO
                    back in CIN);
  wile PH1 do AL:=A; BL:=B; endwile;
  wile PH2 do LOUT:=FL$0_FO; endwile;
  terminal AO.=LOUT; PPH1.=PH1; PPH2.=PH2;
  terminal V_CAT_0.=AL . not(AL);
  terminal V_CAT_1.=BL . not(BL);
end (* of ALU_REG_1 *).
```

Fig. 5.6 - Textual form of cell 'alu_reg_1'.

Just to stress the correspondence between the textual and the graphic form, the following main constructs of both are here listed. They are exactly those allowing the user to specify geometrical and topological information at Register Transfer level:

- declaration of the side location (front, right, back and left) for ports (fig.

5.7);
- abutment operators (fig. 5.8);
- mirroring operators (fig. 5.9);
- rotation operators (fig. 5.10).

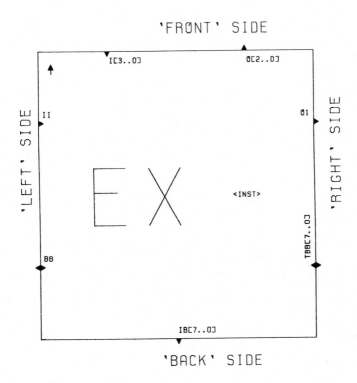

CELL EX (FRONT IN I[3..0] OUT O[2..0] RIGHT OUT O1 BI TBB[7..0]

        BACK OUT IB[7..0] LEFT IN II BI BB);

Fig. 5.7 - Side declaration in ABLED/KARL III user-defined cells.

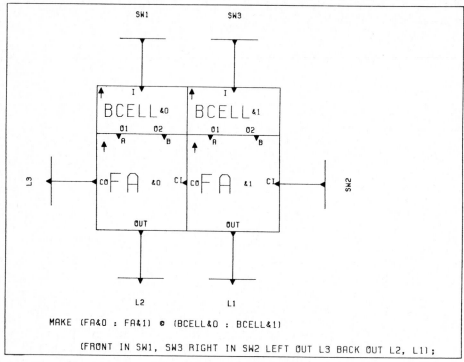

Fig. 5.8 - Abutment operators in ABLED/KARL III user-defined cells.

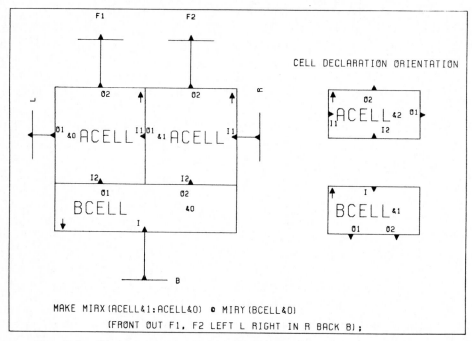

Fig. 5.9 - Mirroring operators in ABLED/KARL III user-defined cells.

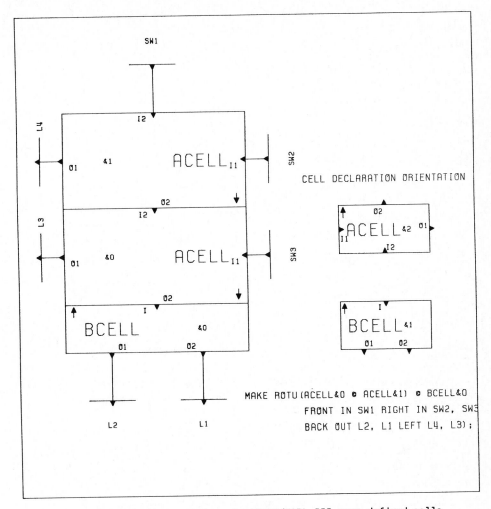

Fig. 5.10 - Rotation operators in ABLED/KARL III user-defined cells.

The objects handled by ABLED are:

- user-defined rectangular cells with ports of type input, output and bi-directional; they can be nested for describing a hierarchy;
- multiplexers and demultiplexers (fully decoded and non-fully decoded);
- logic, arithmetic and relational operators;
- standard functions for code conversion, bit test, wiring (shift, shuffle and mirror);
- operators changing data path width (catenation and subscripting);
- technology independent bus (tri-state, pull-up and pull-down);
- registers with different clocking mechanism (edge triggered, master-slave and latch);

- RAMs;
- constants and ROMs;
- PLAs;
- clock generators;
- delayers with min-max delay time;
- user-defined functions.

Moreover ABLED helps the users when the values of the parameters of the cell being instantiated are specified by automatic computation of data path width, semantic checks on texts associated to the structure, etc....

5.4.2 Behavioural descriptions with KARENE

KARENE is the result of a composite effort, mainly based on the work of the groups at IMAG of Grenoble, where the IRENE language has been developed, and at University of Kaiserslautern, where the KARL language has been originated.

KARENE keeps from the former the constructs for describing behaviours and from the latter the way of describing structures.

Using the behavioural component of KARENE, which is directly derived from IRENE [8], the designer textually enters his high level description. The supported style is exemplified in fig. 5.11.

The example is taken from a larger description (see paragraph 5.5) carried out to test tools we are illustrating

In this example, at the same time, the user has specified:

- a set of operative resources embedded in the description (that will belong to
  an operative part, if the standard operative/control unit decomposition method
  is followed):
  - latch type registers like RDEC, RDECSL (18 bits), etc...;
  - a set of constant values like ZERO, ALLZERO, etc...;
  - logical ('and', 'or'), arithmetic ('+'), relational ('=') and shift ('dshr')
    operators;
- a set of states: $RESET, $DECPAR[2..0] declared within the module and $CKH,
  $CKL inherited from father modules and specifying the finest timing;
- a control flow connecting the previous resources together and with the
  inp/out/bidirectional ports, actions being attached to each (sub)state and executed in a non-procedural way.

For verification purposes, a pseudo-hardware generator has been implemented in order to produce a structural description of the module, acceptable by the standard KARL environment. To this end, a purely behavioural KARENE module is canonically and automatically transformed into a KARL III cell, composed of two sub-

```
(*****************************************************************)
module DECPAR (left in DR; INTACK out FINTRAS; LOADSR in RESET
               back in I_DATA[7..0] right out SHS; DATA[9..0]);
constant ZERO.=0; ALLZERO[7..0].=0; ONE[3..0].=1;
         TWO[3..0].=2; EIGHT[3..0].=8;
reg RDEC[17..0] latch; RDECSL[17..0] latch;
    CDEC[3..0] latch;  CDECMS[3..0] latch;
    BN[3..0] latch;  BNSL[3..0] latch;
    CF[1..0] latch;  CFMS[1..0] latch;
sequence STATE[4]: $RESET,$DECPAR[2..0]; general
         eachtime (RESET or INTACK): $RESET;
endseq
begin
 $RESET:$CKH: CFMS:=ALLZERO[1..0] / CDEC:=ALLZERO[3..0] /
              BN:=ALLZERO[3..0];
 $RESET:$CKL: BNSL:=ALLZERO[3..0] / goto $DECPAR[0];
 $DECPAR[0]:$CKH: BN:=BNSL / CDEC:=ALLZERO[3..0];
 $DECPAR[0]:$CKL: if DR then begin
                             LOADSR.=ONE /
                             RDEC:=IDATA.RDECSL[9..0] /
                             CDECMS:=CDEC + ONE /
                             goto $DECPAR[1];
                           end
                     else goto $DECPAR[0]
                 endif;
 $DECPAR[1]:$CKH: RDECSL:=RDEC / CDEC:=CDECMS;
 $DECPAR[1]:$CKL: RDEC:=dshr&1(RDECSL) /
                  CDECMS:=CDEC + ONE /
                  FLAG1.=(CDEC=BN) /
                  FLAG2.=(CDEC=EIGHT) /
                  if FLAG1 then goto $DECPAR[2]
                        else if FLAG2 then begin
                                        BNSL:=BN + TWO /
                                        goto $DECPAR[0]
                                      end
                                 else goto $DECPAR[1]
                             endif
                  endif;
 $DECPAR[2]:$CKH: SHS.=ONE / DATA.=RDEC[9..0];
 $DECPAR[2]:$CKL: FLAG3.=(BN=EIGHT) /
                  if FLAG3 then begin
                            BNSL:=ALLZERO[3..0] /
                            FINTRAS.=((CF + ONE)=ALLZERO[1..0]) /
                            CF.=CFMS + ONE /
                            goto $DECPAR[0]
                          end
                     else goto $DECPAR[1]
                  endif;
end
(*****************************************************************)
```

Fig. 5.11 - Behavioural description of module DECPAR.

cells: the first for the operative part, the second for the control part (see
fig. 5.12).

```
cell decpar_ou (front out DATA[9..0] left out FINTRAS in INTACK
               back in I_DATA[7..0]; RESET; CKH; CKL
               right out FLAG1; FLAG2; FLAG3; RST in COM[6..0]);
  constant ZERO.=0; ALLZERO[7..0].=0; ONE[3..0].=1;
           TWO[3..0].=2; EIGHT[3..0].=8;
  register RDEC[17..0]; RDECSL[17..0]; CDEC[3..0]; CDECMS[3..0];
           BN[3..0];    BNSL[3..0];    CF[1..0];   CFMS[1..0];
  node FLAG1; FLAG2; FLAG3; DATA[9..0]; FINTRAS; RST;
  begin (* decpar_ou: operative unit *)
  if COM[5] then wile CKL do RDEC:=if COM[4] then I_DATA.RDECSL[9..0]
                                             else dshr&1(RDECSL)
                                   endif;
             endwile;
  endif;
  if COM[5] then wile CKH do RDECSL:=RDEC; endwile; endif;
  if COM[3] then wile CKH do CDEC:=if COM[4] then ALLZERO[3..0]
                                             else CDECMS
                                   endif;
             endwile;
  endif;
  if COM[5] then wile CKL do CDECMS:=CDEC + ONE; endwile; endif;
  if COM[4] then wile CKH do BN:=if COM[2] then ALLZERO[3..0]
                                           else BNSL
                                 endif;
             endwile;
  endif;
  if COM[2] then wile CKL do BNSL:=if COM[1] then ALLZERO[3..0]
                                             else BN+TWO
                                   endif;
             endwile;
  endif;
  if COM[0] then wile CKL do CF:=CFMS + ONE[1..0]; endwile; endif;
  if COM[0] then wile CKH do CFMS:=if COM[2] then ALLZERO[1..0]
                                             else CF
                                   endif;
             endwile;
  endif;
  terminal FLAG1.=(BN=CDEC); FLAG2.=not(FLAG1) and (CDEC=EIGHT);
           FLAG3.=(BN=EIGHT); DATA.=RDEC[9..0]; RST.=RESET or INTACK;
           FINTRAS.=(CF=not(ALLZERO[1..0])) and COM[6];
  end; (* decpar_ou: operative unit *)
```

Fig. 5.12.a - Textual representation of cell DECPAR (operative unit).

```
cell decpar_cu (front out LOADSR in DR
                left in FLAG1; FLAG2; FLAG3; RST out COM[6..0]
                back in CKH; CKL right out SHS);
  cell TBL (back out OUTPUT[10..0] front in INPUT[6..0]); external;
  register PRES_S[1..0]; NEXT_S[1..0];
  node LOADSR; SHS; OUTPUT[10..0]; COM[6..0];
begin (* decpar_cu: control unit *)
make TBL(back out OUTPUT front in (PRES_S.RST.DR.FLAG1.FLAG2.FLAG3));
  terminal SHS.=OUTPUT[8]; LOADSR.=OUTPUT[7]; COM.=OUTPUT[6..0];
  wile CKL do NEXT_S:=OUTPUT[10..9]; endwile;
  wile CKH do PRES_S:=NEXT_S; endwile;
end; (* decpar_cu: control unit *)
```

Fig. 5.12.b - Textual representation of cell DECPAR (control unit).

```
(********************************************************************)
cell DECPAR(left in DR,INTACK out FINTRAS,LOADSR in RESET
            back in I_DATA[7..0]; CKH; CKL right out SHS; DATA[9..0]);
(**)
  cell decpar_ou(front out DATA[9..0] left out FINTRAS in INTACK
                back in I_DATA[7..0]; RESET; CKH; CKL
                right out FLAG1; FLAG2; FLAG3; RST in COM[6..0]); external;
cell decpar_cu(front out LOADSR in DR
               left in FLAG1; FLAG2; FLAG3; RST out COM[6..0]
               back in CKH; CKL right out SHS); external;
(**)
 node FINTRAS; LOADSR; SHS; DATA[9..0]; FLAG1; FLAG2; FLAG3; RST; COM[6..0];
begin (* decpar cell *****************)
 make decpar_ou(front out DATA left out FINTRAS in INTACK
                back in I_DATA; RESET; CKH; CKL
                right out FLAG1; FLAG2; FLAG3; RST in COM);
 make decpar_cu(front out LOADSR in DR
                left in FLAG1; FLAG2; FLAG3; RST out COM
                back in CKH; CKL right out SHS);
end    (* decpar cell *****************)
```

Fig. 5.12.c - Textual representation of cell DECPAR.

This synthesis procedure is oriented to the generation of texts executable by the RT structural simulator. Nevertheless, it can be used as a preliminary study of the synthesis problems connected to hardware generation.

Actually, in our system, the partition of the behavioural description into an operative part and into a control part is accomplished manually, using either the textual editor or the ABL editor.

Relevant aspects in the behavioural component of KARENE are:

- user-defined modules with ports of type input, output and bi-directional; they can be nested for describing a hierarchy of sequencement;
- hierarchical (nested) definition of states and substates [8], like in

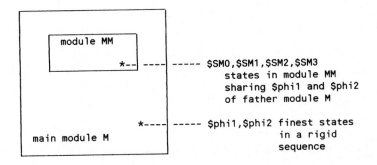

where states within the module MM are specified by:

```
$SM0:$phi1:..../..../....;
$SM0:$phi2:..../..../....;
```

- parallel actions (loading of memory elements, selection of data paths) within
  states;
- jumps to other states using conditional or unconditional 'goto's;
- hierarchical and nested definition of clocks to feed pure structural parts (see
  mixed-mode descriptions);
- definition and specification of asynchronous signals (the 'eachtime'
  construct);
- possibility of using subroutines in describing the automaton.

5.4.3 Mixed-mode descriptions

In our system, main motivations for offering the user the opportunity to have
mixed descriptions are twofold.

Firstly, the designer of telecommunication hardware, usually conceives the circuit to be developed in term of structures and behaviours connected together and
concurrently working. These share the finest timing definition (a single clock, a
two phase or multi-phase clocks) and correspond to the structural cells and to
the behavioural modules in the sense just explained.

Secondly, when some structural blocks, described either in ABLED or in KARL III
textual form, already tested and general enough to be used in different projects
and perhaps already laid-out, is merged with a new block described only at behavioural level, it is necessary to have a mixed-mode capability of describing the
whole circuit to simulate the resulting module.

The basic idea is to consider cells (structural blocks) in KARL III at the same
level as KARENE modules (behavioural blocks).

Therefore the HDL subsystems manage two kinds of objects:

KARL_cell: it is a structural, sequential block, with parameters (ports) of
               input/output/bidirectional type. Moreover these are the only
               external variables seen by the block.

KARENE_module: it is a behavioural block (a finite state automaton controlling a
               non-univocally determined structure) with parameters (ports) of
               input/output/bidirectional type. It shares the sequencing (states
               and clocks) specified in the module in which it is declared and to
               which it belongs.

Both cells and modules are machines working in parallel, independently from each
other and exchanging the values of the declared parameters with their external
world.

A main module can hold both cells and modules in its internal description. Cells
and modules are strictly synchronized; the main module assures synchronization

(correct parameter passing, management of its own sequencement, which is shared by son modules and inputted to son cells) among them.

Obviously this main module assures, at the same time, the links with the external environment (assignment to output and bidirectional variables, reading of input and bidirectional parameters).

In the following example (fig. 5.13) a module M (main) is hierarchically decomposed into cells S1 and S2 (structurally described in KARL III/ABLED) and modules M1 and M2 (behaviourally described in KARENE).

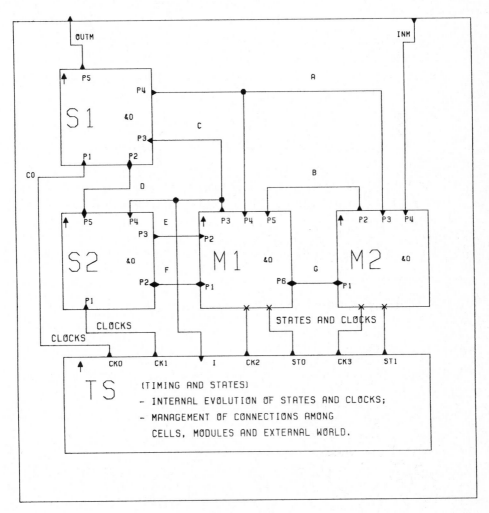

Fig. 5.13 - Modules and Cells in mixed descriptions.

Notice that:

- M1 and M2 modules share the clock variables CK2 and CK3 and the states ST0 and

ST1 of the main module M;

- S1 and S2 cells share only the clock variables CK0 and CK1 but as input parameters;
- TS block (Timing and States) corresponds to the sequencement declaration and generates clocks and states for son cells and modules.

The corresponding textual form follows:

```
module M (front out P0 in P1);

    <decl_part>        (****** variables and functions in M ******)
    <sequencing>       (****** states and clocks in M       ******)

  (****** start structural external cell list:         ******)
  (****** actual names and actual parameters.          ******)
    cell S1&0 (front out OUTM right out A; C back in C0 bi D); external;
    cell S2&0 (front bi D in C right out E bi F back in C1); external;

  (****** start behavioural modules list:              ******)
  (****** actual names and actual parameters.          ******)
    module M1 (front out C in A; B right bi G left in E bi F);
       .
       .         (****** internal description of module M1 ******)
       .
    end; (****** module M1 ******)
    module M2 (front out B in A; INM left bi G);
       .
       .         (****** internal description of module M2 ******)
       .
    end; (****** module M2 ******)

  begin (****** module M sequencement description ******)
     .
     .
     .
  end; (****** module M ******)
```

5.4.4 RT level simulator

The simulator, developed at University of Kaiserslautern in 1981 [10], manages four value signals:

- logical '0' and '1',
- undefined '?' to model delays and conflicts,
- high impedance '*' to model bidirectional buses.

It is composed by a LOADER, that loads the main cell being simulated and is abled to dynamically link the RTCODEd descriptions of the referenced cells, and a command interpreter SCIL. This allows the user to describe, in a procedural way, the stimuli fed into the circuit under test by commands on hardware resources (set, reset, assign, print, plot,...) and structured programming constructs (repeat ... until, while ... endwhile, loop ... endloop, if ... then ... else ... endif, def

... enddef).

5.4.5 Floor plan editor and testing tools

ABLED entered graphical descriptions can be used to specify geometrical/topological characteristics of the intented floor plan:

- cell orientation (rotation and mirroring),
- proximity of abutted cells,
- interconnection among cells.

Starting from this information, the ARIANNA chip floor plan editor enables the user to evaluate, place and route in a symbolic way the physical components of the cell. The main advantage, being the floor plan derived from an RT specification, is that its evaluation is done at a very early stage of the design process, and as a consequence, if the layout evaluation is not satisfactory, critical structural parts can be modified and re-edited, and a new floor plan evaluation started.

The testing tools work on the RTCODEd representations (output of KARL compiler), allowing the user to study the circuit at functional level [32].

A fault model generator is provided [31], [33], that extracts the models directly from the layout and connects them to the RT level. Moreover, a fault generator, a fault simulator, a test pattern generator and a testability analyzer are integrated at KARL level.

The main advantage is that the test engineer, working at RT level, has at his disposal a faster simulation environment. The advanced concept regards the fault model generator, that overcomes the traditional stuck-at logic model and starts directly from the layout, providing more realistic models for a given technology.

5.4.6 Some Specific Architectural Tools

Among the specialized tools of our HDL subsystem we shortly recall here:

- the micro-program compiler [18]: starting from a high level description, it produces an optimized microcode for a user defined micro-machine. To debug and validate the implemented microprograms the RT level simulator is used;
- a verifier of communication protocols between control units [19]. The tool is designed according to the following scheme:
  - the behaviour of the interfaces and communication protocols is described by means of a Petri net formalism;
  - this formal description is used to validate the protocol, to derived descriptions at different levels (e. g. timing diagrams) and to extract a RT level description of the ideal machine implementing the described protocol;

- using KARENE and the associated tools, a real machine is obtained by manual
transformations. Using the RT simulator, the real and the ideal machines are
compared for the equality of the responses under the same sequences of stimuli.

5.5 AN EXAMPLE DESIGNING AT ARCHITECTURAL LEVEL

To test the HDL subsystem we present here, a real, complex VLSI custom circuit
has been completely (re)designed. We have chosen a speech synthesizer integrated
circuit, designed in CSELT in 1983 [5], as a significant example among digital
systems typically encountered in the telecommunication field.

We report here only some parts of the whole work [6], just to show how the
hierarchical decomposition method is supported within our system.

The methodology followed in the 1983 design was based on functional specification
in Fortran, RT level description and simulation with RTS1/A [7], gate level
description, simulation and timing verification using LOGCAP.

Instead, the redesign experiment steps through the following stages: a purely
behavioural specification using KARENE, a number of mixed-mode decompositions,
with simulations performed to validate the performed transformations, a purely
structural description with ABLED/KARL III assembling library cells, symbolic
floor planning using the ARIANNA editor, RT level fault simulation and testability analysis.

Fig. 5.14 gives a global view of the speech synthesizer chip.

The blocks IIL, ECCSAMPLE, TR_LOGIC, SSRL, LLSS and SIR are directly at structural level (KARL cells), while ACQPAR and SYNTH_BLK are behaviours (KARENE
modules). The block TIMING is part of the mixed description and specifies the
sequencement (two phase clock) shared by every module/cell. The description
hierarchy is partially shown below:

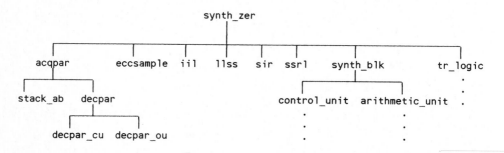

In particular the 'synth_blk' module contains, in a textual form, the description
of the speech synthesis algorithm and is composed by two modules 'arithmetic_

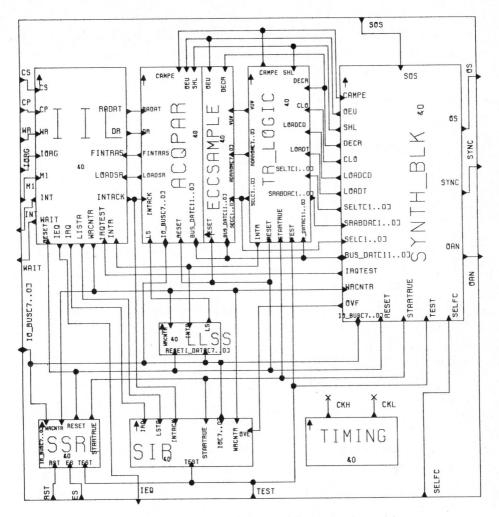

Fig. 5.14 - Global view of the speech synthesizer chip.

unit' and 'control_unit' that are further expanded at lower levels. The 'acqpar' module, devoted to data exchange between an external micro-processor and the internal 'synth_blk', has been decomposed by the designer, staying in mixed mode (see fig. 5.15), into a structural KARL cell 'stack_ab' (a double stack of registers, that can be directly implemented, as a set of structural parts (see fig. 5.16), and the behavioural module 'decpar', previously described in paragraph 5.4.2.

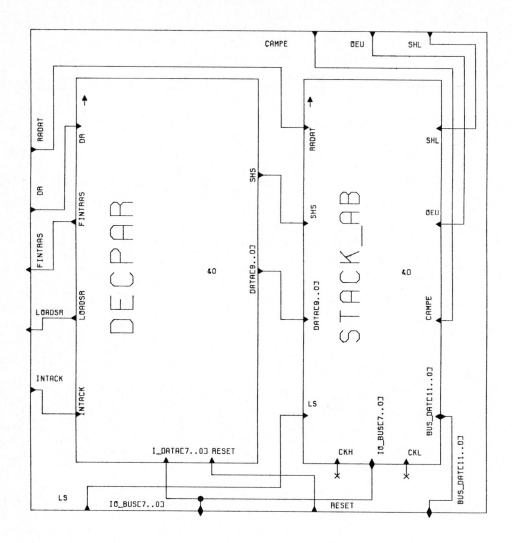

Fig. 5.15 - Decomposition of module 'acqpar' at mixed level.

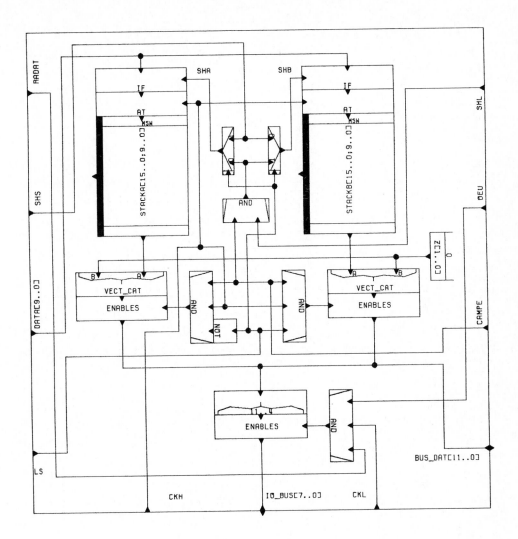

Fig. 5.16 - Structural description of cell 'stack_ab'.

5.6 CONCLUSION

We have presented the HDL subsystem of a VLSI circuit CAD system, whose main feature, in our view, consists in being open to host a diversity of design methodologies. On one hand this is meant to be an answer to a particular need of the telecommunication area, for which the system has been conceived; on the other, it guarantees capability of evolving, as new tools become available and modify the existing methodologies.

The last point is better exemplified with reference to the kind of high level descriptions that we suppose to be entered in our subsystem, as initial specifications: each design entity is defined as a set of interconnected (i. e. a structure of) behaviours and structures, working synchronously in parallel. When silicon compilers based on structural decomposition (into operative and control units) become available, they could be consistently applied to the behavioural blocks of such a specification, while other, more specialized, synthesis tools (e. g. for digital filters) could be applied directly to the structural blocks.

Particular emphasis has been given to a graphic representation being the counterpart of the textual RT language. That is not only because our experience shows that usually engineers use a form of 'spatial thinking', but also because this allowed us to annotate structural blocks with topological/geometrical information interpretable by floor plan tools, which in modern methodologies go in parallel with the design at RT level. Prerequisite of this approach is the availability of a technology independent graphical language (ABL) directly corresponding to the concepts of the underlying RT language (KARL III) and therefore automatically translatable in textual form.

Using the system to redesign a real complex chip, we observed a definite improvement over commercially available systems, where the only way to express behaviours is an interface from the structural or logic level to a conventional programming language.

As a whole, the presented work is an example of the kind of industrial and academic cooperation achieved within the CVT project, in aiding well founded and sufficiently stabilized ideas to gain industrial maturity and wider acceptance. We would like to explicitly remind here the work of IMAG, Grenoble, and of University of Kaiserslautern, on top of which the HDL system was built.

Possible directions of further work include: a study on the automatic generation of layout from behavioural descriptions through a stepwise transformational approach (structural synthesis plus interactive optimization); extension to lower levels, by incorporating a more flexible model of time within the simulator; considering more general communication and concurrency semantics for mixed behavioural/structural descriptions.

5.7 REFERENCES

[1] L. I. Maissel, H. Ofek, Hardware Design and Description Languages in IBM, IBM Journal of Research and Development, September 1984, pp. 557-563.

[2] R. W. Hartenstein, E. von Puttkamer, KARL, a hardware description language as part of a CAD tool for LSI, Proceedings of the 3rd CHDL Conference, Palo Alto, 1979, pp. 155-161.

[3] A. M. Beyls, B. Hennion, J. Lecourvoisier, G. Mazare, A. Puissochet, A Design Methodology Based Upon Symbolic Layout and Integrated CAD tools, Proceedings of 19th Design Automation Conference, 1982, pp. 872-875.

[4] C. Jullien, A. Leblond, CVT-Data Base Interface, CVT Report, CNET, Grenoble, July 1985.

[5] G. Capizzi, C. Cianci, M. Melgara, An easily testable speech synthesizer, Euromicro 83, Madrid, September 1983, pp. 45-52.

[6] G. Girardi, A. Zappalorto, An experiment on designing with KARENE, CVT Report, CSELT, Torino, 1984.

[7] H. J. Knobloch, Description and simulation of complex digital systems by means of the register transfer language RTS1/A, in Computer Design Aids for VLSI Circuits, P. Antognetti, D. O. Pederson, H. De Man eds., Sijthoff & Noordhoff 1981, pp. 285-319.

[8] F. Anceau, S. Marine, K. Jahidi, IRENE: un language de description de circuits integres logiques, IMAG R. R. n. 356, March 1983.

[9] G. Girardi, R. Hartenstein, U. Welters, ABLED: a RT level Schematic Editor and Simulator Interface, Euromicro 1985, Brussels, pp. 193-200.

[10] B. Weber, PASCAL-Implementierung eines Simulators auf KARL-2 Basis, Diplomarbeit, University of Kaiserslautern, 1981.

[11] S. Nacthsheim, The INTEL Design Automation System, Proceedings of 21st Design Automation Conference, 1984, pp. 459-465.

[12] K. Tham, R. Willoner, D. Wimp, Functional Design Verification by Multi-level Simulation, Proceedings of 21st Design Automation Conference, 1984, pp. 473-478.

[13] D. Ackley, J. Carnegie, E. B. Hassler, Jr., Hardware Description System Development, Proceedings of ICCC 82, New York 1982, pp. 608-611.

[14] F. Rubin, P. W. Horstmann, A Logic Design Front-End for Improved Engineering Productivity, Proceedings of 20th Design Automation Conference, 1983, pp. 239-245.

[15] J. B. Bendas, Design through Transformation, Proceedings of 20th Design Automation Conference, 1983, pp. 253-256.

[16] M. Shahdad, R. Lipsett, E. Marschner, K. Sheehan, R. Waxman and D. Ackley: VHSIC Hardware Description Language IEEE Computer, Feb. 1985, pp. 94-103.

[17] G. L. Smith, R. J. Bahnsen, H. Halliwell, Boolean Comparison of Hardware and Flowcharts, IBM Journal of Research and Development, Jan. 82, pp. 106-116.

[18] A. Giordano, M. Maresca, T. Vernazza, An Architecture Driven Microcode Optimizer, CVT Report, University of Genova, July 1985.

406 G. Girardi, S. Giorcelli and G. Giandonato

[19] P. Civera, C. Conte, D. Del Corso, F. Maddaleno, Petri net Models for The Description of Parallel Bus Cmmunication Protocols, CVT Report, Politecnico di Torino, April 1984.

[20] R. Hartenstein: Fundamentals of structured Hardware Design, North Holland Publishing Co., Amsterdam/New York, 1977.

[21] G. Girardi, ABL editor: user manual, CVT report, CSELT Torino, 1984.

[22] U. Welters, ABL2KARL translator: algorithm description, CVT report, University of Kaiserslautern, F.R.G., 1985.

[23] K. Lemmert, et al., KARL-III reference manual, CVT report, University of Kaiserslautern, 1984.

[24] C. Mead, L. Conway, Introduction to VLSI Systems, Addison-Wesley, Reading, Mass., 1980.

[25] A. Di Janni, M. Italiano, A Unified User Interface for a Cad System, Proceedings of 22nd Design Automation Conference, 1985, pp. 9-15.

[26] A. Di Janni, A Monitor for Complex CAD Systems, CVT Report, CSELT Torino, November 1985.

[27] G. Arato, O. Gaiotto, P. Antognetti, A. De Gloria: Arianna: a floor-planning tool, Proceeding of 11th European Solid State Circuits Conference, Toulouse, 1985, pp. 193, 200.

[28] G. De Micheli, A. Sangiovanni-Vincentelli, T. Villa, Computer-aided synthesis of PLA-based Finite State Machines, Proceedings of ICCAD, Santa Clara, Ca., September 1983, pp. 154-156.

[29] L. Lavagno, ASMA: an algorithmic state machine description language and preprocessor, CVT Report, CSELT Torino, 1984.

[30] H. M. Lipp, Methodical Aspects of Logic Synthesis, Proceedings of the IEEE, January 1983, pp. 88-97.

[31] M. Melgara, M. Paolini, R. Roncella, An algorithm for asynchronous NMOS/CMOS network analysis in a CAD tool for physical Fault Simulation, Microprocessing and Microprogramming, October 1984, pp 117-123.

[32] S. Morpurgo, A. Hunger, M. Melgara, C. Segre, RTL test generation and validation for VLSI: an integrated set of tools for KARL, Proceedings of 6th CHDL Conference, Tokyo 1985, pp. 261-271.

[33] M. Melgara, M. Paolini, R. Roncella, S. Morpurgo, CVT-FERT: automatic generator of analytical faults at RT-level from electrical and topological descriptions, Proceedings of the International Test Conference, Philadelphia, October 1984, pp . 250-256.

6. Implementation Techniques for Multi-level Hardware Description Languages

6. IMPLEMENTATION TECHNIQUES FOR MULTI-LEVEL HARDWARE DESCRIPTION LANGUAGES

Dominique BORRIONE* and Claude LE FAOU

Laboratoire IMAG-ARTEMIS
BP 68, 38402 St. Martin d'Hères Cedex, France

* Professor at Université de Provence, Marseille

Languages which cover several levels of abstraction in hardware design are characterized by a set of kernel notions common to all levels, and a set of level dependant primitives. The architecture of a CAD system based on such languages can be divided into two processing phases, which are presented in this paper with an emphasis on the problems raised by the simulation of multi-level descriptions. All the algorithms and scheduling modes adapted to each modelling level must be taken into account, ensuring conversions between language levels, and communication between the various scheduling modes.

1 MULTI-LEVEL HARDWARE DESCRIPTION LANGUAGES (HDL'S)

This chapter places itself in the context of languages which cover several levels of hardware design, with an emphasis on the possibility given to the user to mix levels within a single description of his design.

1.1 NEED FOR MULTIPLE-LEVEL MODELLING

It is now well recognized that designing a digital circuit involves a number of steps. From the functional specifications of the circuit, which state the required behaviour in a more or less formal way, to the verified fabrication documents (layout and interconnection of primitive modules, whether ICs on a card or transistors on a chip), the circuit undergoes several descriptions at various abstraction levels. Research on silicon compilers attempts at providing tools which directly and automatically produce the fabrication documents from the functional specifications; however, the state of the art for the eighties is not the usage of silicon compilers, except for well identified functional blocks in selected technologies. For some years, the circuit designer will have to live with the necessity to refine his/her design in a stepwise manner, and thus manipulate several models, possibly involving various modelling principles.

Between the functional specifications of a digital circuit and the masks for that circuit, the following design levels have been widely recognized, although some of them may be stepped over in a particular design task, depending on its complexity and the availability of ready made building blocks :

```
- system level
- architectural level
- register transfer level
- gate level
- switch level (for MOS technology)
- circuit level.
```

During the design process, the designer often wishes to describe parts of the circuit under consideration at various levels. Abstracts levels allow for the manipulation of synthetic informations, while levels closer to the technological implementation allow for precise and detailed verifications. In order to avoid an excessive quantity of data, when only a small section of a circuit is being zoomed at, it is highly desirable to mix description levels within a single model.

Until recently, each modelling level had to use one (among several dozens) single level HDL, and no or little relation existed between languages geared to neighbouring levels. A considerable progress is brought by the emergence of multi-level languages, built to cover several modelling levels, with a common kernel of primitives for all levels. Three such languages have been reported in the past years: CASCADE[6], CONLAN[23], and VHDL[26].

1.2 MIXED LEVEL MODELLING

Based on the existence of a multi-level language, a stronger requirement is the availability of mixed level capabilities, that is the possibility of interconnecting parts written at various modelling levels within a single description. This implies the existence of interface functions, whether explicit or implicit, to translate the informations exchanged by parts of the model belonging to different levels, and knowing incompatible types of objects.

Two categories of mixed levels may be sought :

- spacial mixed level corresponds to a circuit where different parts are described at different levels.

- temporal mixed level corresponds to the dynamic replacement of the model for a part of the circuit by another model, during a simulation experiment.

One may imagine to combine the two categories into a "spacio-temporal mixed level" model; yet, to our knowledge, no simulator supports temporal mixed level, nor a fortiori the spacio-temporal combination.

Mixed level modelling is simplified if a model is written as a hierarchy of nested modules. Each module may be written at a particular level. A given module may have several alternatives, each written at a particular level. In all cases, a module is seen as a whole, and may be processed and checked separately, which greatly simplifies comprehension and design. With modular models, the problem of translating informations between levels only happens at well identified points : module interfaces.

1.3 CHARACTERISTICS OF A MULTI-LEVEL HARDWARE DESCRIPTION LANGUAGE

A multi-level HDL can be divided into :

- a kernel of notions, data types, operators and constructors, common to several or all levels of description;

- for each level, specific value and carrier types, operators, and primitive components.

1.3.1 Kernel Notions

1. Structural decomposition

   This decomposition corresponds to the identification of physically disjoint parts. The design is described as a network of interconnected boxes, each box in turn containing a network of interconnected boxes etc.... until leaf blocks of the hierarchy are reached.

   Structural decomposition is modelled by some notion of module (e.g. "description segment" in CASCADE or CONLAN) and the fact that a module contains interconnected units which are themselves instances of more elementary modules.

   In the following, we shall distinguish between :

   . the model of an arbitrarily complex hardware "module", possibly parameterized, described at a particular level ; such module corresponds to a circuit generator ;

   . an instance of module, corresponding to a particular circuit, with all parameters fixed ; such instance is the unit of compilation and verification, and shall later be referred to as "unit".

2. Communication between units

   Units are declared and interconnected in the body of an enclosing module, thus expressing structural decomposition . If a unit is to be checked independantly of any model in which it is instanciated, it must communicate with its enclosing environnement only through its interface carriers. In a multi-level HDL, each interface element should be typed, and its direction specified as, for instance: "in" (input), "out" (output), "inout" (alternatively input or output), or "nd" (non-directional, reserved for switch and circuit levels).

3. Functional decomposition

   This decomposition corresponds to the identification, within a given box, of various functions which orerate on the same physical resources. Their grouping or sequencing is usually described by a control part, thus expressing the behaviour of the box.

The functional decomposition does not necessarily imply a particular implementation. It may be expressed in terms of statements close to those of programming languages (assignments, functions, procedures, conditionals). The primitives available for writing it depend strongly on the description level.

4. Value and carrier types

A value type is a finite or infinite set of elements, identified by their constant denotation, together with a set of operations defined on these elements. Typical examples are:

. type INTEGER, or REAL, with arithmetic and comparison operators,

. type BOOL, or multi-valued LOGIC, with boolean operators.

Carriers are value containers; their contents may vary in time. Carriers correspond to abstract (variables), or physical (registers, terminals, electric wires) devices, whose state is characterized by their content value. In addition, a carrier type is characterized by its time behaviour and operators for accessing and modifying the value, which may be independant of the value type and initial value, considered as parameters to a generic type definition[23].

Each language level fixes the available value and carrier types, and the admissible combinations of generic carrier type and value type.

Value and carrier objects may be structured into arrays and records, as in most Algol-like modern programming languages.

5. Operational statements

Operational statements are used to describe behaviour, in the body of a function, procedure or hardware module description. They are based on the concepts of procedure and function invocations. They perform computations based on past and present values of a model carrier, and modify the values of model carriers. The basic operational statements are the assignment of the result of an expression to a carrier, and the procedure call.

Operational statements may be conditioned ("if" and "case" conditionals), or repeated (usually, "while" and "for" suggest sequential repetition, and "over" indicates repetition without any specification of sequence). Conditionals and repetitions may be nested to any depth.

6. Sequencing and parallelism

In most HDL's, operational statements are concurrent, and their lexical order has no effect on the result of a simulation run. Attempts to simultaneously modify the same variable with different values are detected during simulation, and reported as an error.

In a multi-level HDL, sequentiality as in programming languages, if provided, will appear at system level only. At other levels, operational statements, or groups of concurrent statements may have a sequential evaluation only if preceded by conditions which successively take value "true", thus inducing the order of evaluation. Some language levels provide special statements to describe sequencing control according to an automaton model or to a control graph model.

1.3.2 Characteristics Of The Various Levels

1. System level

   At system level, a designer specifies the timing and functional behaviour of digital systems, and the interactions between system components, without specifying any implementation detail. Routing of messages in a network, or synchronizing communicating parallel processes, are typical examples.

   The operative part is expressed as assignments to abstract variables (carrier type VARIABLE). Operations may be permanently valid, or their execution may depend on the model control part.

   The control part is described using primitives which are often equivalent to a directed control graph model, such as Macro E-Nets [22] or LOGOS nets [24]. A comparison of graph models can be found in [14]. A typical interpreted control graph interconnects two kinds of nodes:

   - "Places" represent events, or conditions for starting a computation. In models where places hold boolean values, we shall call them "control signals". A control signal holds value 1 if the event it stands for has occurred and has not yet been taken into account; otherwise, it holds 0.

   - Transitions modify the value of one or several places to which they are directly connected (transition "firing"). This corresponds to taking into account one or more events (its input places) thus creating other events (its output places).

     Operation statements may be attached to a transition. When the transition fires, these statements are executed.

2. Hardware architecture and register transfer levels

   Modules written at these levels specify state changes of hardware elements, but not the details of their data paths. A description may be:

   - synchronous: time is expressed in terms of clock cycles; one or several independent clocks are provided from the outside, through the interface. The system state may change only on a clock pulse. Combinational circuits are assumed to stabilize within the cycle of the fastest clock.

.   asynchronous: delays are expressed in terms of units of time, and
past values of carriers may be referenced in expressions. A
synchronization pulse may be the rising or falling edge of a local
boolean carrier and not only an external clock pulse.

The behaviour part of a description may represent:

.   Combinational circuits, which are memoryless. All statements are
connections of possibly complex expressions to pins or connection
wires, later on called TERMINALs. All statements are permanently
active.

.   Sequential circuits, which contain memory elements.   Carriers of
type LATCH or REGISTER are loaded under the condition of a boolean
level or rising/falling edge.

Typically, the control is modelled by an automaton.   A set of
concurrent statements that depend on a given state are labelled with
the state identifier. Sequencing is modelled by changing the automaton
state, which takes one clock period.

3.   Gate Level

At gate level, a description is a network of interconnected gates and
busses. The edges are statically oriented.

The primitive components correspond to the usual logic gates (NOT,
NAND, AND, NOR, OR, XOR, XNOR), the transfer gate (TG), tri-state gates
(TNOT, TAND, TOR, TXOR, TXNOR, TNAND, TNOR), and simple models of BUS.

Their interface elements are directional. They are usually carriers
taking values in 3 or 4 state logic.

The various kinds of timing attributes (typical, minimum, maximum
values of rise and fall delays) are parameters for which default values
may be fixed.

4.   Switch level

At switch level, a description is a network of interconnected
transistors and nodes; the edges are non-oriented. Switch level is
specially adapted to MOS technology[8].   The three types of MOS
transistors (N-channel, P-channel and Depletion) are given as primitive
components, with a strength parameter: NTRANS, DTRANS, PTRANS.

Their three interface elements are non-directional(GATE,   DRAIN,
SOURCE), but their value is abstracted to dicrete logic (usually 3 or
more states).

Other primitive components are the NODE, which models an
interconnection point, with a variable number of non-directional
interface PINS, and the INPUT, a special case of NODE which is
connected to primary inputs or ground (GND), or power (VDD).

5. Circuit level

At the electrical level, one is interested in determining the current
and voltage at selected circuit wires, as a function of time. The
model is written in terms of the laws of physics, using the MKSA unit
system. The main characteristics of circuit level, compared to the
other levels, is that all carriers take values of type REAL, and that a
description represents a continuous model of the circuit.

Behaviour is induced by the interconnection of primitive and
user-defined components. The primitive components are, for instance,
the models of various transistors, and of the elementary dipoles which
may be found in an electrical circuit: resistor (R), conductance (G),
voltage source (E), current soure (J), capacity (C) and self inductance
(L). Their interface is made of non-directional pins.

Two functions "i" and "v" are attached to each pin P, and return the
intensity of the current in P, and the voltage of P. Each primitive
component, with interface pins P1,P2...Pn defines an equation on i(Pj),
v(Pj), according to the electrical laws for the component. In
addition, the compiler automatically generates (Kirchhoff law):
$$i(P1) + i(P2) + \ldots + i(Pn) = 0$$

2   GLOBAL SYSTEM ARCHITECTURE

A description written in a multi-level HDL may be the primary input to a number
of CAD software tools, including graphical manipulation, good and fault
simulation, synthesis, formal verification, test pattern generation etc... The
trend today is to integrate all these tools as a hierarchical network of
environments, within a single CAD system. One such hierarchy, shown on Figure
1, and reproduced from [20], displays the partial structure of the CASCADE
system.

Whatever the application, a description will undergo two distinct processing
phases, as shown on Figure 2.

2.1   PHASE 1

Phase 1 is language oriented. In a multi-level HDL, it is parameterized with
the language level being used. Phase 1 is the construction and the verification
of a model, independantly of the CAD tool which will be applied to that model.

It performs lexical, syntactic and semantic checks, in order to verify that the
description is correct with respect to the language definition, and more
important, meets the constraints imposed on the hardware design for that level
(such constraints may be part of the HDL level semantics). It corresponds to
the "Modeling" environment of figure 1.

D. Borrione and C. Le Faou

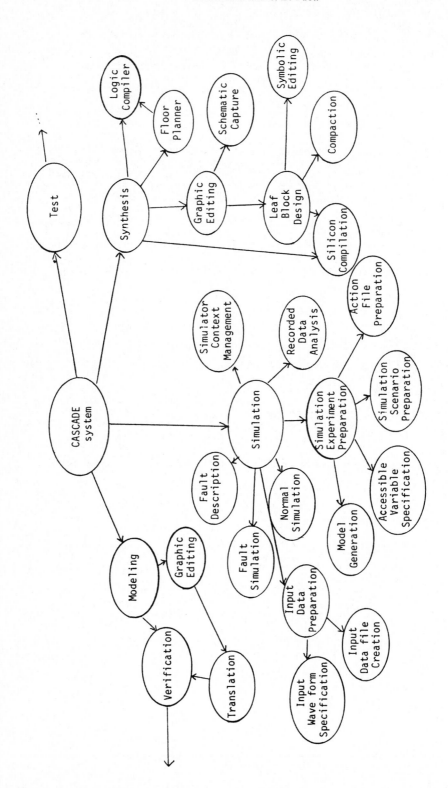

Figure 1 :   Part of Cascade Environments   (HEN)

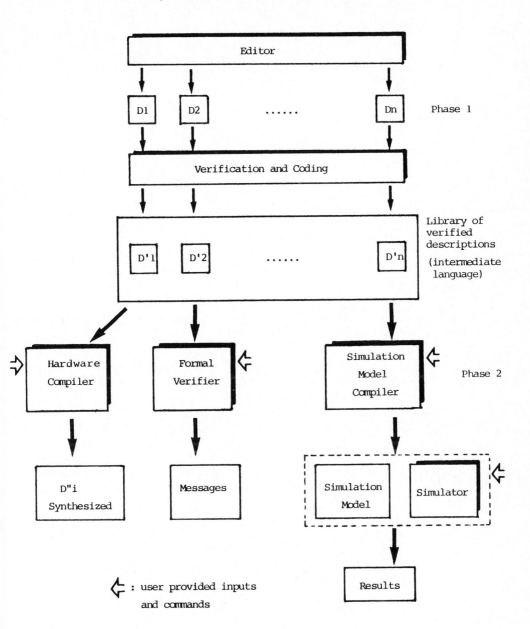

Figure 2 : Processing a description

2.1.1 Need For A Mixed Textual And Graphical Editor

The description of a user-defined component must first be edited into a computer file.   A general purpose text editor may of course be used; but such editors have no information about the meaning of the character string being manipulated, and bring no help to the user for the construction of his/her model.

Special purpose graphical editors have been in use for years, and many are commercially available for gate or electrical level descriptions. In this case, the user does not even see a textual description. The model is built as an interconnection of primitive or previously defined non-primitive modules. However, these tools are attached to a single level of description, where only one type of carrier and of value is available. Another strong limitation lies in the fact that only structural descriptions are supported; the user has no mean to define the direct behaviour of a module: he/she may only interconnect existing components.

Multi-level HDL's require more elaborate editors, parameterized with the language level, with mixed graphical and textual capabilities. Stating the reference language level makes immediately available all the primitive components for that level, as for the single-level editors discussed previously. In addition, for RT level and more abstract, the user should be able to specify additional informations such as dimensions and value type of carriers other than the default scalar interconnection wire. A switch to a syntax directed textual editor should, as is already done in programming environments [12], help the user input syntactically correct behavioural descriptions.

Although no editor has been produced that fulfills all these requirements for multi-level HDL's, recent progresses in software engineering [4] should encourage CAD teams to build more user-friendly modelling environments.

2.1.2 Verification And Coding

The verification and coding package corresponds to a compiler front-end, and should be common for all applications. Quite naturally, the verification of each module description should be done separately, and should produce an output internal form for each one. When a module is described as a network of units, only the interfaces of nested units need be provided to the verifier.   The principles of the verifier are shown on Figure 3.

In addition to the usual lexical and syntactic analysis, the maximum number of static semantic checks should be performed at this point, including the following non exhaustive list of verifications:

  - type and dimension compatibility in expressions;

  - actual interconnections of nested units with respect to the declared formal interfaces;

  - context in which carriers are being modified, according to their declared type;

  - fanout limitations; etc...

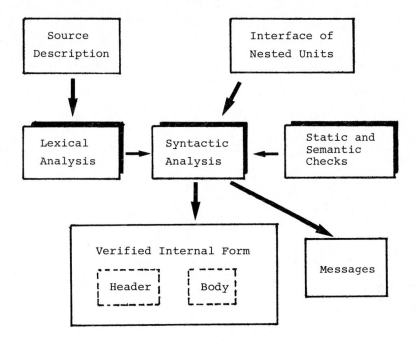

Figure 3 : Verification of a source description

The verified internal form can be split into a coding of the description header (reference language level, identifier, interface), and a coding of the description body (structural/behavioural description). Depending upon the subsequent application, only one aspect might be used. Yet the union of the two should:

- lose no information with respect to the source HDL description;

- allow a translation back into source language;

- be general enough for a variety of applications;

- allow easy manipulations, transformations, and visualizations.

In general, the verified internal form is a combination of table and list structures. Proposals have been made for standard internal forms, in order to allow transfers between different CAD systems[13].

2.1.3 Use Of Compiler-writing Tools

Once the semantics specific to a HDL have been clearly identified, the implementation of Phase 1 is very similar in nature to the writing of a programming environment and of a compiler. For a multi-level HDL, the complexity is of the same order of magnitude than that of an ADA compiler, and calls upon classical compiler-writing techniques [1,3].

The utilization of the best packages from software engineering technology is the key to efficient production of reliable and powerful modelling environments. For instance, it will be advantageous to use a meta editor, parameterized with the syntax of the source language, to build special purpose editors for each HDL level [16].

The usage of compiler-compilers is now widely reported [15,17,19]. Figure 4 shows the principles of the construction of a syntax directed verifier, using a typical parser generator such as [11].

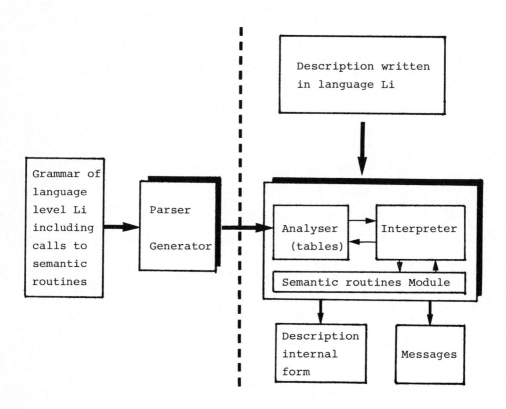

Figure 4 : Constructing the verifier using a
            Compiler-compiler

On the left hand side of Figure 4, the HDL syntax is provided under an appropriate format, and must fulfill constraints imposed by the parser generator (Europeans seem to prefer predictive parsers based on LL(1) grammars, while Americans are more inclined to LR(1) or LALR based bottom-up techniques [2]). Calls to semantic routines are included in the syntax.

The parser generator produces a set of tables which are linked to a standard interpreter, and to the module of verification and coding semantic routines provided by the compiler writer. The result is a syntax directed verifier which accepts as input a user description written in reference language Li, and outputs the description internal form and possible error messages (right hand side of Figure 4).

2.2 PHASE 2

Phase 2 is application oriented. It takes the output of phase 1 in standard internal form, and performs one or several of the following tasks:

- A link-edit between all the internal forms for the various units, when a model is described as an interconnected network of sub-modules.

- Direct transformations on the internal form to produce a modified model of the same circuit (optimizations, synthesis).

- A translation into another code, that is required as input by an existing application package, or that can be more efficiently processed by some algorithm (test, formal proofs, true value and fault simulation). This is quite similar to code generation in optimizing compilers.

For reasons of space, this paper cannot cover all the possible phase 2 packages for the various application environments. We shall limit ouserlves to the problems raised by the production of a multi-level mixed mode simulation model, and shall insist on the cohabitation problems that must be taken into account

- between abstraction levels,

- between simulation algorithms and scheduling modes, which will therefore be first reviewed.

2.3 EXAMPLE

A small micro-processor based system, called PODSYSTEM, is built on the MC6800 family parts [21]. The system is organized around an 8-bit parallel microprocessor POD, a 16-bit address bus BUSA and an 8-bit bidirectionnal data bus BUSD. As shown on Figure 5, a 128 byte RAM for data, a 1024 byte ROM for program, and a parallel I/O interface adapter (PIA) are tied on the two busses.

At the architectural level, the PODSYSTEM description is naturally decomposed into 4 units, one per actual integrated circuit. Each unit is an instance of an

external description segment which describes its behaviour: unit PIA821 is an
instance of PIA, and POD is an instance of MICRO. Description segments RAM and
ROM could be parameterized with an integer attribute SIZE.   Unit RAM10 is an
instance of RAM with actual SIZE=128, and unit ROM708 is an instance of ROM with
actual SIZE=1024.   The two busses are local carriers of the   enclosing
description PODSYSTEM.

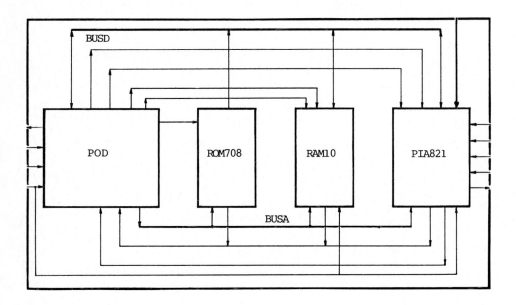

Figure 5 : PODSYSTEM example

Building a simulation model for the PODSYSTEM is shown on Figure 6. Each module
description is verified separately by phase 1, and stored in a catalogue of
verified standard internal forms.

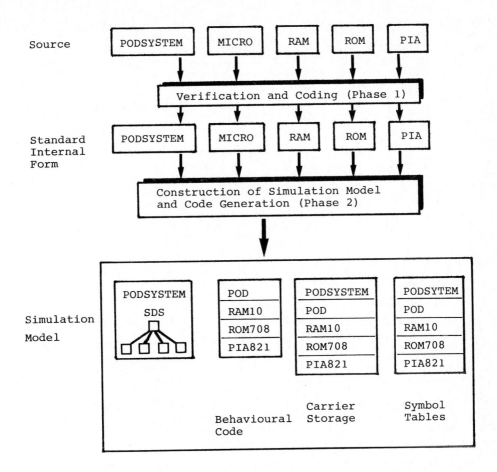

Figure 6 : Production of a simulation model

Starting from the outer unit, phase 2 contructs a tree corresponding to the nested structure, layer after layer. Every node in the tree will be treated differently according to its particular level of abstraction and its associated scheduling mode. The result of these treatments can be split into:

- an interpreted part, the Scheduler Data Structure (SDS), which stands for a (possibly re-organized) structural decomposition. For the PODSYSTEM example, the SDS will be represented by the tree:

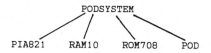

- a compiled part, which stands for the translation of behavioural descriptions.

3 SCHEDULING AND SIMULATION ORIENTED DATA STRUCTURES

The scheduler is in charge of the simulator time management. At each point in time, the scheduler determines if all, or parts, of the model is active, and invokes the evaluation procedures for the active parts. When all active parts have been computed, the scheduler is given control back, and either increases the simulated time in order to perform simulation for a subsequent time point, or returns control to the user if the circuit has become passive, or if the requested maximum simulation time has been reached.

When several parts of a model are not always simultaneously active, the scheduler is provided a hierarchical Scheduler Data Structure (SDS). Each non-terminal node in the SDS corresponds to a model part which has been further divided into sections which may not be simultaneously active.

A (compiled or interpreted) piece of algorithm, called "Simulation Block" (SB), is associated with each terminal node in the SDS. It represents the behaviour of the model part which the terminal node stands for, and is executed whenever that node is found active.

In the PODSYSTEM example (Figure 5), a first level of nodes in the SDS would correspond to the POD, ROM708, RAM10 and PIA821 units; the node associated with POD could probably be further decomposed, leading to more nodes in the SDS, while the node for RAM10 might be a terminal one.

Several scheduling modes are defined, according to the criteria taken for time progression, and the algorithms used for evaluationg the model. Each scheduling mode attempts at optimizing the simulator computation cost, according to the model of time (discrete/continuous) and the synchronization hypothesis for the described circuit.

The scheduling modes generally adopted for typical HDL abstraction levels are summarized in the following table. Although other choices can be made, we shall take this correspondance as a working hypothesis in the following paragraphs.

Abstraction level	Scheduling mode
System level	Discrete event driven
Architectural level	Discrete compiled
Register transfer level	Discrete compiled
Gate level	Discrete event driven
Switch level	Discrete event driven
Circuit level	Continuous

3.1 DISCRETE EVENT DRIVEN MODE

This mode is the one chosen in asynchronous models, when at any given time a very small part of the circuit is active - only some transitions in the control graph of a system level description, only some gates or some transistors at gate or switch level. Consequently, it would be very costly to verify the whole circuit, at each cycle of computation, just to know which part has to be executed (for example, the execution time for a gate is much smaller than the time necessary to know whether it is active or not).

This event driven mode has been explained, at logical gate level, in many papers [7]. Each gate corresponds to a terminal node in the SDS. The future events are linked to a "time wheel" [27]. For each gate activation, the "time wheel" is updated taking into account the delay of the gate and its fanout.

At system level, each transition of the control graph corresponds to a node in the SDS. If there are instructions controlled by the transition, these instructions are translated during phase2 into a procedure which is executed whenever a transition firing is simulated.

The "time management" here is also performed by a "time wheel", or a linked event list, which makes it possible to point towards the transition node of the SDS. For every activation of such a node, simulation actions will be launched, and the scheduler will update the "time wheel" taking into account the delay associated to the transition.

From now on, we shall call "dynamic scheduler" the program which realizes this scheduling mode.

3.2 DISCRETE COMPILED SCHEDULING MODE

This mode is generally used at architectural and register transfer levels. We shall call static scheduler the program which realizes this scheduling.

Let us take the example of an 8-bit ripple carry adder (Figure 7), made up of 8 one-bit full adders (Figure 8). The associated SDS is represented on Figure 9. The horizontal arrows represent the scheduling order of the boxes, which has been statically computed during the second phase of compilation. The SDS builder has tried to preserve the initial nesting until simulation.

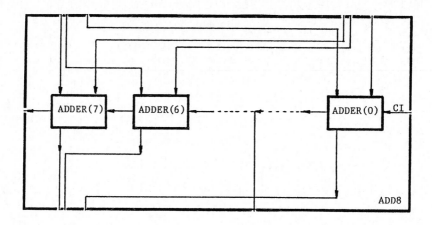

Figure 7 : 8-bit adder

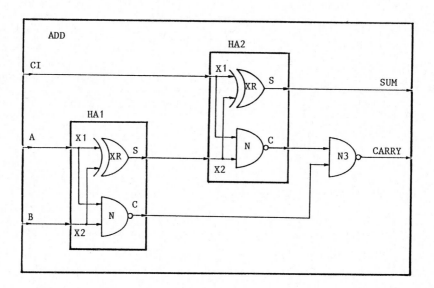

Figure 8 : One bit adder

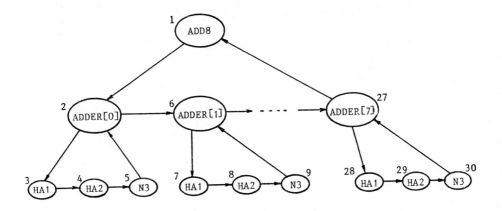

Figure 9 : 8-bit adder SDS

In the SDS, we distinguish two kinds of nodes: the "composition nodes" (ADD8, ADDER[i]) which are used to transfer control to the lower level, and the "terminal nodes" which will allow the execution of the SB when activated.

The boxes associated to the "composition nodes" are purely structural (they contain only interconnexions of boxes). The boxes associated to the "terminal nodes" are purely functional, they contain only instructions.

The indexing of the nodes on Figure 9 corresponds to the order followed by the scheduler for distributing control.

3.2.1 Examination Of A "terminal Node"

Let us consider node HA2 scheduled at time t. It contains the following scheduling information:

-   an activity flag which is set to true if one of its inputs has been modified at time t (either by an output of box HA1 or by a primary input CI);

-   a next event time which is the first time in the time wheel attached to the terminal node HA2.

If the activity flag is true or if the next event time is equal to t, HA2 is activated. The associated SB is evaluated, and sends back to the scheduler two pieces of information which will be stored in node HA2:

-   the new next event time (which is NIL if the node is latent);

-   an output modification flag (when this flag is true, the scheduler scans the fanout list in order to position the activity flag of the successor boxes).

3.2.2 Examination Of A "composition Node"

Let us consider node ADDER[0] at time t. It contains the same information items: activity flag and next event time.

Assume that node ADDER[0] is activated at time t. Its activity flag is true if one of its inputs has been modified. In this case, the first work of the scheduler consists in propagating activity through the fanout of this input. For example, if input CI of ADDER[0] has been modified at time t, its fanout within ADDER[0] leads to HA2 and node HA2 has its activity flag set to true.

When the scheduler has given control to ADDER[0], it will examine the nodes of the lower layer (HA1, HA2, N3) and possibly activate them. After that, coming back to ADDER[0], the scheduler stores two pieces of information in this node:

-   the new next event time
    $$tn \ (ADDER[0]) = MIN \ (tn \ (HA1), \ tn \ (HA2), \ tn \ (N3))$$
    possibly NIL if HA1, HA2 and N3 have been flagged latent;

-   an output modification flag.

For example, if output S of HA2 has been modified, its fanout leads to output SUM of ADDER[0], and thus the output modification flag of ADDER[0] is set. Then the scheduler will transmit activity at time t to the above layer in the nesting.

3.3 CONTINUOUS SCHEDULING MODE

This mode is the one used at circuit level. Time is measured in seconds, in the standard MKSA unit system, and is represented as a real number. The mathematical problem consists in solving a set of non-linear algebro-differential equations with initial conditions between two dates t1 and t2. In that case, the scheduler is linked with the simulation algorithms. The division of time between these two dates will be performed automatically and the integration step will generally vary, according to numerical difficulties and precision.

The scheduler will know dates t1 and t2 only. If in a model only the "continuous mode" is used during simulation, the nesting of boxes is flattened down and the SDS contains only a single node.

4 BUILDING THE SIMULATION ORIENTED DATA STRUCTURE

4.1 BUILDING OF THE "SMT", A SCHEDULING MODE GUIDED PROCESS

Phase 2 has to build Scheduler Data Structures (SDS) which are optimized for a specific scheduling mode, and may greatly differ from one scheduling mode to the other. In a multi-level model, it is therefore necessary to represent what parts of the model are associated with each scheduling mode. This information can be represented by a Scheduling Mode Tree (SMT).

On the example of Figure 5, let us assume that PODSYSTEM, RAM10, ROM708 and PIA821 are defined at system level, while POD is defined at RT level. POD contains an adder described partly at gate and partly at circuit level. The associated SMT is shown on Figure 10.

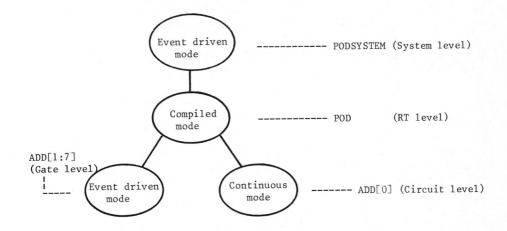

Figure 10 : POD SYSTEM Scheduling Mode Tree

Each SMT node corresponds to a part of the circuit combined with a particular scheduler. The Scheduler Data Structure associated with each circuit part is specific of each scheduling mode, and will each be interpreted by its associated scheduler.

4.2 TRANSLATION OF A "COMPILED SCHEDULING MODE" TYPE NODE

The SDS construction mechanism is top-down. Starting from the outer module, a tree corresponding to the nesting of units is built layer after layer.

For instance, the nested structure for Figures 7 and 8 is represented by the tree of Figure 9. At each layer, static levelizing is performed, and the result corresponds to the horizontal arrows of Figure 9.

If a zero delay loop is found, at a particular level of the hierarchy, one of the following two strategies may be applied:

- The user-defined nesting is preserved, and the presence of the loop is indicated in the SDS. The simulator will have to apply a stabilization algorithm on the looped blocks. This is the only possible strategy if the loop includes blocks described at different language levels.

- The user-defined structure, at this level, is modified by a restructuring algorithm. In this case, the loop

  . either will be suppressed

  . or will be pushed down to a terminal node of the hierarchy.

The terminal nodes of the SDS are associated to purely functional boxes, which contain only behavioural statements. For optimization reasons, these a priori unordered parallel statements can be ordered to obtain stable values in the carriers in one pass of computation. If a loop is found during the ordering process, a stabilization sequence can be added. The ordered set of instructions is then compiled at the end of phase 2, to make a Simulation Block (SB).

4.3 TRANSLATION OF AN "EVENT DRIVEN MODE" TYPE NODE

At gate and switch levels, the nested structure of units, if any, is always flattened down. The SDS produced is thus a one level hierarchy, where each node other than the root corresponds to a primitive component of the circuit (gate, transistor or node).

Static levelizing, for zero delay parts of the circuit, can be performed during phase 2 in order to detect zero-delay loops, and to optimize the updating of the time wheel and the gate (or transistor) evaluation.

At system level, where the events to be scheduled are the firing of transitions in the control graph, one of these two strategies may be chosen:

- to flatten down the part of the circuit described at system level,

- to preserve the initial nesting.

In the second case, the scheduling has to be hierarchical, and is a little more complex. An SDS node corresponds

- either to a transition of the control graph, which in turn points to its associated operative part instructions;

- or to a special purpose descriptor which points to the inner box to be activated.

4.4 TRANSLATION OF A "CONTINUOUS SCHEDULING MODE" TYPE NODE

Continuous scheduling corresponds to circuit level, and the description is always flattened down. The scheduler is strongly interleaved with the simulation algorithms. These algorithms, in the general case, have to solve a non-linear algebro-differential system with initial conditions.

The system F of algebrodifferential equations is built up, using for example as variables the node voltages, pin voltages and pin currents [18]. If some parts of the circuit are weakly coupled, the variables and the system can be automatically reordered, and this property will be exploited during simulation [5,25].

The result of this treatment is an Electrical Simulation Block (ESB). It possibly contains computation of system F or of parts of system F, computation of the Jacobian matrix or of parts of the Jacobian matrix, choice of pivots (LU transform), resolution of triangular systems . All these computations can be translated into machine code directly executable by the algorithms of the simulator.

A one bit full adder at circuit level is shown on Figure 11. The circuit level compiler builds up an algebrodifferential system, which is not represented here, whose dimension is approximately 600 corresponding to 42 transistors. This dimension depends on the choice of variables (nodal method, modified nodal method, tableau method, etc...).

5 SCHEDULER - SIMULATOR

5.1 GLOBAL SIMULATOR ARCHITECTURE

The principles of a global multi-level simulator architecture are represented on Figure 12. A scheduler supervisor explores the SMT. For each SMT node, corresponding to a particular scheduling mode, a SDS has been created during phase 2; this SDS and the specific scheduler are invoked by the supervisor. This scheduler, in turn, explores its SDS, and, upon reaching a terminal SDS node, either calls a specific simulator to perform the evaluation of a simulation block, or returns control back to the scheduler supervisor which will go down in the SMT.

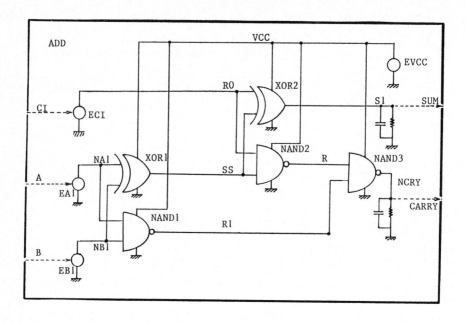

Figure 11 : one bit adder at circuit level

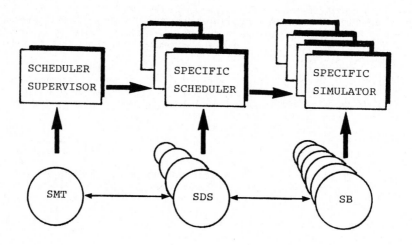

Figure 12 : Principles of a Multi-level simulator architecture

5.2 MIXED MODE SCHEDULING

Cohabitation mechanisms have to be defined for the simultaneous evaluation of multi-level models that require the presence of several scheduling modes.

5.2.1 Continuous Mode Inside Discrete Mode

This is the case when a circuit level module is contained within a module described at any other level. The box at circuit level needs to be given two dates t1 and t2. The electrical simulation algorithms will themselves perform the subdivision of time, which they consider as a real variable. At logical level, time is considered as an integer variable and is expressed in multiples of an elementary duration.

In this case the user is requested to give the ratio between the discrete and the continuous time units (i.e. express the discrete time unit in seconds). The scheduler will then perform the time conversions when handling the continous mode.

5.2.1.1 Continuous Mode Inside Discrete Compiled Mode

Let us consider the example of Figure 7 written at RT level except for description ADD written at circuit level.

The scheduler will treat the nodes ADDER[i] in the way described in paragraph 3.2 (such node is activated if the activity flag is true or if the next event time is equal to t1). But the scheduler supervisor has to transmit to the continuous mode specific scheduler, and then to the Electrical Simulation algorithms, another time t2 which is necessary for the continuous mode. This time interval corresponds for the scheduler to a "quiet time interval" during which it guarantees to the electrical box that its inputs will not be modified [18]. The static ordering allows the scheduler to know that. In this example, t2 corresponds to the closer next event.

When control has been given to the node ADDER[3] for instance, at time t1, then the continuous specific scheduler calls from the algorithm base the Electrical Simulation algorithm and the associated ESB during the interval t1-t2. Several cases can occur:

1. A modification is detected in an output pin of ADDER[3] before time t2. This means that a modification of a sufficient magnitude to be interpreted as a different logical value, at logical level outside ADDER[3], has been detected. Let t3 be the time of this modification. The control is returned to the scheduler together with t3 as next event time.

2. No output variation occurs before t2 but ADDER[3] continues to work. Then control is returned to the scheduler together with t2 as next event time.

3. No output variation occurs until t2 but ADDER[3] stops working. It is latent. Control is returned to the scheduler together with NIL as next event time.

5.2.1.2 Continuous Mode Inside Discrete Event Driven Mode

The activating process is the same as above. The only difference is on the evaluation of time t2. As the scheduler works in the event driven mode, it cannot know this time. It will consequently choose the following time (t2=t1+1).

The control is then given to the node from t1 to t1+1 (in discrete time units). Therefore, only cases 2 and 3 of previous paragraph can occur when the control is returned to the scheduler. This mechanism remains efficient as long as the discrete time unit is big compared to the time unit used at circuit level.

5.2.2 Discrete Compiled Mode Inside Event Driven Mode

This mode can occur for example with boxes described at RT level within a box at system or gate level.

Let us consider the example of box ADD8 of Figure 7 inside a module described at system level. The mechanism is very simple. The dynamic scheduler looks at the time wheel and finds at time t1 an event associated to node ADD8.   The control is given for this node to the static scheduler (via the scheduler supervisor). At t1, the static scheduler interprets the SDS of Figure 9 as explained above. When this examination is finished, the node ADD8 will contain the next event time t2 and will be flagged for output modification. These two pieces of information are then exploited by the dynamic scheduler in order to update the time wheel.

5.2.3 Discrete Event Driven Mode Inside Discrete Compiled Mode

This mode can occur for example for a module described at gate level within a module described at RT level. For the static scheduler, the mechanism is the same as when it examines a node working in continuous mode (see paragraph 5.2.1.1).

Let us consider the example of Figure 7 with ADD described at gate level.   When node ADDER[i] is to be evaluated, the static scheduler gives control,via the scheduler supervisor, to the dynamic scheduler between t1 and t2.   The three cases described in paragraph 5.2.1.1 can then occur:

1. output modification at time t3 before t2: the control is returned to the static scheduler with time t3 as next event;

2. no output variation before t2 but the time wheel is not empty and the control is returned with time t2 as next event;

3. no output variation until t2 and no more event stored in the time wheel: the node is latent. The control is returned with next event time = NIL.

For this mechanism to be efficient, the dynamic scheduler must have a large number of events to manage during the time interval [t1:t2].

5.3 COMMUNICATION BETWEEN LEVELS

5.3.1 Time Units

Apart from the circuit level, in which a physical unit of time is being used, the various discrete modelling levels work on an abstract unit of time. When simulating a multi-level model, a different unit of time may correspond to each modelling level.

A possible way to manage time communication between the various levels may consist in

1. stating, for each modelling level, the ratio between the local time unit and the "basic logical time unit",

2. stating the correspondance between the "basic logical time unit" and the physical time unit (i.e. giving the value of the basic logical time unit in seconds).

5.3.2 Interface Functions

As we have seen on the ADD example, a circuit may be modelled at various levels, under the same description segment identifier. A data management system should be responsible for storing and retrieving the various representations of the circuit. Therefore, a source description may be independent of the description levels of its nested units; it may be simulated several times without modification, at quite different levels, by specifying to the simulation compiler which representation to take for each individual unit.

In order for this property to be available, all representations must have compatible interfaces: same number of interface elements, declared with same direction, same identifier, same dimension. The type assumed by an interface element may depend on the language level. Compilation of the unit interconnection must thus invoke a conversion function whenever the value type of the unit actual interface carrier does not match the value type of the formal interface. An interface conversion function has to return either the converted value, or an error message for non permitted translations, for all combinations of language levels and primitive value types. This function might depend on the technology used by the designer, and should be part of the compiler for efficiency reasons.

REFERENCES

[1] Aho A.V., Ullman J.D. "Principles of Compiler Design", Addison Wesley
    Publishing Company, 1977

[2] Backhouse R.C. "Syntax of Programming Languages - Theory and Practice"
    Prentice Hall International, London, 1979

[3] Bauer F.L., Eickel J. "Compiler Construction: an advanced course",
    Springer Verlag, New York, N.Y., 1974

[4] Beaudoin-Lafon M., Gresse C. "Caty: un environnement de programmation pour
    une construction graphique et interactive de programmes" TSI, Vol. 3 No 4,
    Dunod, July-Aug. 1984

[5] Bona M. "Quelques outils numeriques pour la resolution de systemes
    algebrodifferentiels de grande dimension. Application au projet CASCADE."
    These Docteur Ingenieur, USMG, Grenoble 1983.

[6] D. Borrione: "The CASCADE Multi-Level Hardware Description Language",
    IMAG/ARTEMIS Research Report No 514, Feb. 1985.

[7] Breuer M.A., Friedman A.D. "Diagnosis and Reliable Design of Digital
    Systems." Potomac, MD: Comp.Sc.press, 1976.

[8] Bryant R.E. "Switch level modeling of MOS digital circuits." Proc. IEEE
    Int. Symp. on circuits and systems. Rome, 1982, pp 68-71.

[9] Collection of Proceedings of the "International Conference on Computer
    Hardware Description Languages and their Applications", Sept. 1975, Oct.
    1979, Sept. 1981, May 1983, Aug. 1985.

[10] Collection of Proceedings of the IEEE-ACM annual "Design Automation
     Conference" series, June yearly.

[11] Delaunay M., Grabowiecki J.F. "Note technique d'utilisation du
     Transformateur de Grammaire Ll(1)", IMAG Research Report No. 234,
     Grenoble, Sept. 1980.

[12] Donzeau-Gouge V., Huet G., Kahn G., Lang B. "Programming Environments
     based on structured editors: the Mentor experience", INRIA Research Report
     No 26, Rocquencourt, July 1980

[13] "EDIF Specification - Electronic Design Interchange Format", Version 095,
     EDIF Steering Committee, Nov. 1984

[14] Foo S.Y., Musgrave G. "Comparison of Graph Models for Parallel Computation
     and their Extension" Proc. International Conference on Computer Hardware
     Description Languages and their Applications, New York, Sept. 1975

[15] Johnson S.C. "YACC - Yet Another Compiler Compiler", CSTR 32, Bell
     Laboratories, Murray Hill, N.J., 1974

[16] Kahn G., Lang B., Melese B., Morcos E. "Metal : a formalism to specify
     formalisms", Science of Computer Programming, North Holland, Sept. 1983

[17] Koster C.H.A. "Using the CDL compiler-compiler", in [3]

[18] C. Le Faou: "Hierarchical Multilevel Mixed-Mode Simulation in CASCADE", IMAG/ARTEMIS Research Report No 513, Feb. 1985.

[19] Lesk M.E. "LEX - a lexical analyser generator", CSTR 39, Bell Laboratories, Murray Hill, N.J., 1975

[20] J. Mermet: "Several Steps Towards a Circuits Integrated CAD System: CASCADE", Proc. CHDL'85 International Conference, Tokyo, Aug. 1985,

[21] MOTOROLA Inc.: "M6800 Micro Processor Programming Manual", and "M6800 Application Manual", 1975.

[22] Noe J.D., Nutt G.J. "Macro E-Nets for representation of parallel systems" IEEE Trans on Computers, Vol. C-22, No 8, Aug. 1973

[23] R.Piloty, M.Barbacci, D.Borrione, D.Dietmeyer, F.Hill and P.Skelly: "CONLAN Report", Lecture notes in Computer Science 151, Springer-Verlag, Berlin, 1983.

[24] C.W. Rose: "LOGOS and the Software Engineer", Proc. Fall Joint Computer Conference, 1972, pp. 311-323.

[25] Rabbat N.B., Sangiovanni-Vincentelli A.L., Hsuej Y. Hsieh "A multilevel Newton algorithm with macro modeling and latency for the analysis of large scale non-linear circuits in the time domain." IEEE Circuits and Systems, Vol. CAS-26, No 9, 1979.

[26] Shahdad M., Lipsett R., Marschner E., Sheehan K., Cohen H., Waxman R., Ackley D. "VHSIC Hardware Description Language" IEEE Computer, Vol 18, No 2, February 1985

[27] Ulrich E.G., Herbert D. "Speed and accuracy in digital network simulation based in structural modeling." Proc. 19th DAC, Las Vegas 1979.

7. Standardization Efforts

7.1. THE CONLAN PROJECT:
Concepts, Implementations, and Applications

Robert PILOTY

Technische Hochschule Darmstadt, F.R.G.

Dominique BORRIONE

Université de Provence, Marseille, France (*)

The Consensus Language project has produced a construction
mechanism that describes system behavior and structures.
It has already yielded derivative hardware description
languages and gate-array synthesis software and promises
further design tools for the future.

Conlan -CONsensus LANguage- is a general, formal language construction mechanism
for the description of hardware and firmware at different levels of abstraction.
Its basic object types and operations describe the behavior and the structure of
digital systems. First results on the global Conlan approach [5,6] or on
special aspects [7,8] were published starting in 1980. They were developed by
an international working group established in 1975 and consisting of R. Piloty,
chairman; M. Barbacci, CMU Pittsburg; D. Borrione, IMAG Grenoble, France;
D.Dietmeyer, Univ. of Wisconsin; F. Hill, Univ. of Arizona; and P. Skelly,
Honeywell Phoenix.

The Conlan construction mechanism uses the notion of language derivation; it
expresses objects, operational primitives, and syntax of a new member of the
Conlan family in terms of the corresponding entities of a previously defined
language, Base ConLan. Its properties and the formal rules for the derivation
of new languages from BCL are rigorously specified in the Conlan Report [1].

A Conlan hardware description language called Wislan was defined, and gate-array
synthesis software based on Wislan was implemented at the University of
Wisconsin [2,3]. Other efforts to apply the Conlan system to the derivation of
specific languages from BCL and to develop appropriate software tools to support
the derivation process and the construction of simulators are taking place in
Darmstadt and Grenoble [4].

In providing an informal introduction to Base Consensus Language, this article
outlines work on the derivation of languages from BCL and describes the status
and plans for software tools supporting language derivation and implementation.

(*) The work reported here was performed for D. Borrione, at the IMAG/ARTEMIS
Laboratory in Grenoble, France.

1 GENERAL CONLAN CONCEPTS

Conlan assumes a universe of primitive objects, divided into values -data objects such as integer constants or strings of characters- and carriers, which are containers of varying values. Objects sharing common properties are grouped into types; types that play a similar role are grouped into classes. We distinguish between the classes of value types and of carrier types.

Type and class are two of the six structuring concepts available as basic building blocks, called segments, to define Conlan entities. Another two are operations on objects: a function returns an object as result; an activity modifies the content of one or more carriers. For instance, the operation that returns the XOR results of two Boolean values is defined as a function; conversely, the operation that assigns to a parameter variable the XOR of two Boolean values is defined as an activity.

The last two structuring concepts are directly related to the two complementary aims of Conlan. A description is a model of a digital system, written in a particular Conlan language. A language is a set of type, class, operation, and description definitions to be used as primitives for a given modeling level of abstraction.

To a large extent different groups or people are expected to write these two types of segments: the toolmakers who prepare new members of the Conlan family and software to support their new members and the users who use those new languages and supporting software to record their hardware design efforts. In many organizations toolmakers are members of the design automation department while users are members of engineering design description departments. A toolmaker defines languages; a user writes hardware description segments in terms of the primitive segments of one of these languages and his own user-defined segments.

A valid Conlan text consists of properly nested blocks, the outermost being a segment prefaced with a reference language statement (key-word REFLAN). The reference language in the Reflan statement provides the syntax and semantics available for the body of the text.

Segments also start with a keyword, which is followed by an identifier and terminated by keyword END, which may be followed by the segment identifier. Initiating keywords for segments are TYPE, CLASS, FUNCTION, ACTIVITY, DESCRIPTION and CONLAN. CONLAN segments are intended for use by toolmakers to define new languages. DESCRIPTION segments are intended mainly for users to describe hardware.

Blocks that refer to or invoke an instance of a segment are called statements. Most begin with a keyword and terminate with END optionally postfixed to the opening keyword. Primitive statements are identical in all Conlan languages. Some may be used for language definition only, so must be written in a text where the outermost segment is a CONLAN segment. Their keyword terminates with symbol "@". Likewise, identifiers for toolmaker entities (such as type identifiers) are terminated by symbol "@". They can be referenced only in a language definition segment, not in a description segment; user entities, however, are referenceable in both.

Statements IF, CASE, and DECLARE are self-explanatory. The OVER statement indicates repetition without any sequence specification. The USE statement expresses the structural decomposition of a description into nested instances of smaller descriptions and assigns an identifier to each instance. Behavioral description is expressed in terms of operation invocations in a form that resembles procedure calls in a programming language. However, in Conlan there is no implied or explicit division between data structure and control structure for describing behavior by operation invocation. Lists of operation invocations do not imply a sequence of execution. Also, no sequencing operations, like GOTO, are provided as primitives. All operations invocations are a priori executed concurrently within one block; conditional invocations of operations (IF and CASE statements) permit users to describe time sequencing by allowing them to specify under which conditions an operation is to be executed.

ASSERT statements are especially valuable in hardware description. They allow toolmakers and users to specify predicates and other Boolean relations between inputs and outputs of operations or descriptions that they expect to be true: an error condition exists if an assertion turns out false. Setup and hold-time specifications on flip-flops are naturally checked against assertions, as are size constraints on array parameters.

Other statements are reserved for language definition. The FORMAT@ statement provides the toolmaker with a means of deriving the syntax of a new language from the syntax of its reference language by permitting additions or deletions. Productions may receive additional meaning through the FORMAT@ statement, but with limits set by a system of rights on productions. By carefully placing rights on BCL grammar, the Conlan working group intends to ensure the consistency of all member languages. Its efforts should also permit toolmakers to keep the syntax of a new language simple and, at the same time, incorporate new constructs to denote features, such as an infix for a new operation. The core syntax is indelible, however, so it remains common to all Conlan members.

2 BASE CONLAN

BCL is defined by a set of object types and operations, the syntax, and an associated computation model.

2.1 VALUE TYPES

The elements of value types are used primarily to express the instantaneous system state of the described hardware units. BCL provides the basic scalar value types int, bool, and string with their corresponding constant denotations and the usual operations. In addition, the types pint, nnint, bint(m,n:int) are introduced as subtypes of int, designating positive, non-negative, and bounded integers; the latter has left and right bounds m and n as type parameters. As subtypes, they retain all integer operations.

2.2 SIGNAL TYPES

Because of the delay and inertial properties of such hardware components as gates, combinational networks, and storage elements, the description of signal types must, in general, permit the computation of their behavior as a sequence of system states in real time, measured in multiples of some elementary time unit. The system state, during some time units, is normally a function of one or more past system states. This function requires that the language provide a means to refer to past values of the system state. On the other hand, the next real-time state of a system may be a complex mathematical function of preceding states. Its evaluation may require many computational steps to which no elapsed time can be attributed. These considerations lead to the BCL model of time and signals.

2.3 THE BCL TIME MODEL

BCL provides a discrete time model that breaks real-time into discrete instants
separated by a single time unit. The beginning of each new time unit contains
an indefinite number of computation "steps" identified with integers greater
than zero. Successive steps provide only a before/after relation. Values
obtained at the last step of computation are then associated with that instant.
When operation invocations are nested, the inner operation uses "local steps" to
complete its computation before the outer operation uses its result.

2.4 THE BCL SIGNAL MODEL

The signal is a set of values mapped onto the BCL time model; that is, it is a
sequence of values. One generic signal type is

TYPE signal(V: val_type)

A specific signal type designator is obtained by binding its parameter v to one
value type, for example, "signal(bool)", "signal(pint)", etc.

Figure 1a shows an example of a signal x of type signal(bool) and duration five.
It is defined as a nested structure of tuples, or ordered sets, and is denoted
with tuple delimiters "(." and ".)". The first level is a tuple that holds the
default value zero for initialization and the signal itself. The signal is a
tuple of five elements corresponding to the five instants, while the rightmost
element corresponding to the present time unit contains all computation step
values obtained. Computation at the step level is terminated as soon as all
signals in the system have stabilized, so at least the last two step values of
the current time unit are the same in all signals.

Upon termination of the current time unit, the algorithm of the BCL computation
model discards all step values except the last, appends a new time unit tuple,
and initializes it with the last value of the preceding instant (Figure 1c).
Two counters carrying positive integers are maintained by the algorithm: the
time counter t@ and the step counter s@. Upon termination of the current time
unit, t@ is incremented by one and s@ is reset to 1.

Two user functions are provided by TYPE signal(v).

FUNCTION delay (x: signal(v); d: pint): signal(v)
shifts a signal x by d time units towards the present (Figure 1b), fills the
first d units with default values, and sets all computation steps of the current
interval to the value at t@-d. FORMAT@ statement provides the infix notation
x%d for convenience.

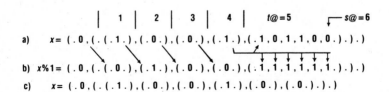

**Figure 1. BCL signals: a) five-time unit signal with computation steps in cur-
rent interval; b) same signal one-time unit delayed; and c) same signal after
shrinking and initialization of next interval.**

INTERPRETER@ FUNCTION val(x: signal(v)): v
returns the rightmost computation step value. Keyword INTERPRETER@ indicates
that BCL automatically invokes this function whenever a signal appears as an
actual parameter of an operation or description and whenever the corresponding
formal parameter expects a value type. Thus, by combination of x%d and val(x),
the BCL user can directly refer to any past real-time value of a signal.

Example:

Let "stable" be a function that tests whether or not a Boolean signal x has been
stable for the last n time units. The Conlan function definition segment
defines a function with this property recursively as

```
FUNCTION stable(x: signal(bool); n(pint): bool
  RETURN
    IF n = 1 THEN 1
    ELSE x%n = x%(n-1) & stable(x, n-1)
    ENDIF
ENDstable
```

2.5 CARRIER TYPES

BCL carriers contain a BCL signal. Three generic carrier types are

```
TYPE terminal(v: val_type; def: v)
TYPE variable(v: val_type; def: v)
TYPE rt_variable(v: val_type; def: v)
```

Instances of these carriers may be created and named in the DECLARE statement.
Parameter v representing the value type of the signal in the carrier and
parameter def representing a default value chosen from the domain of v must be
bound at the point of declaration.

```
DECLARE a, b: terminal(int, 1);
        c, d: rt_variable(bool; 0) END
```

Declared carriers are initially empty; they contain no signal. Frequently used
carrier types may be abbreviated by subtype definition, for example, by type
segment:

```
SUBTYPE btm0 BODY terminal(bool, 0) ENDbtm0
```

2.5.1 User Activities For Carriers.

In each carrier type, there is one user activity. It appends a value y to the
signal of carrier x at every step where it is invoked. Infix notations are
provided for convenience:

```
x1 .= y connects y to terminal x1.
x2 := y assigns y to variable x2.
x3 <- y transfers y into rt_variable x3.
```

If two or more connections to the same terminal are invoked with different
source values during the same step, a "collision" is reported. Likewise, if two
different values are assigned to the same variable or a rt_variable, a collision
is also reported.

2.5.2 Time Properties Of Carriers.

Additional operations are defined with key-word INTERPRETER@ in each carrier
TYPE definition segment; they are automatically invoked by the evaluation
environment and determine the memory and time characteristics of a carrier type.

For instance,

            INTERPRETER@ ACTIVITY finstep(W x: terminal(v, def))

is automatically invoked by the environment whenever no connect activity on x is
active. For example, if in the single conditional invocation

                    IF c THEN x .= y ENDIF

condition c is off during one step, it appends default value "def" to the signal
of x. Therefore, terminals lose their values during off-periods (see Figure 2).

Activity "finstep", when applied to a variable, extends the signal during "off"
steps with the last value assigned in the preceding "on" step. Variables retain
their values during off periods on the step level (see Figure 3).

Figure 4 shows the evaluation of a conditional invocation of a rt_variable.   It
retains the value from the preceding instant when the condition is off.

Other activities are invoked upon stabilization of an operation or description
segment to delete intermediate step values and extend carrier signals by one
interval before the initialization of the computations for the next time unit
[1].

Functions "sig" and "val" return the signal and the current instant value of a
carrier.   They are provided in each carrier type for automatic dereferencing.
The result of val applied to a rt_variable is shown on Figure 4:   the
rt_variable is a unit delay carrier.

Figure 2. Connection to a terminal.

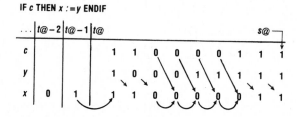

Figure 3. Assignment to a variable.

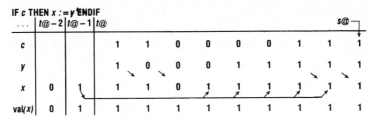

Figure 4. Transfer to a *rt*__variable.

2.6 CONSTRUCTOR TYPES

In addition to scalar user types outlined in the preceding sections BCL provides
toolmakers with powerful array and record constructors for spatial composition
of elements taken from the scalar user types.

2.6.1 Array Types

One master constructor

TYPE array@(u: any@)

is provided as a generic type. Individual array types are obtained by binding
its type parameter u to any type in particular from the scalar value-, signal-
or carrier types defined by BCL or its user. Arrays may have an arbitrary
number of dimensions.

Arrays with explicitly stated array dimensions can be formed.

SUBTYPE fixed_array@ (u: any@; d: array_dimension@)

is a subconstructor derived from master constructor array and specifies the
dimensions as second parameter d. After FORMAT@ statement to suppress
designator "fixed_array" a typical fixed_array declaration in BCL reads as
follows:

DECLARE a, b, c: btm0[0:7; 1:16] END

Many functions are provided for testing the attributes of an actual array
parameter and for indexing, transforming, comparing, and catenating arrays. For
indexing, the usual bracket notation is provided.

Example:

Assume DECLARE a: btm0[0:31; 1:8] END
then

      a[3;7] selects an element.
      a[3,4; 2:5] selects a slice.
      a[1:31, 0; 1:8] rotates left one position in the first dimension.
      a[1, 1, 5, 28:20; 1:8] is also permissible.

2.6.2 Records Types

A generic type

        TYPE record@ (f: field_descriptor_list@)

allows the declaration of records, e.g,

        DECLARE a, b: record@('f1': bvar0; 'f2': btm0)

with field selection written "!", e.g.,

        'f2'!a .= x

3   BASIC FORMS OF BEHAVIOR DESCRIPTION

BCL supports general forms of hardware behavior description. In designing a computer hardware description language, the three fundamental modes of behavior modeling must be considered:

1. signal flow,

2. conditioned state transition, and

3. controlled state transition.

Signal flow specifies the system behavior by a list of permanent activity invocations that continuously and concurrently compute the signals in a selected set of output and internal carriers as a function of a set of input and internal carrier signals. No retention properties for the system carriers are required in this mode.

In conditioned state transition, behavior is described by a list of conditional invocations of activities driving the system carriers. The conditions are expressed in terms of present or past values of potentially any carrier in the system. In a given carrier, new signal values are computed only if one of the activities driving it is invoked. Otherwise the carrier retains its state or is assigned a default value. This mode is particularly useful when only a small part of the system carriers change state at a given moment. Then, a designer must determine where and when something happens in the carriers. Computation efficiency is gained when only a few conditions are true at any time.

In the controlled state transition, system carriers are split into control carriers and data carriers. Only the state of a small number of control carriers (in case of a purely sequential procedure only the state of one, the "locus of control") determines if and when an activity may be invoked. Languages supporting procedure description normally provide special constant denotations for control states (e.g., labels), special sequencing activities (e.g., GOTO, FORK, NEXT). Programming languages, or automaton- and control-graph-based languages are of this type.

BCL supports only signal flow and conditioned state transition. As the root language of an extensible family, BCL covers only the basic situations in hardware description. Nonetheless, languages for controlled state transition can be defined from BCL. Although it does not directly provide multivalued logic types normally required at low-level logic simulation, convenient facilities are available to construct them from the BCL types based on TYPE, ACTIVITY, and FUNCTION segments.

3.1 SIGNAL FLOW SPECIFICATION

The following signal flow description illustrates an rs-flipflop with unit delay NOR connectives:

```
DESCRIPTION rsff (IN r,s: signal(bool); OUT q, nq: btm0)
   BODY
      nq .= nor(r%1, q%1),   q .= nor(s%1, nq%1)
ENDrsff
```

It executes a transition of nq = 1, q = 0 to nq = 0, q =1 in two time units, if r is kept at zero and s at 1 for two intervals. The signals in nq and q accumulate three step values in the current time unit during transition.   The first value in nq and q is carried over from the preceding time unit, the next corresponds to the new state, the third is identical to the second and establishes termination of that time unit.

3.2 CONDITIONED STATE TRANSITION SPECIFICATIONS

Consider

```
FUNCTION count_ones (x: array(bool)): pint
   ASSERT dim_count(x) = 1& 1b(x,1)<rb(x,1) ENDASSERT
   BODY
      DECLARE i,r: ivar(0) ENDDECLARE
      IF i < dimsize(x,1) THEN i := i+1 ENDIF;
      IF i < dimsize(x,1) & x[1b(x,1) + 1-i] = 1 THEN r := r+1 ENDIF
ENDcount_ones
```

This function computes the number of ones contained in a Boolean vector of arbitrary range on the step level.   The assert statement assures that only one-dimensional arrays with ascending indices are accepted as actuals. The two auxiliary integer variables, i and r, which are declared in the body, are initialized with value 0. Variable i scans the input vector x.   Variable r counts the ones.   For a vector of size 8, 9 local steps are required to terminate.

Specifications in terms of conditioned state transitions are particularly convenient to specify the input/output behavior of modules or building blocks (grey boxes) in real time, as the following example of an edge triggered d-flipflop should demonstrate:

```
DESCRIPTION dff (tsu, th,tp: pint)
                (IN d, ck: signal(bool); OUT q, nq: bvar(0))
   ASSERT tp>th ENDASSERT
   BODY
   ASSERT IF ck%th & ~ck%(th+1)
          THEN stable(d, tsu + th)
             & stable0(ck%th, tsu)
             & stable1(ck, th)
          ELSE 1 ENDIF
   ENDASSERT
      IF ck%(tp-1) & ~ck%tp THEN q:=d, nq:=~d ENDIF
ENDdff
```

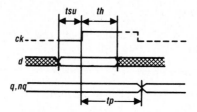

Figure 5. Timing diagram of a d-flip-flop.

This dff-template has an attribute parameter list and an I/O parameter list. The first contains setup, hold, and propagation time as attributes. Attributes must be bound at the point of instantiation; for example,

USE ff1, ff2: dff(3,4,7) ENDUSE

The IN list expects a Boolean signal for data and clock. The OUT list states Boolean variables q and nq as output carriers [1,5]. One attribute assertion requires a longer propagation delay than hold time. The next assertion monitors data and clock inputs for proper shape, as shown in the timing diagram in Figure 5. Finally, the action of the flipflop is specified in terms of conditional assignments to q and nq.

4 DERIVATION OF LANGUAGES

The Conlan construction mechanisms and Base ConLan types used to define existing HDLs assured accurate results. We proceeded then to derive the semantics of the Cassandre, Lascar, and Lasso languages from BCL [9]. We selected these languages because

- One of the authors had all information about them;

- The languages cover design levels from RT to system level;

- Cassandre and Lascar have been used in industrial environmemts for years; and

- Lasso offers useful features, such as timed specifications.

A discussion of Lascar, together with portions of its formal Conlan derivation was published in 1982 [4].

Defining the primitive operators and types of these languages proved to be easy. Some syntactic aspects of these languages could also be defined, with the FORMAT@ statement of Conlan, particularly the various kinds of conditional and control statements. Other syntactic rules were incompatible with Conlan syntax; interface, formal parameter, and local object declarations were among the most obvious. However, BCL offers more powerful constructs for array indexing and type definition than the other three languages.

After being validated on existing HDLs, the Conlan definition principles were applied to derive new languages from BCL. The first was Wislan, followed by TPDL, the Time Profile Description Language [10]. Strictly speaking, TPDL is not a full HDL, but a general set of types and operators that describe logical and temporal conditions to be used in the Cascade HDL [11] as well as in the Cascade simulator command language. TPDL offers a set of primitives to express

- some relations among variables of the model that may hold only during certain periods of time or may depend on other relations (model specification by temporal assertions);

- some observations, measurements, or input assignments that may depend on the occurrence of an event or be valid only for some time intervals (commands of the simulation experiment).

In both cases, statements are permanently valid, unless preceded by conditions that limit their validity. These conditions are expressed with a variety of operators as combinations of sets of events and sets of intervals.

4.1 AN EXAMPLE OF TYPE DERIVATION

In the definition of a new Conlan language, one of the key features is the derivation of new types from existing types of the reference language. The BCL-type "signal(bool)" provides the mechanism for deriving the TPDL notion of a set of events.

The value of one or more carriers of a description is tested by a Boolean expression. An event is defined by a 1 to 0 or 0 to 1 transition of such an expression. It has no duration and occurs at a specific instant in time, called its date. There is a one-to-one correspondence between the occurrence of an event and a date. A set of events is defined when all transitions of a Boolean expression are considered.

The primitive "any_time" generates an event each time a Boolean expression becomes 1, while the primitive "any_change" generates an event each time a Boolean expression changes its value (from 1 to 0 or from 0 to 1). These two primitives, therefore, produce a set of events from a Boolean expression.

A particular expression can be selected to produce a unique event. The primitive "as_soon_as" produces an event the first time a Boolean expression becomes 1, whereas the "number(n,e)" produces an event at the nth occurrence of an event within a set of events e. "At" generates an event at a given date. As an example, let "be" be a Boolean expression with the time behavior shown in Figure 6a, with respect to the time scale of Figure 6b. Here the basic time unit is 1 ns. The function count counts the number of occurrences of the event parameter and produces an integer result. The every primitive allows the generation of a series of events with a given periodicity (a number of time units), starting from an event or a date that can be either included or excluded. The "delay" primitive delays the occurrence of all the events of a specified set, for a fixed period. Additional primitives are provided to generate sets of events such as union and ordered or unordered sequences of events.

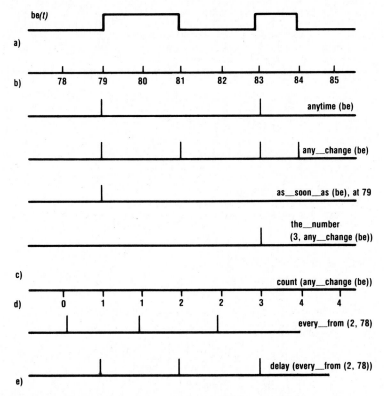

Figure 6. Typical sets of events returned by TPDL expressions.

4.2 DEFINITION OF TYPE EVENT IN BCL

The previous informal definition of events, although easy to comprehend, fails to provide precise information to the user and is even less adequate for the implementor. A precise semantic definition is given by the TYPE segment "set_of_events" written in BCL. Figure 7 shows portions of this segment. Space constraints have eliminated many functions, as well as FORMAT@ statements that specify the infix notation for the functions. Line numbers are provided to ease the explanation, but they do not belong to the Conlan text.

Sets of events are derived from Boolean signals with default value 0 (line 5). Five functions are carried from type signal (bool) to type set_of_events (line 7); the first three, identified by the end symbol @, are included for definition only and are not available to the user:

- "slength@" returns the length of the signal; for example, the number of time points, for which a value exists;

- "inst_value@" and "select_rts@" return the value of a signal x at time i, denoted x{i,} and x{i} respectively, considered either as one instantaneous value at time i, or as a sequence of instantaneous values due to several steps of computations at time i;

```
 1  TYPE set__of__ events
 2    BODY
 3       ''/sets of events are derived from boolean signals with default value 0 / ''
 4
 5    ALL e: signal(bool) WITH dpart@(e) = 0 ENDALL
 6
 7    CARRY slength@, inst__value@, select__rts@, pack, delay ENDCARRY
 8
 9       ''/ returns a set__of__events containing a 1 whenever x rises /''
10    FUNCTION anytime(x: signal(bool)) : set__of__events
11       RETURN THE@ z:set__of__events WITH
12         slength@(z) = slength@(x) AND
13         FORALL@ i: pint IS
14           i > slength@(z) OR z{i} = pack__css@( ¯(x%1{i,}) & x{i,})
15         ENDFORALL
16       ENDTHE
17    ENDanytime
18
19       ''/ Functions anychange, at: similar to anytime /''
20    STATIC FUNCTION the__number(ATT i: pint; x: set__of__events): set__of__events
21
22       BODY
23         DECLARE a, t: variable(nnint, 0) ENDDECLARE
24         IF a < i & x{ } THEN a: = a + 1 ENDIF,
25         IF a = i − 1 & x{ } THEN t: = t@ ENDIF
26         RETURN at(t)
27    ENDthe__number
28
29    STATIC FUNCTION as__soon__as(x: signal(bool)): set__of__events
30       RETURN the__number(1, anytime(x))
31    ENDas__soon__as
32
33    STATIC FUNCTION count(x: set__of__events): nnint
34       BODY
35         DECLARE a: variable(nnint, 0) ENDDECLARE
36         IF x{ } THEN a: = a + 1 ENDIF
37       RETURN a
38    ENDcount
39
40       ''/ periodic event starting at a specified origin, included /''
41    FUNCTION every__from(ATT period: pint; origin: pint): set__of__events
42       RETURN
43         IF origin > t@ THEN pack(0)
44         ELSE THE@ z: set__of__events WITH
45             slength@(z) = t@ AND
46             FORALL@ i: pint IS i > t@ OR
47               i < origin AND z{i} = pack css@(0) OR
48               IF (i − origin) MOD period = 0
49                 THEN z{i} = pack__css@(1)
50                 ELSE z{i} = pack__css@(0)
51               ENDIF
52             ENDFORALL
53           ENDTHE
54         ENDIF
55    ENDevery__from
56
57  ENDset__of__events
58
59  SUBTYPE event
60       BODY ALL e: set__of__events WITH
61             FORALL@ i: pint IS
62             FORALL@ j: pint IS
63               i > slength@(e) OR j > slength@(e) OR
64               i = j OR  ¯ (e{i,} & e{j,})
65             ENDFORALL ENDFORALL ENDALL
66  ENDevent
```

Figure 7. Derivation of TYPE "set__of__events."

- "pack" transforms a value v into a signal of length t@ with present and all past values equal to v;

- "delay" denoted x%i in infix Conlan notation, returns a signal, displacing values i time units toward the present.

Then, the functions specific to type set_of_events are defined.

As a general rule, all functions (except count) return a set_of_events of length equal to the value of t@ at each invocation. These functions can be defined in two ways:

- by computing the resulting set_of_events from their parameters and finding the unique set that fulfills a predicate (see anytime: lines 10-17, every_from: lines 41-55);

- by computing at each point in time the present value of the set_of_events and accumulating the full history in a local variable of a static function, where previous values of local objects are retained (see the_number_that: lines 20-27).

Some functions are easily written by invocation of previously defined functions (see as_soon_as: lines 29-31).

A single event is considered to be a special case of a set_of_events, which occurs only once. Thus, type event is defined as subtype of set_of_events (lines 59-66).

5 SOFTWARE DEVELOPMENTS

Researchers at the Institut fuer Datentechnik in Darmstadt developed the software system Conlit as a basic set of Conlan implementation tools. It contains Cocofe, a Conlan Compiler Front-end generator [12], and Pargen [13], a parser generator complementing Cocofe. Both are written in standard Pascal.

Cocofe is table-driven. It implements all languages derived formally in the ~Conlan framework. It performs lexical and static semantic analysis and generates an intermediate form of a Conlan text. This intermediate form, called Ireen [14], is independent of any particular Conlan language, It is an abstract data type on a domain of Pascal record tree.

Each node of an Ireen tree represents an instance of some Conlan construct; for example, identifier, description segment, operation invocation, etc. Each edge represents a nesting relation specified in the original Conlan text. The static, semantic analysis [15] covers name analysis, type and class checking, check and evaluation of compile-time expressions (including subtype checking), check of access rights, data-flow directions, processing of syntax modification, overloading resolution with automatic parameter matching, and preparatory actions for run-time subtype checking.

Cocofe can be used as a front-end generator for a specific Conlan language definition or as the front end itself for this member language (figure 8). In the first mode, it generates a Reflan table for a specific new member language L2 from the table of an existing member language L1 and from the CONLAN segment that defines L2 in terms of L1. The Reflan table contains the headers of all segments specified in the language definition segment.

In the second mode, Cocofe uses the Reflan table for L2 to process hardware description segments written with L2 as reference language. A description is transformed into a corresponding Ireen structure, which is then used as input to compiler backends to implement specific tools (simulators, timing verifiers, hardware generators, etc.).

Pargen [13] complements Cocofe by generating parsers tailored to Conlan languages. A generated parser may then replace Cocofe's buit-in interpretative, language-independent parser to gain speed when processing descriptions written in a specific language.

Both Cocofe and Pargen are operational. They have been installed for exploratory work at several sites in Europe and the United States.

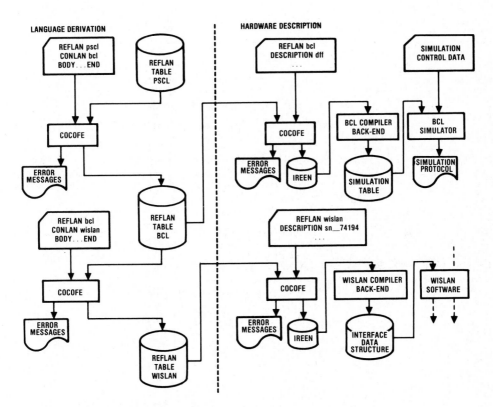

Figure 8. The application of the Cocofe system.

5.1 REGLAN SIMULATOR

Conlan software activities in Darmstadt are shifting toward development of compiler back ends to implement specific simulators and formal verification tools at different levels of abstraction. The first step in our work has been to derive a suitable description language from BCL. Then a corresponding compiler front end is generated by Conlit. Next, we either connect existing tools to the Ireen data structure through suitable interface software or directly implement new tools.

Recently, the Reglan project to test the entire implementation cycle was set up. It involves language definition through the Conlan report rule, front-end generation with Conlit, and back-end development from the Ireen interface. Reglan covers the register-transfer and system level. It provides most features of languages at these levels of description, including

- real-time models for operations on edge-triggered and latch registers; on open collector, tristate unidirectional and bidirectional buses; and on single and multiport RAM and ROM memories as carrier types with time parameters for zero and unit delay register and bus transfer modelling;

- abstract variables for the basic value types (bool, four-valued logic, integers, character strings) with time delayed assignment for functional I/O description of modules (grey boxes) and for building control structures, (for example, tokens in petri nets) as in CAP [16];

- a "change" function to detect signal transitions in a given time window; a "delay" function with parameters for maximal and minimal delay yielding "U" (undefined) during this tolerance after an input transition. Both functions are used typically to write assertions on I/O signal time relations or to model delay properties of combinational networks; and

- an ON statement, complementing the IF statement. These statements permit efficient modeling for real-time automata with data-dependent operation time.

Although intended for industrial use, Reglan and the simulator will be tested by students in courses and projects on computer design. It will provide extended capabilities (structured description, modeling of different register types, and time relations) compared to the RTS1a simulator which is currently in use.

5.2 FUTURE PLANS

An ideal integrated design environment supports a set of design stations and permits design specification at arbitrary levels of spatial, temporal, and operational abstraction. It provides access to a common database holding previously specified design and a common set of tools operating on them. This database configuration is feasible only with a uniform, or level-independent, interface connecting the database, design stations (user interface), and toolbox to make user interface (language design), and tool design independent of the database.

In its present state, Conlan with its implementation tools has only two of the six properties required to support such an integrated design system: it provides a uniform user interface for a large variety of languages, and Conlit software provides a common interface between user descriptions and tools operating on them.

Before Conlan can be integrated into a design system, it needs

1. a database to handle design objects in Ireen;

2. constructs for expressing structural decomposition, including metric and technological attributes of modules and their ports to support placement and floor-planning, routing and abutment;

3. the means to specify and handle a description as a layout; and

4. a graphic user interface.

The Icode project, recently started, is designed to satisfy these requirements. It will be based on an extended version of Ireen, which will include

. a uniform Conlan port concept for description interfaces with metric and technological attributes of a port (position, width, material, etc.), in addition to operational semantics;

. standard metric attributes for module instances (position, size, orientation, etc.);

. representations of several versions of the same object;

. layout by composition of rectangles; and

. user-defined compatibility rules on cell/module contours to support consistency checks during compositions of cell nets or composite cells.

The basic software components of the planned Icode system are shown in Figure 9. It will consist of the front-end, design database, and toolbox.

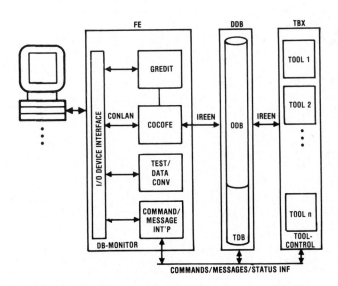

Figure 9. Integrated Conlan Design Environment.

The front end acts as an interface between workstations and the database, It provides facilities for design object generation:

1. a multilevel graphic/interactive editor Gredit for schematic entry, cell placement and routing, composite cell construction, and symbolic/metric layout; and

2. the Conlan compiler front end Cocofe system for translation for graphic and textual object descriptions in Ireen form.

In addition, conversion and display functions for I/O test data as well as command and message interpretation for the database and toolbox will be performed in the front end.

The database stores, retrieves, and changes Ireen objects under control of a monitor executing commands from the front end and toolbox. It also contains design description and test data.

The toolbox provides a uniform tool-control environment for an open set of tools, which include design and compatibility rule checkers, simulators, extractors, and generative tools with Ireen as a standard interface to the database.

6 THE CERES PROJECT

In 1980 the CAD research team of the IMAG institute in Grenoble initiated an integrated CAD system project for VLSI circuit design. The first phase of the project lasted three years. Researchers defined the multi-level hardware description language Cascade, based on the concepts of Conlan [5]. They completed specifications for the compilation and mixed-mode simulation [10] mechanisms for Cascade and defined the architecture of an integrated set of software design tools built around Cascade.

Since 1983, the implementation of a prototype Cascade Environment for the Realisation of Electronic Systems, or Ceres, has been supported by the European Economic Council. Ceres has become a multi-national project involving academic institutions -IMAG in France and Politecnico di Torino in Italy- and industrial firms in France, Italy, Great Britain, Belgium, and the Netherlands.

Figure 10 shows an overview of Ceres. The box in dotted lined encloses the Cascade description and verification subsystem, which offers a unique description language framework with a common set of constructions, segments, and operators with unique semantics at all design levels. Ceres also provides

1. a hardware-free user interface with graphic dialogues and results based on the general purpose, graphic system Clovis, developed in Grenoble: the sophistication of man-machine interaction will depend essentially on the particular work station being used;

2. a comprehensive, interactive, floor-planner, supporting a top-down methodology for translating behavior into topology and geometry, and linking Cascade at selected description levels;

3. a common design data structure and management system, but this aspect is probably the least original; and

4. a multilevel parallel simulator that works on a single internal data structure produced by the Cascade compiler.

It can activate blocks at various abstraction levels, using a variety of algorithms during the same simulation run; for example, explicit, discrete and continuous methods can be used together with implicit methods inside the same model. We are now studying gate-level, concurrent fault simulation for blocks with parallel simulation at other levels. A first, incomplete prototype of Ceres was demonstrated in July 1985.

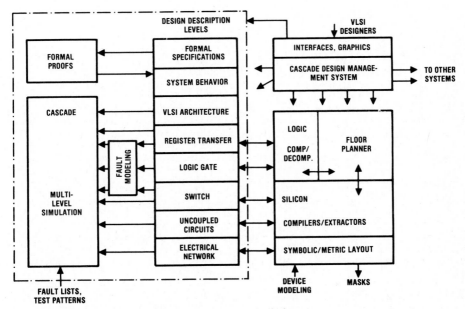

Figure 10. Ceres, a Cascade environment for electronic system design.

THE CASCADE DESCRIPTION LANGUAGE

Cascade covers all hardware modelling levels from electrical components networks to system behavior. It can be divided into

. core Cascade, a kernel of notions common to several or all levels of description, with unique syntax and semantics;

. six predefined languages levels, which include value and carrier types, operations and description segments, and control primitives directly available at each design level.

The relation of BCL to the semantics of core Cascade is obvious. Its value and generic carrier types, operation segments, description segments with attributes and description instantiation, and the array constructor have been taken over. Likewise, DECLARE, IF, CASE, OVER, and ASSERT statements have been reproduced in their full semantics.

Extensions to Conlan have been necessary to incorporate electrical and mixed discrete-continuous modeling. Cascade knows the "real" value type, the electrical-wire and electrical-node carrier types, and the derivation operator for differential equations as primitive entities. The USE statement has also been extended to allow for data conversion when modules described at different Cascade levels are interconnected. Default conversions are provided, which may be overriden by user-defined functions. On the other hand, none of the toolmaker specific statements of Conlan have been incorporated in Cascade. Language definition segments are restricted to a set of segment definitions without any syntax alteration and only subtypes of existing types can be defined by the user.

7 CONCLUSION

Defining and implementing hardware description languages and associated software tools within the Conlan framework offers significant benefits to CAD toolmakers. All design levels can be covered by a family of languages derived from a common root and using a powerful set of constructs that provide a clear semantic definition. Coherence between these languages is enforced by their sharing a kernel of types, operators, segment constructors, and syntax rules. Considerable freedom is, nonetheless, left to the language designer to extend and remove primitive segments and statements in a member language, thus adjusting it to a particular technology or design environment. Software is now available to automatically check static semantics and consistency rules in a language definition derived from BCL and produce the associated compiler front end. Any such compiler front end transforms a user description into a standard data structure, which provides a well defined interface between a variety of back-end tools. First Conlan implementation tools are available from the Technische Hochschule Darmstadt and are presently being used in several sites in Europe and the US. In Darmstadt and Grenoble, work continues on projects to construct full CAD systems in which hardware description is based on the Conlan approach.

REFERENCES

[1] R. Piloty, et al., "CONLAN Report", Lecture Notes in Computer Science No 151, Springer Verlag, New York, 1983.

[2] M. K. Engh, A. K. Vaidya, and D. L. Dietmeyer, "WISLAN - A CONLAN Member for Gate Array Design", Proc. Sixth Int'l Symp. CHDL, North Holland Publishing Co., New York, 1983, pp. 31-42.

[3] A. K. Vaidya, D. L. Dietmeyer, and M. K. Engh, "WISLAN - Technology Transformation and Optimization", Proc. Sixth Int'l Symp. CHDL, North Holland Publishing Co., New York, 1983, pp. 43-54.

[4] R. Piloty and D. Borrione, "The CONLAN Project: Status and Future Plans", Proc. 19th Design Automation Conf., June 1982, pp. 202-212.

[5] R. Piloty, et al., "CONLAN - A Formal Construction Method for Hardware Description Languages: Basic Principles, Language Derivation, Language Application", NCC 1980 Proc., Vol. 49, pp. 209-236.

[6] R.   Piloty, et al., "An Overview of CONLAN - A Formal Construction Method
     for Hardware Description Languages", Proc. IFIP Congress 1980, Oct. 1980,
     pp. 199-204.

[7] D. Borrione, "The Worker Model of Evaluation for Computer Description
     Languages", Proc.   Fifth Int'l Symp.   CHDL, Sept. 1981, North Holland
     Publishing Co., New York, 1981, pp. 3-21.

[8] H.   Eveking, "The Application of CONLAN Assertions to the   Correct
     Description of Hardware", Proc. Fifth Int'l Symp. CHDL, Sept. 1981, North
     Holland Publishing Co., New York, 1981, pp. 37-50.

[9] D. Borrione, "Langages de Description de Systemes Logiques - Propositions
     pour une methode formelle de definition", These d'etat, INPG, Grenoble,
     France, July 1, 1981.

[10] D. Borrione, et al., "A Time-Profile Description Language for System
     Specification and Simulation", IMAG Research Report No.406, Grenoble, Nov.
     1983.

[11] J. Mermet, "Circuits and Systems Computer Aided Design and Engineering:
     CASCADE:, Proc.   IFIP Int'l Conf. Computer Applications in Production and
     Engineering CAPE '83, North Holland Publishing Co., New York, 1983, pp.
     245-262.

[12] B. Kreling, "CONLAN Implementation Tools: COCOFE - CONLAN Compiler
     Front-End, User Manual", Report RO 84/3, Institut fuer Datentechnik,
     Technische Hochschule Darmstadt, Mar. 1984.

[13] A, Kappas, "CONLAN Implementation Tools: PARGEN - The CONLIT
     Parser-generator, User Manual (Ver.   01)", Report 83/7, Institut fuer
     Datentechnik, Technische Hochschule Darmstadt, Sept. 1983.

[14] B. Kreling, "IREEN - An Intermediate Form of CONLAN, Reference Manual",
     Report RO 84/2, Institut fuer Datentechnik, Technische Hochschule Darmstadt,
     Mar. 1984.

[15] B. Kreling, "Static Semantics of CONLAN - Rules and Algorithms", Report RO
     84/1, Institut fuer Datentechnik, Technische Hochschule Darmstadt, Mar.
     1984.

[16] F.   J.   Ramming, "CAP/DSDL: Preliminary Language Reference Manual",
     Internal Report, Institut fuer Informatik, University of Dortmund, 1981.

7.2. ELECTRONIC DESIGN INTERCHANGE FORMAT (EDIF)
Proposed Standard

Paul H. STANFORD

1.0 Disclaimer

It should be noted that this description of the Electronic Design
Interchange Format does not in any way purport to be a formal definition of that format.

Should there be any discrepancies between this account and the
official standard document, the latter is to be taken as the correct
definition. This summary is based on version 1.0.0 of the EDIF
standard, which was the result of efforts from several people from
many different organisations.

There is not sufficient detail included here to fully define the
syntax or semantics of the format: this is intended to be an overview to show the capabilities of the format, and its methods.

2.0   Introduction

EDIF is a standard format for the exchange of electronic data, with particular emphasis on integrated circuit descriptions and semicustom data. The principle goal of the format designers was to facilitate the exchange of such data between companies, including foundries, vendors, workstations, and customers. The need for such an unambiguous medium has become increasingly acute in recent years, especially due to the volume and complexity of data needed to describe or use semicustom devices (such as gate arrays and standard cells), and the diversity of available workstations.

It should be noted that EDIF is not claimed to be a cure for all difficulties which can occur in moving circuit designs. The normal marketing and technical discussions prior to such exchange are just as important as ever, including decisions as to what type of data should be exchanged, what actions are expected, and so on. Also, the existence of EDIF will not affect any severe incompatibilities between different parties. The task of the format is to faithfully express the information which is appropriate for the exchange, and thus relieve the problems which are fairly attributable to the means of expressing information. The format can be thought of as a bridge between design databases, and as such it is expected to be generated and processed by software rather than by engineers.

3.0   History

The original cooperative efforts to design such a format began in late 1983, with the encouragement and manpower of companies which believed that such an endeavour could succeed. These included Texas Instruments, Motorola, Tektronix, National Semiconductor, Daisy Systems and Mentor Graphics.

During 1984, there was a press release to announce these efforts, three public meetings (both in America and Europe), and the publication of two preliminary drafts. By mid-year, the number of companies actively involved in various aspects of the development of EDIF rose to several dozen, with many more participating to the extent of requesting information and submitting comments. It is thus fair so say that the format is the result of widespread cooperation amongst the companies and individuals who are most concerned with using it. This will continue to be the case for future enhancements, which should reflect the needs of all interested parties.

Version 1.0.0 was published in February 1985, little over a year after work began. This version was intended to be used for exchange, as distinct from the review status of previous versions. Changes in the format which would not be compatible with this version will be avoided as far as possible, so that software and libraries can be constructed with greater confidence of its future value.

4.0   Overall format

The following description is intended to give an idea of the nature
(both syntax and semantics) of an EDIF file during its transmission
between different sites. Any further applications within either company
is beyond the scope of this description.

4.1 Medium

Currently, there is no particular physical medium associated with
EDIF files.   However, these are expected to be magnetic tape,
communication links, or any other suitable medium. It is stated that the
data consists of a single physical file.

The format has been defined in terms of a text file, as this is
thought   to   reduce   the   possibilities   for   machine-to-machine
incompatibilities. It will be assumed that the information has been
decoded from the particular physical medium into a series of ASCII text
characters, including the "end of line" characters linefeed and carriage-
return (or can be treated as if this has been done). Any considerations
of storage or transmission medium such as block or record boundaries will
be of no further concern.

4.2 Structure

The syntax used in EDIF is closely related to (and borrowed from) the
LISP programming language. This was done to simplify the task of machine

compilation of EDIF files, and to allow easy design of the inherently nested structure of any modern circuit design database.

At the lowest scanning level, an EDIF file can be fragmented into a series of tokens as follows. Both left and right parentheses () are special one-character tokens. Any combination of one or more space characters (blank, tab, linefeed or return) also act as token separators. A string contains all characters from a double-quote (") to the following double-quote, and is considered to be a single token. Except within strings, there is no distinction between upper and lower case characters.

The basic structure of EDIF is that of a list, formed by enclosing a series of primitive data items and nested lists within a pair of left and right parentheses: (a b c d). EDIF goes beyond LISP in that every such list is required to begin with a keyword, which is associated with a definite meaning in the standard. Further, the items which may (or must) be found following a particular keyword are also carefully specified.

4.3 Primitive Data

The primitive data items in EDIF are identifiers (names), numbers and strings. An entire file consists of only these items, keywords, parentheses and space. Other more complex items such as points are also described here, as they occur in a wide variety of places within the format.

4.3.1 Identifiers

The names in EDIF are restricted to contain between 1 and 255 alphanumeric or underscore characters. The initial character must be a letter; upper and lower case variants are considered to be the same. In the event that more relaxed character sets are used within a design system, the "rename" facility is available to record any name mappings used in creating the file.

4.3.2 Numbers

Only integers can be directly represented as tokens in EDIF, although real numbers can be expressed using the "E format". An integer is represented as a sequence of decimal digits, possibly preceded by a sign character (without internal spaces). The absolute value should be less than two to the power thirty one, to ensure that the value can be stored as a thirty two bit integer in either one or two complement form.

4.3.2.1 E Format

An expression of the form (E m p) where E is a keyword (indicating exponentiation), and m and p are integers, represents the number m times ten to the power p. Thus (E 1 6) represents a million; (E 2 -3) represents two thousandths. Again for machine transportability, the absolute value of such a form should be zero or lie between ten to the power plus or minus thirty five.

4.3.2.2 Minomax

In some cases, such as timing data, a range of values is required in place of a single value. The expression (Minomax a b c), where a, b and c are numbers, indicates a minimum value a, nominal value b, and maximum value c. Where there is no restriction intended, the form (unconstrained) is used in place of a number, including the parentheses; where a field is not desired, the form (undefined) is used.

4.3.3 Strings

Strings are occasionally used within an EDIF file. They consist of a double-quote, a series of characters, and a matching double-quote ("). Within the string, unprintable characters, "new-line" and tab characters are ignored. Special characters (including " and %) are encoded as a four character group of the form %ddd, where ddd is a three digit decimal representation of the ASCII code of the desired character.

4.4 Points

A cartesian point with x-coordinate x and y co-ordinate y is represented as (Point x y), where x and y are restricted to be integers.

The interpretation of this representation depends in part on the location of the expression. For example, within a cell description, the coordinates are relative to the (unplaced) cell origin. Also, the relationship between coordinate values and physical dimensions is governed by scale statements corresponding to the appropriate library.

4.5   Comment

Comments within EDIF should not be viewed in the same way as comments within programming languages or design languages. Their function is to preserve and communicate design comments which may be available within the originating database.

These take the form (Comment string ...); for example (comment "derived from adder abcd"). To simplify the description of the rest of the syntax, we shall use phrases such as "the comment form contains one or more strings" to mean the same thing. The precise contexts in which a comment may appear are defined in the official specification.

4.6   Member

The member form contains an identifier, followed by integers which function as array indices. For example, (member p 3) represents the element of an array "p" with index 3.

4.7   Qualify

The qualify form contains any number of identifiers or members, and is used in a variety of contexts to express a "qualified name". For example, (qualify nand23 out) would indicate a component port on an instance named "nand23" which correspond to port "out" on the lower cell.

4.8 Multiple

In various parts of the format, as defined in the formal specification, a single value could be replaced by a Multiple form, which may contain more than one value. This acts as an abbreviation in the sense that the same meaning could be expressed by using several individual statements. For example, declarations of input ports A, B, C could be compressed into a single declaration of (multiple A B C); both methods of expression mean exactly the same thing.

4.9 UserData

The userdata form contains an identifier, followed by any number of unrestricted primitive items or lists. It may be found within most other forms (after any "positional" items), as defined in the official format specification.

This is intended to be a mechanism whereby consenting parties can incorporate "private" data within the structure of an EDIF file without resorting to using variants of the format itself.

The contents and meaning of such a construct must be agreed by the participants. If no such pre-arangement has occured, the use of userdata within a file should be considered a serious error (unless the sender warns the recipient to ignore them). The format itself defined no meaning to any userdata constructs, and their use should be avoided, or limited as much as possible, as they severely limit the transportability of the resulting file.

5.0   EDIF syntax overview

The initial keyword in an EDIF file is the word "EDIF", so that the
entire description is a single, large list of the form (EDIF ....).   The
EDIF form contains a name (to help identify particular transmissions), a
status form, followed by any number of design, external and library forms.

The design gives a "starting point" into a design hierarchy, the
library contains cell descriptions, and the external form declares
libraries which are used by, but not defined within, this file.   Such
libraries should have been transferred between the participants on another
occasion.

5.1 Status

The status form corresponds to any data management information which
may be recorded within a database. It contains details of the version of
EDIF that was used to encode the data, and dates of creation and update of
the associated object.   It is intended to assist in the task of
maintaining libraries, and in verifying that correct versions of data are
being used.

It can occur within the top level of the EDIF file, and also within
individual libraries, design forms, technology forms, cells and views.
For brevity, it will not be mentioned for each of these.

The status form may contain an Edifversion, an Ediflevel, and any
number of Written forms.

5.1.1 EdifVersion

This consists of the numbers which are specified as the format version on the official EDIF specification; for example, (edifversion 1 0 0). The first number corresponds to major revisions or additions, the second to minor revisions, and the third to preliminary or review versions.

5.1.2 EdifLevel

The level of an EDIF description (or part of a description) indicates the degree of complexity or sophistication required to interpret or process the data. There are three levels defined: level zero contains only simple, static descriptions; level one allows the definition of cell parameters, and the use of functions; level two will include constructs needed for the description of procedural cells. Each level is an extension of the lower levels. A cell described at one level may contain instances of cells described at higher levels; thus a level zero cell may use level one cells, provided it assigns explicit values to its parameters.

Version 1 0 0 allows both level zero and level one descriptions; the level two constructs are (or were) under development. We shall concentrate on the description of the level zero format for most of this account.

The Ediflevel form contains a single integer giving the (highest) level of the format used within the appropriate object (file, library, cell or view). It thus acts as a warning to processors of the format that may not be able to translate all levels. For example, (ediflevel 0).

5.1.3 Written

The Written form is used to convey some of the "history" of an object, including its creation, last update and intermediate updates (if available).

5.1.3.1 Timestamp

This contains integers to fix the year, month, day, hour, minute and second of creation or update. These are declared to be expressed in terms of Greenwich Mean time, which may be important if frequent revisions are transmitted between different time zones. For example, (timestamp 1985 5 8 21 10 0).

5.1.3.2 Accounting

Other possibly helpful information documenting the update can be encoded into the accounting form, which contains a name followed by an optional string. The name has no intrinsic meaning to the format, and is intended to be used to assist in various "trouble shooting" activities, such as determining the author of the cell. It could also be used to record the program used to translate the data, or the internal versions of the originating system.

For example, (accounting author "A.B. Smith")

5.2 Design

In a hierarchical design of cells placed within cells, there needs to be a "starting point" to the structure. The design form serves this function; it contains a name (to externally distinguish between designs, if more than one is defined), followed by an indication of the top-most cell of the design. As this requires the name of a library and a cell, a qualified name is used.

5.3 External

The External form simply contains the names of one or more libraries which are required to complete the full description. Such an omission should be agreed prior to the transfer. It can be used to avoid resending large libraries, whilst alerting the recipient to the fact that certain libraries are required.

5.4 Library

The bulk of information in an EDIF file is contained within its libraries. It begins with the name of the library (as used within the rest of the file), and may contain a Technology form and any number of Cell descriptions. The technology information, such as scaling factors, is relevant to each of the cells in the same library. Cells from different libraries may have the same name without implying any connection. Different libraries must have different library names.

5.4.1 Technology

Each library may contain an optional Technology form. It contains a
name (for external reference), and a variety of forms used to collect
information relevant to the whole library. This includes coordinate
scaling,   logic   value   definitions,   graphic   display options, and
declarations of figure layers. It is expected to hold geometric spacing
rules and general electrical information in later releases.

5.4.1.1 NumberDefinition

Distance and time scales may be established by this form, which
contains the name of the unit system followed by a number of scale forms.
The name SI implies that the reference units are Meter, Farad, Ampere,
Ohm, degree Celsius, Second and Volt for types distance, capacitance,
current, resistance, temperature, time and voltage respectively. Any
other unit system would need to be agreed beforehand by the participants.

5.4.1.1.1 Scale

The relationship between values used within the library and external
units for one type of value is specified by a Scale form. It contains the
type of value (for example, distance or time), a value in the "internal"
scale, and the corresponding value in terms of the external scale (for
example, meters). Thus a library which expresses coordinates in microns
should contain the statement (scale distance 1 (E 1 -6)), or equivalently
(scale distance 1000000 1).

5.4.1.2 Define

Within the technology block, the major use of Define is to express legal figure layer names. For example, (define output figuregroup metal2) establishes "metal2" as the name of a layer, or group of figures. For layers which are "internal", not intended for pattern generation, the word "local" is used in place of "output".

The Define form is used to define other types of object in other parts of the format, such as ports; this will be discussed in that context.

5.4.1.3 Rename

In the event that the name for an object in the originating database violates the EDIF restrictions on identifiers, it should be replaced by a valid name. The Rename form allows the original name to be preserved, giving the option for its reinstatement by the recipient if their system can handle unrestricted names.

The Rename form simply contains the defined name used within the description, followed by a string which records the original name. For example, (rename new12 "1.(*bar").

5.4.1.4 FigureGroupDefault

Each figure group name (layer name) can be associated with a number of display and expansion options. The form begins with the previously defined name, and may contain forms to define its default path width, pathtype, color, fillpattern or borderpattern. Any of these can be overridden on individual geometries, and will be described after the figures on which they act are described.

5.4.1.5 GridMap

The gridmap is used to establish an extra scaling function to be performed on the points of each cell definition, and enables cell data to be recorded in terms of x and y pitches instead of uniform distances.

For example, if a general distance scale of microns has been established, then (gridmap 5 7) means that cell data in that library is expressed in terms of 5 micron steps in the x direction and 7 micron steps in the y direction. Thus a point coded as (point 2 1) would have "real" coordinates 10 microns by 7 microns.

5.4.1.6 SimulationInfo

In an effort to allow simulation information to be exchanged without making an arbitrary choice of the "correct" logic values to use, EDIF allows the declaration of the values appropriate for the sender, and certain relationships between them.

The form contains a name (used to identify this simulation environment), followed by SimulationValue, Isolated and Arbitrate forms.

SimulationValue contains a list of names, which correspond to the particular range of logic values of interest. At this time, no electrical "meaning" is attached to these names, which serve as a link between model definitions, timing descriptions, and test patterns.

Isolated contains one of these logic values, which is the desired default value to be assigned to an unused input port, or other isolated circuit node. This could indicate an error condition in terms of externally communicated semantics for the chosen logic names.

Arbitrate contains a single Pairwise form, which contains a sequence of logic values. This double nesting is intended to allow further expansion if more sophisticated arbitration is required. The purpose of this item is to define the effect of "wired functions"; that is, the desired result when more than one signal output is joined to the same net.

An assumption is made in Pairwise that it is sufficient to arbitrate the wiring of just two values together. Pairwise contains a list of logic values obtained by reading the upper diagonal of the truth table for the "wired function", using the same ordering of values as used in the SimulationValue statement. This is the only place where that ordering is used. Thus the list contains the result of wiring first with first, first with second, and so on, then second with second, second with third, and eventually last with last.

The effect of Isolated and Arbitrate is that a net can be assigned a logic value (during simulation), no mater how many outputs drive the net.

5.4.1.7   SimulationMap

Each SimulationMap provides a mapping between different environments
created by a previous SimulationInfo. This enables the correspondence
between different logic value schemes to be defined. This would be
necessary, for example, if cells described in terms of different values
occured in the same netlist.

The form contains the simulation names involved in the mapping, say
S1 and S2, followed by any number of Statemap forms: (simulationmap S1 S2
...). The mapping will provide a function from values in S2 to values
in S1. The simulation names should be qualified by a library name if they
are defined in different libraries.

Statemap contains a single value from S1 followed by all values from
S2 which should map to that value. For example, (statemap true tw tf tr
ts tp).

5.4.2   Cell

The major constituent of libraries are cells. Each cell begins with
a name, which must be different from any other cell name in the same
library. After the name, a cell may contain a Status block, a ViewMap and
any number of named Views, each of which is intended to contain a
description of the same entity (the cell) in different terms. This
contrasts with the library structure, which is not "described" by each of
its cells.   For example, if a cell contains a mask layout view and a

netlist, then the layout should faithfully implement the function defined
by the netlist.

5.4.2.1 View

Each view contains a view type, a view name (which is unique within
the cell), and may contain Status, Interface and/or Contents forms.

The view type is one of: netlist, schematic, symbolic, masklayout,
behavior, document or stranger. These limit the forms allowed within the
view, and may color their exact meaning. Thus, a metal figure would have
no place in a netlist description, and a logic symbol in a schematic view
would not be expected to be used for pattern generation.

The different viewtypes will be explained in more detail later.   It
should be noted that the range of descriptions includes pure netlists
(without graphics),   logic/schematic   diagrams   (with   graphics   and
connectivity), full layout, model definition, textual documentation and
the emergency escape stranger view. Also, a cell may contain more than
one view of the same type.

5.4.2.1.1 Interface

The data contained within the Interface roughly corresponds to a
datasheet. It contains information required to make use of a cell such as
its outline, port definitions and timing constraints.   In many exchange
scenarios, only interface descriptions would be required, as their

contents may be irrelevant, or proprietary, or both.

5.4.2.1.2 Contents

The contents part of a view constitutes the detailed description of
the cell in terms of other cells and specific objects appropriate to the
view type. It may contain a variety of constructs, which depend to some
extent on the view type. This is where a logic diagram may be stored, or
a detailed mask layout, or a model description.

5.4.2.2 ViewMap

The viewmap provides the linkage between different views of the same
cell, such as the correspondence between logical ports in one view and
actual geometric ports in another, or components (cell instances) in
different views.   This should be interpreted as a mapping amongst these
styles of description rather than as additional information about any
single view.

The viewmap may contain one or more portmap or instancemap. These
consist of two or more qualified names of ports or instances which
"correspond" in the above sense; the qualification involves the view name,
one or more instance names (where appropriate) followed by the actual port
or instance name. The portmap is restricted to ports declared within the
containing cell.

6.0   Cell Interface

We now discuss the objects which may be found within an interface description in more detail.   Recall that an interface occurs within a view, which is part of a cell, in a library, within the EDIF file itself.

It is important to note that the interface forms a "summary" of the detailed description of the cell, not an addition to it. For example, figures associated with a port in a masklayout interface define the location and nature of connection points to the cell, but do not generate mask patterns: the contents should be used for this function.

6.1 Body

This contains the cell's logic symbol in a schematic view, or geometric outlines (boundaries) in a layout view. The latter act as protection frames in other cells which use this cell.

6.2 Define

This is used to declare ports, signals, and (for level one), parameters.

6.2.1 ArrayDefinition

Within a definition, this allows the declaration of arrays of ports, signals or other types of data. These can be multidimensional; individual elements can be referenced with the Member form.

6.3 Rename

Used to record any name mappings which were performed in the
translation from the senders database into EDIF.

6.4 PortImplementation

This records the graphics or actual geometries, and positions, of
cell ports. Where appropriate, it can also contain data needed to express
fanin and fanout restrictions, and factors used in load induced delay
calculations, with a well defined property form.

6.5 Permutable

The fact that certain ports may be interchanged during netlist layout
can be expressed with the Permutable form. For example, the inputs of a
nand gate could be declared to be permutable, thus allowing extra
flexibility in routing.

More complex situations, such as sets of ports which may be exchanged
as a whole whilst preserving order within each set, can be expressed by
nesting Permutable and NonPermutable statements.

6.6 Joined

Ports in an interface can be Joined, which means that they are
internally connected. This allows for the fact that either of two (or
more) ports may be chosen to make a connection, and that the signal could

pass through the cell to assist external routing.

6.7 MustJoin

Conversely, MustJoin declares that two or more ports need to be connected externally whenever the cell is used, possibly as a result of an internal routing failure. More complex situations of partial routing failures can be expressed by nesting joins and mustjoins.

6.8 WeakJoined

This expresses an intermediate condition, in which any of a number of ports may be chosen as a valid connection to the cell, but that the external routing cannot use the ports as a feed through. In other words, the ports are connected as far as the operation of the cell is concerned, but should not be treated as being joined for the purposes of routing any containing cell.

Such a condition could arise when the internal connection used are implemented on a material that is not suitable for extensive routing.

6.9 Unused

There may be situations where it is sensible to declare a port which is not actually used within the cell. The Unused form can record this fact.

6.10   Timing

Signal delay between ports can be expressed in the Timing section,
plus any required setup or hold constraint. It contains the name of the
appropriate simulation name (see SimulationInfo) and any number of Delay
or Stable forms.

6.10.1   Delay

Each   Delay   contains   a   number   of   Transition   statements,   which
expresses a range of numbers associated with the signal delay from a named
port to another named port, resulting in a transition between named logic
values.   It   thus   allows   pin-to-pin timing descriptions, separately for
"rising", "falling", and any other type of transition.

The times given can include minimal, nominal and maximum delays; they
can also include both intrinsic values and load-induced incremental
values.

6.10.2   Stable

The Stable form is used to demand that an input signal be stable
throughout a time interval, defined in relationship to a specific logic
transition of a port.   For example, (stable data -10 5 clock low high)
would demand a setup time of 10ns and a hold time of 5ns in "data",
relative to the rising edge of port "clock" (assuming appropriate
declarations within the library's technology block).

6.11   ArrayRelatedInfo

The relationship between gate array "macro" cells, which may contain just metal patterns, and the underlying array master, is expressed with Plug and Socket attributes. Cells which may only be placed in certain locations can be declared to be "plug" cells, and list compatible socket cell names. Cells which are marked as being "sockets" may be placed within the gate array master description, and thus restrict placement of corresponding plugs to those locations.

Features such as socket symmetry, and multiple socket requirements, can also be expressed.

6.12   Simulate

The Simulate form contains logical test pattern sequences. It can be used to express simulation exercises for the cell, or both stimulus and response data for verification purposes. The information can be expressed in a simple list of test vectors, or may include explicit descriptions of "scanned" input and output techniques.

There may be more than one simulate in a cell; they are distinguished by name, and contain a reference to the appropriate simulation name.

6.12.1   IgnoreValue

This allows the temporary creation of a new logic "value", which is used to indicate that an output should not be checked.

6.12.2 WaveValue

WaveValue allows the definition of names which have a defined logic
behavior throughout a given time interval, such as a single clock pulse.
These augment the static logic values within the test patterns.

6.12.2.1 LogicWaveform

LogicWaveform is used within wavevalue, logicinput and logicoutput.
It contains a sequence of previously defined values which, with the given
timing information, define signal behavior over a length of time.

6.12.3 PortListAlias

This is used to abbreviate a list of port names, for use within Apply
statements.

6.12.4 Apply

The Apply statements form the major part of the Simulate description.
They are interpreted as acting sequentially on the   cell,   during
simulation. Each apply specifies the number of "cycles" of activity which
are described, and may specify the duration of each cycle, and any number
of logicinput and logicoutput statements.

6.12.4.1 LogicInput

This form indicates a subset of the cell's ports, followed by a logicwaveform description of their inputs for the duration of the Apply. This list may be truncated after no further changes are required on any port.

6.12.4.2 LogicOutput

This is similar to logicinput, except that expected outputs may be specified. Again, there is some freedom to express the values associated with several ports for a short time, or few ports for several cycles, as appropriate.

All logic inputs and outputs within an Apply are described relative to the start of operation of the Apply.

7.0   Cell Contents

The contents of a cell view contains the detailed implementation of the cell according to that view. For example, in a MaskLayout view this would hold the explicit geometries and cell placements needed for pattern generation; in a Schematic view it would hold the logic or schematic diagram. Many different constructs are available for the contents, with some restrictions based on the view type, as documented in the specification.

Brief details will be given on the major forms, to give an overview of the various types of information that may be expressed.

7.1 FigureGroup

This is used to convey basic geometric items, and is appropriate for MaskLayout and Schematic views. It contains the name of a figuregroup, as defined in the technology; this provides default display characteristics, and a link into external processing through the name. These defaults may be overridden for particular geometries, which is most appropriate for features such as path width.

The basic types of figure supported are rectangles, polygons, paths, circles, shapes and dots. Shapes are similar to polygons, but may include both straight and curved edges; they are expected to be most appropriate in describing logic symbols. Dots are only appropriate in schematics, as they have no specified extent. There is an Annotate form which may be used for adding text to geometric displays, but not for real mask generation.

7.1.1 SignalGroup

A signal may be attached to one or more figures by collecting them within a SignalGroup list, which contains the desired signal name. This enables the association between Schematic or Netlist descriptions and MaskLayout descriptions to be preserved.

7.2 Define

Local signals may be defined in the contents. In level one EDIF, local parameters may also be defined. The link with original invalid names may be preserved with the Rename form, as in the interface.

7.3 Global

A global declaration of a port may only be made within the contents. It is an abbreviation, and implies the joining of the named port with any port of the same name in lower cell instances.

7.4 CriticalSignal

Signals may be given a "criticality value" to assist routers prioritise the task of synthesising interconnect for each net.

7.5 Unused

Deliberately unused component ports may be named within the Unused form to avoid false diagnostics from netlist completion checks.

7.6 Joined

This is the basic form for recording netlist topology. It may contain defined local signal names, port names, or qualified port names (indicating ports from cell instances). All items collected in the same Joined form are declared to belong to the same net; however, a net need not be defined by a single Joined form.

MustJoin is a variation on Joined, and corresponds to the case of an incomplete routing attempt. For simulation purposes they mean the same thing.

7.7 Wire

Wires may be used in Schematic views, and combine the joining of ports with a graphical representation (which must be in the form of a path).

7.8 Instance

The Instance form is used to establish cell hierarchy. It creates a named instance of another cell as part of the description of a view. Where appropriate, it may contain coordinate transformation data such as scaling, rotation and translation. Multiple translations may be used to achieve more than one instance of a cell in a regular geometric array.

If the lower cell contains parameters, values for these may be provided within the instance.

7.9 Timing data

Both required and measures path delays can be specified within the contents. All the flexibility of the Delay form is available, as described for the interface. However, in the contents delays may be given for explicit pathways through or within the cell.

7.10 LogicModel

This form is only used within a Behavior view. It may be used to give a very low level description of the logical behavior of the cell, and is expected to be appropriate for very simple model descriptions.

Such a model description consists of a number of implications, which give the logical value and time interval for an output port conditional on similar information for a number of input ports. It thus combines a simple truth table description with detailed timing relationships.

7.11 Section

This is the structural element used in Documentation views, and allows a very simple level of formatting to be indicated with the text. It may contain strings (for the actual text), nested sections, or instances of other documentation views.

8.0 Parametric EDIF

All the constructs described for EDIF take an extra power in level one, through the use of parameters and simple expressions. Values may be passed into cells, and functions used to define their operation on the cell. In most cases, any explicit values appropriate to a level zero description can be replaced by parameters or expressions in level one.

This is clearly appropriate in describing the interface to simulation models, even if only their interface can be described. It is essential for the description of complex parametric layout cells, such as in a parametric cell library.

The motive for separating EDIF into levels is to provide a smooth path into higher level descriptions, without requiring unnecessary sophistication to exchange static cell descriptions.

9.0   The Future

Plans to extend the format include continued development of functions for level one, algorithmic forms for level two, and further efforts in the areas of modelling, hardware test descriptions, printed circuit board applications, and methods of abbreviating the format.

AUTHOR INDEX

Borrione, D., 409, 441
Fujita, M., 283
Giandonato, G., 395
Giorcelli, S., 375
Girardi, G., 375
Glesner, M., 227
Hartenstein, R.W., 3, 15
Joepen, H., 227
Le Faou, C., 409
Lemmert, K., 51, 197
Lin, T., 253
Mavridis, A., 163

Melgara, M., 337
Nebel, W., 197
Park, N., 111
Parker, A., 111
Piloty, R., 441
Prinetto, P., 75
Schuck, J., 227
Stanford, P., 463
Su, S.Y.H., 253
Wehn, N., 227
Welters, U., 137
Wodtko, A., 313